RESEARCH METHODOLOGY FOR SCIENTIFIC RESEARCH

RESEARCH METHODOLOGY FOR SCIENTIFIC RESEARCH

SECOND EDITION

K. PRATHAPAN

Assistant Professor
Department of Physics
Govt. Brennen College
Kannur, Kerala, India

TECHSAR™

NEW DELHI

Published and Printed by

Techsar Pvt. Ltd.
4435-36/7, Ansari Road, Daryaganj
New Delhi–110 002 (India)
E-mail: tech_info@techsarworld.com
Website: www.techsarworld.com

ISBN: 978-93-90620-17-3

Reprint 2025

Dedicated to

My Parents and Teachers

Preface to the Second Edition

I am very happy with the wide acceptance received for the first edition of *Research Methodology for Scientific Research* among researchers and academicians. In recent years, the book has been included in the syllabi of various MSc, MPhil, and PhD programmes of various Indian universities. The first edition was published in 2014 and since then the research field has changed considerably due to the developments in technology and other related aspects. The knowledge is accumulating day by day.

The acceptance of the book among the scientific community encouraged me to consider a revised second edition of the book. I got many responses from various people in teaching and scientific communities and I considered all those responses while preparing the revised manuscript. A large number of the most relevant topics were included in every chapter. Most of the additions are related to the academic research prevailing in Indian universities. Philosophies like Reductionism and Holism, Modernism, Postmodernism, etc., are also discussed in this edition in chapter 2. Various components of a research proposal, submission methods, sources of financial support, etc., are included in chapter 6. Discussion on the features of scientific conferences and presentation of research papers along with various available videoconferencing tools are also provided. A new chapter on the *Preparation of Research Thesis* is added in this edition as chapter 8. This chapter provides a detailed description of the various components of a research thesis, their details, and guidelines for preparing each component. The chapter on Research Ethics includes discussion on new topics like intellectual property rights, royalty, and research ethics committees. A separate chapter on publication ethics is included in the new edition as chapter 10.

I hope that the revised edition will satisfy the needs of researchers and research students in the field of science and mathematics. I also take this opportunity to thank all researchers and academicians for their valuable suggestions. Suggestions for improvement are greatly appreciated.

K. Prathapan

Preface to the First Edition

An understanding of the methodology of scientific research is essential for maintaining the quality of research programmes. Unfortunately, majority of the science students are least concerned about the method of acquiring scientific knowledge. Realizing this fact, research methodology is made a part of curriculum in many of the universities. This book is intended to provide explanation of various philosophies of science and the scientific methods and meets the requirements of the MPhil and PhD students in science stream who are preparing for their examination in Research Methodology under various Indian universities. Every aspect is discussed in detail and a sequential development of the topic is adopted for maintaining the continuity. Illustrations and examples are provided wherever necessary.

First chapter provides an overview of the research process and its various aspects. Second chapter deals with the various philosophies of scientific research. Philosophies like induction, deduction, positivism, falsificationism, anti-positivism, post-positivism, relativism, realism, etc., are discussed.

Third chapter deals with the planning and designing of research. Selection of a research problem, criteria of good research, sources of research ideas, etc., are discussed. Validity and reliability of research, artifacts and bias, managerialism and scientific research, leadership in research are also discussed in detail.

Fourth chapter presents various components of scientific method. The formation of hypothesis and its testing are discussed in detail. Chapter five is dedicated to the various data analysis methods, presentation of data and error analysis.

Chapter six gives a detailed information of the various types of scientific literature, journals and various journal indices. In chapter seven the art of scientific writing is discussed in detail. The various components of a research paper are discussed. This chapter also includes various requirements for communicating to a journal.

Chapter eight discusses the various ethical issues in scientific research and guidelines for good research practice.

While preparing the manuscript I consulted various books, research articles, magazines, blogs and websites of many educational and research institutions. I am grateful to the authors and publishers of those materials. A detailed list of such references is given at the end of each chapter. Suggestions for further reading are also given at the end of each chapter.

The support and encouragement provided by the members of my family, friends and colleagues is highly appreciated. I wish to express my grateful thanks to Dr. V P N Nampoori, Professor, International School of Photonics, CUSAT, Dr. K G Gopuchandran and other faculty members of the Department of Optoelectronics, Kerala University for introducing me to the world of scientific research.

K. Prathapan

Contents

Chapter 1

Perspectives of Scientific Research

Religion is a culture of faith; science is a culture of doubt.

—**Richard Feynman**

There is much that science doesn't understand, many mysteries are still to be resolved. In a Universe tens of billions of light years across and some ten or fifteen billion years old, this may be the case forever. We are constantly stumbling on surprises.

—**Carl Sagan**

1.1 SCIENCE AND TECHNOLOGY

"Science is all about models of the real world, whether natural (basic science) or man made (applied science, or technology). These models are not discovered in nature, for nature has no numbers, no coordinate systems, no parameters, no equations, no logic, no predictions, neither linearity nor non-linearity, nor many of the other attributes of science. Models are man's creations, written in the languages of science: natural language, logic, and mathematics. They are built upon the structure of a specified factual domain". —**Jeff Glassman**

Science and technology (S&T) has emerged as an important tool for strengthening the economic and social development. Science and technology provide people with the knowledge and tools to understand and address many of the challenges. A significant progress in various spheres of science and technology can be achieved through a strong network of research institutions, universities and trained manpower. During the past few decades majority of the countries focused on application-oriented research and development (R & D) for technology generation and human resource development. As a part of government policy bright and gifted students are encouraged to take up science as a career and to integrate the developments in science and technology with all spheres of national and international activities for improving the life standards, employment generation, environment protection and ecological security.

1.1.1 Science

Science is a system of knowledge based on principles of reason and empirical proof that is built up over time through the dedicated hard work of thousands of scientists. Science aims at a systematic and logical reconstruction of empirical reality. Even the very definition of science was a matter of debate for centuries and even now it is. Many philosophers and scientists attempted various definitions of science. Some definitions are given below:

1. The systematic observation of natural events and conditions in order to discover facts about them and to formulate laws and principles based on these facts.
2. The organized body of knowledge that is derived from such observations and that can be verified or tested by further investigation.
3. Any specific branch of this general body of knowledge, such as biology, physics, geology, or astronomy.

—*Academic Press Dictionary of Science & Technology*

Science is an intellectual activity carried on by humans that is designed to discover information about the natural world in which humans live and to discover the ways in which this information can be organized into meaningful patterns. A primary aim of science is to collect facts (data). An ultimate purpose of science is to discern the order that exists between and amongst the various facts. —**Dr. Sheldon Gottlieb**

I think that we shall have to get accustomed to the idea that we must not look upon science as a "body of knowledge", but rather as a system of hypotheses, or as a system of guesses or anticipations that in principle cannot be justified, but with which we work as long as they stand up to tests, and of which we are never justified in saying that we know they are "true"... —**Karl R. Popper**

From a general point of view science is a form of knowledge that describes and explains the natural and physical world and its place in the universe. Scientific knowledge is often understood as pure knowledge and is unaffected by human passions. Scientific knowledge is by nature abstract and theoretical. It is often developed through controlled experiments in artificial and idealized laboratory settings. The *Compact Oxford English Dictionary defines science as:*

"a systematically organized body of knowledge on any subject"

In fact, science is not only a body of knowledge but "a way of knowing". Scientific investigation involves exploration, experimentation, observation and measurement, and analysis and interpretation of data. These activities require specific skills and habits of mind; such as, accuracy, discipline, and integrity in the application of scientific principles are fundamental to scientific activity.

Mexican scientist Ruy Perez Tamayo characterized science as a *creative human activity whose aim is the comprehension of nature and whose product is knowledge obtained by reliable methods and which aspires to obtain as much rational consensus as possible*. However these characterizations are epistemological (production of knowledge by reliable methods) but are subjected to bias as the science overlooks the role of the agents generating knowledge. In this context the Oxford Dictionary rightly says that *science is the intellectual and practical activity encompassing the systematic study of the structure and behaviour of the physical and natural world through observation and experiment.*

Science is often divided into basic and applied science. Basic science focuses on understanding the natural world and phenomena. Even though the scientists engaged in basic research are not concerned about the societal applications of their research works; the knowledge that comes from basic research often has anticipated practical outcomes. The science that addresses the societal problems is referred to as applied science.

1.1.1.1 Characteristics of science

Michael P. Clough & Jerrid W. Kruse [1] in their wok *Characteristics of Science: Understanding Scientists and their Work* describe the following characteristics of science. These are all self-explanatory.

- Science is not the same as technology.
- Scientists will employ whatever methods they find useful for understanding the natural world.
- Doing science well require imagination and creativity.
- The generation and acceptance of scientific knowledge takes much time.
- Science has a subjective aspect.
- Well-established science knowledge is durable, but always open to revision.
- Well-established science ideas are not easily abandoned.
- Hypotheses, laws and theories are different, yet related kinds of knowledge.
- Doing science is a social and collaborative process.

The statement that "science aims at a systematic and logical reconstruction of empirical reality" itself reveals some characteristics of science (Robert Lafaille and Hans Wildeboer, 1995). These are:

- *Systematic:* Fundamentally, science is systematic in nature. It concerns the way of gaining insight, the exploitation of knowledge, the way in which the empirical cycle is carried out as well as the way in which the reconstruction takes place.
- *Logic:* Besides logical ones, also literary and pictorial reconstructions of reality are possible. They are not considered to be scientific, although they may have a deep impact on science. Rational formulation is a formal criterion of science.
- *Reconstruction:* This implies a difference between image and reality. They are not simply identical. Science is directed towards theory construction and uses this in further investigations.

- *Empirical reality:* Science limits itself to observational realities. Empirical methods are necessary tools to attune facts and theories as much as possible.

1.1.2 Technology

Technology is a body of knowledge separate from; but related to the sciences. Technology is the application of tools, materials, processes and systems to solve problems and provide benefits to humanity. In other words, technology can be regarded as an attempt to improve our environment and the standard of living. The relationship between science and technology is one where science builds principles or theories and technology provides the practical application of those principles or theories. In a simple sense technology is the output of applied research. Most of the definitions on technology reveal the same idea; for the *Compact Oxford English Dictionary* defines technology as: "the application of scientific knowledge for practical purposes"

Knowledge of content, process and skills in technology include these components [2]:

- Methods of designing and developing solutions.
- Standards for selecting and using appropriate materials, tools and processes.
- Experimental and design specifications for testing and evaluating solutions.
- Criteria for judging the performance and impact of the solutions.
- Evaluating the impact of modifying a system to improve performance.

Another common idea regarding the technology is that it is composed mainly of artefacts and techniques [3].

Techniques are systems of abilities and rules that are useful in solving problems. Various numerical techniques for solving differential equations, cloning technique, etc., are examples.

Artefacts are objects that result from the transformation of other objects through the application of techniques. Artefacts are created through intentional human activities. The transformation of a polished sharp stone into a knife through intentional actions results in the artefact *knife*. We are surrounded by artefacts: television, clock, calculator, computer, vehicles, etc., are artefacts.

1.1.3 Techno-science

Twentieth century is marked by the emergence and growth of a new kind of system which is often regarded as the *techno-science* where the science and technology is indissolubly merged together. Such a system requires the collaboration of scientists, technologists, managers and investors. As Javier Echeverria, (2003) observed these systems require enormous amount of money for their financing, and therefore are usually associated with economically powerful groups and institutions [4]. Examples for techno-scientific systems can be found in the areas of nuclear research, space exploration, etc.

Leon Olive in his book *History and Philosophy of Science and Technology, Vol. III: An Introduction to Ethics of Science and Technology* observes that techno-scientific systems are systems of intentional actions that are guided by belief, norms, values and rules and that have scientific and technological foundations. They are oriented towards the achievement of certain aims. They are related to research systems and institutions. Olive further clarifies that techno-scientific systems, like scientific ones, seek to describe, predict or explain facts.

Even though techno-science has undergone enormous growth during the past few decades, now the scenario is changed; we see the coexistence of scientific systems, technical systems, technological systems and techno-scientific systems.

1.2 MEANING OF RESEARCH

Research refers to search for knowledge. Research is organized learning, looking for specific things to add to our store of knowledge. Research is defined as *a scientific investigation of phenomena which includes the collection, presentation, analysis and interpretation of facts that link man's speculation with reality* (Calmorin 1994). Obviously, this can be considered a general statement on research. Research is advanced through many stages like searching of resources, analyzing ides, drawing inferences, etc. The aim of research is to study the unknown and to discover useful things, answer unanswered questions or create that which doesn't currently exists. This results from the curiosity and interest on the part of the scientist, the output being the discovery and creation of knowledge.

According to Clifford Woody "research comprises defining and redefining problems, formulating hypothesis or suggested solutions; collecting, organizing and evaluating data; making deductions and reaching conclusions; and at last carefully testing the conclusions to determine whether they fit the formulating hypothesis". According to Leedy (1989) "research is a *systematic* quest for undiscovered knowledge". The stress is given on "systematic" that reveals the nature of research: planned, organized and has a specific goal. In fact, research is a never-ending process, as many scientists observed, discoveries and creations lead to new discoveries and creations [5]. Some more definitions of research are given below.

"Research is a process of steps used to collect and analyze information to increase our understanding of a topic or issue". —**Creswell**

"In the broadest sense of the word, the definition of research includes any gathering of data, information and facts for the advancement of knowledge".

—**Martyn Shuttleworth**

"Research is an honest exhaustive, intelligent searching for facts and their meanings or implications with reference to a given problem. The product or findings of a given piece of research should be an authentic, verifiable and contribution to knowledge in the field studied." —**P M Cook**

"a studious inquiry or examination; especially: investigation or experimentation aimed at the discovery and interpretation of facts, revision of accepted theories or laws in the light of new facts, or practical application of such new or revised theories or laws". —**The Merriam-Webster Online Dictionary**

Realizing the importance of research for the development of world and humanity various countries have established organizations and launched joined projects. The main objective of such organizations is to provide opportunities, to exchange ideas, information, skills and technologies, and to collaborate on scientific and technological endeavours of mutual interest that can translate the power of science for the benefit of mankind at large. For example, the Indo-US Science and Technology Forum (IUSSTF), established under an agreement between the governments of India and the United States of America in March 2000 is an autonomous society that promotes and catalyzes Indo-US bilateral collaborations in science, technology, engineering and biomedical research through substantive interaction among government, academia and industry.

The activities of such collaborations and forums includes [6]:

Foster scientific excellence by capitalizing on the scientific and technological synergy.

- Disseminate information and create awareness through scientific exchange.
- Build linkages through networking between academia and industry.
- Explore new frontiers by nurturing contacts between young and talented scientists and technologists.
- Pave way to sustainable interactions and build long-term relationships.
- Encourage public-private partnership to foster elements innovation and entrepreneurship.

For undertaking good research works, scientist must be motivated. A number of factors are responsible for creating motivation among the scientific community. They include:

- Curiosity about new things.
- Satisfaction in facing challenging problems.
- Desire to offer service to the society.
- To get a research degree and a good career.
- To get respect from others.
- Interest in practicing the acquired knowledge.

This list is not complete. One can find many other factors that motivate a person to undertake research.

1.3 CHARACTERISTICS OF RESEARCH

The widely agreed upon characteristics of research are:

- Empirical: Research is based on direct experience or observation by the researcher.
- *Logical:* Research is based on valid procedures and principles.
- *Cyclical:* Research is a cyclical process because it starts with a problem and ends with a problem.
- *Analytical:* Research utilizes proven analytical procedures in gathering the data, whether historical, descriptive or experimental and case study.
- *Critical:* Research exhibits careful and precise judgement. A higher level of confidence must be established.
- *Methodical:* Research is conducted in a methodical manner without bias using systematic method and procedures.
- *Replicability:* The research design and procedures are replicated or repeated to enable the researcher to arrive at valid and conclusive results.

Apart from these characteristics Yogesh Kumar Sigh in his book Fundamentals of *Research Methodology* and Statistics points out the characteristics of research that were extracted from various definitions of research. They are [7]:

- It gathers new knowledge or data from primary or first-hand sources.
- It places emphasis upon the discovery of general principles.
- It is an exact systematic and accurate investigation.
- It uses certain valid data gathering devices.
- It is logical and objective.
- The researcher resists the temptation to seek only the data that support his hypotheses.
- The researcher eliminates personal feelings and preferences.
- It endeavours to organize data in quantitative terms.
- Research is patient and unhurried activity.
- The researcher is willing to follow his procedures to the conclusions that may be unpopular and bring social disapproval.
- Research is carefully recorded and reported.
- Conclusions and generalizations are arrived at carefully and cautiously.

1.4 TYPES OF RESEARCH

Research is a scientific study by means of logical and systematized techniques. It aims to discover new facts, new scientific tools, concepts and theories. It consists of identifying the problem, formulating a hypothesis, collecting the data, testing the hypotheses and reaching certain conclusions, either in the form of solution towards the identified problem or in certain generalizations for some theoretical formulation.

In a broad sense research is classified into basic and applied research.

Basic research: Research that is carried out for gathering knowledge is called basic, fundamental or pure research. One can notice the following characteristics of basic research.

- It is mainly concerned with generalizations and with the formulation of a theory.
- It is about finding out what is already there without any greater purpose of research than the explanation itself.
- It is a direct descendent of philosophy, where philosophers and scientists try to understand the underlying principles of existence.

The main motivation of basic research is to expand knowledge and it doesn't have immediate practical use or commercial impact. It forms the foundation for applied research. Basic research often has indirect benefits, which can contribute greatly to the advancement of humanity.

Innumerable examples for basic research can be seen in the history of science:

- Newton's study of motion and gravity.
- Einstein's research on gravity and relativity.
- Studies on optical phenomena.
- Research on the structure of atom.
- Theories on the nuclear structure.
- Scientist's study on human eye.
- Huygens study on electromagnetic waves.
- Darwin's study on the evolution of life on the earth.

Applied research: Applied research focuses on the solution of practical problems rather than the acquisition of knowledge. The problem is addressed using well-known theories and principles. When scientists perform applied research they know that there is immediate practical use for the knowledge. The goal of applied research is to improve the living standard of humanity. This research is generally conducted on large-scale basis, it is expensive also. Such research is often conducted with the support of some financing agency like government, public corporation, World Bank, UNICEF, UGC, DRDO, etc. Hunt defines applied research as "an investigation for ways of using scientific knowledge to solve practical problems". It tends to be descriptive, rather than exploratory and is often based upon pure research. Applied research imparted its influence on every phase of human life. Some examples are:

- Treatment of a specific diseases.
- Development of optical instruments.
- Development of transistor and other electronic devices.
- Development of optical communication systems.
- Nuclear power plants.
- Improvement of agricultural crop production.
- Improvement of energy savings in home appliances.

Applied research depends on the basic research as the basic research provides the underlying ideas to solve a specific problem. For example, the development of a nuclear reactor or a nuclear weapon requires an understanding of nuclear physics. Similarly, the development of a laser instrument for medical applications requires knowledge of the effect of interaction of laser with the skin and other tissues which can only be provided by basic research. Applied research is often carried out by academic or industrial institutions. Often, an academic institution such as a university will have a specific applied research programme funded by an agency such as UGC, DRDO, etc., interested in that programme.

Applied research is further classified as problem oriented research and problem solving research. In problem oriented research a company or an organization undertakes a research programme for solving the problem faced by all the companies or all over the world. For example, WTO does problem oriented research for developing countries. In India agriculture and processed food export development authority (APEDA) conducts regular research for the benefit of agricultural industry. Problem solving research is carried out by an individual company or institution to solve a specific problem faced by them. The research carried out by various companies to increase the performance or customer satisfaction level of their product is problem oriented research.

A further classification of research is possible based on the objectives of research. They are:

- Descriptive research
- Correlational research
- Explanatory research
- Exploratory research
- Expository research

Descriptive research describes systematically a situation, problem or phenomenon or provides information about, say, spreading of a particular disease, or describes attitudes towards an issue. In social science and management research, the term ex post facto research is used for descriptive research. An important characteristic of this method is that it doesn't allow the researcher to control the variable. Survey method is utilized in descriptive research.

Correlational research attempts to discover or establish the existence of a relationship or interdependence between two or more aspects of a situation. The variables can be controlled or modified by the researcher. The studies on the factors affecting the blood pressure or blood sugar, research on the various factors affecting the atmospheric temperature, etc., are examples.

Explanatory research attempts to clarify why and how there is a relationship between two or more aspects of a situation or phenomenon.

Exploratory research is undertaken to explore an area where little is known or to investigate the possibilities of undertaking a particular research. The objective of exploratory research is the development of hypotheses rather than their testing. A good example is the exploration of space.

Expository research is based on the existing information only. The output of such a research is usually a review report. This requires a wide reading of the available information. Then by comparing, analyzing and synthesizing all information about a particular problem, the scientist arrives at new conclusion or insights.

Apart from the above-discussed classification research is classified as: analytical, conceptual, empirical, qualitative, quantitative and historical research. A good description of all these types of research can be seen in Research Methodology: Methods and Techniques, C R Kothari (2004.).

1.5 IMPORTANCE OF RESEARCH ACTIVITIES

The importance of research became evident through the famous Hudson Maxim; which says: “All progress is born of inquiry. Doubt is often better than overconfidence, for it leads to inquiry, and inquiry leads to invention”.

Research is the backbone of the development of every country. It is important in both scientific and non-scientific fields. It can be seen that most of the developed countries utilize a large part of their capital income for the research activities. It is intended to uplift the living standard of humanity. The thrust area of research includes agriculture, natural resources and renewable sources of energy, semiconductor technology, communication, space research and defense. As C R Kothari says; “Research provides the basis for nearly all government policies in our economic system. Through research we can devise alternative policies and can as well examine the consequences of each of these alternatives” [8]. Therefore, it became the duty of the government to chalk out programmes for research activities in various aspects of country’s development. Our government realized the importance of science and technology for the development of the country and it is reflected in the five-year plans. Massive support is provided to basic research, especially to universities, so that our nation can contribute significantly towards science and technology.

The increasingly complex nature of various problems faced by the humanity demands a more dedicated research in all areas. For example, the biggest problem that we face in near future is the energy crisis. Fossil fuels will be depleted within a few decades. The scientific community is behind an alternative to fossil fuels. Large-scale research is going on all over the world in harnessing the solar energy and utilizing hydrogen as a fuel.

S. Rajasekar et al., identify the importance of research activities in various fields through the following avenues [9]:

- A research problem refers to a difficulty which a researcher or a scientific community or an industry or a government organization or a society experiences. It may be a theoretical or a practical situation. It calls for a thorough understanding and possible solution.
- Research on the existing theories and concepts help us identify the range and applications of them.
- It is the fountain of knowledge and provides guidelines for solving problems.
- Research provides basis for many government policies. For example, research on the needs and desires of the people and on the availability of revenues to meet the needs helps a government to prepare a budget.
- It is important in industry and business for higher gain and productivity and to improve the quality of products.
- Mathematical and logical research on business and industry optimizes the problems in them.
- It leads to the identification and characterization of new materials, new living things, new stars, etc.
- Only through research can inventions be made; for example, new and novel phenomena and processes such as superconductivity and cloning have been discovered only through research.
- Social research helps find answers to social problems. They explain social phenomena and seek solution to social problems.
- Research leads to a new style of life and makes it delightful and glorious.

The importance and significance of research can be understood through different viewpoints. An economist is interested in the economic structure of the country and is concerned with the distribution of resources. A social scientist studies the social relationship and addresses various problems in the society. For a nuclear physicist the main concern is the energy production through nuclear reactions. Therefore, it is evident that each and every area is equally important and the research activities in all these areas should be promoted equally.

A different observation on the significance of research is made by C RKothari based on the following viewpoints [10].

- To those students who are to write a master's or Ph.D. thesis, research may mean a careerism or a way to attain a high position in the social structure.
- To professionals in research methodology, research may mean a source of livelihood.
- To philosophers and thinkers, research may mean the outlet for new ideas and insights.
- To literary men and women, research may mean the development of new styles and creative work.
- To analysts and intellectuals, research may mean the generalization of new theories.

1.6 PRINCIPLES OF QUALITY RESEARCH WORK

To ensure quality research work *Research Methodology: A self-Learning Manual* [11] suggests the observation of the following principles.

- *Objectivity:* The purpose of research should be clearly defined and common precept be used
- *Usage of scientific procedure:* The research procedure used should be described in sufficient detail to permit other researchers to repeat the research in a systematic manner.
- *Continuity:* The research should be carried out in such a manner that principle of continuity is ensured.
- *Proper Planning:* The design should be planned in a most scientific way and all aspects of resources, time frame, constraints and procedural aspects be taken into consideration.
- *Integrity:* The researcher should report with complete frankness, flaws in procedural design and evaluate their effect on the findings.
- *Adequacy of Data:* The analysis of data should be sufficiently adequate to reveal its significance and the methods of analysis used should be appropriate.
- *Reliability:* The validity and reliability of data should be checked carefully.
- *Structure:* It means that research is structured with specified sequence in accordance with the well-defined set of rules. Guessing and intuition in arriving at conclusions are rejected.
- *Logic:* This implies that research is guided by the rules of logical reasoning and logical process of induction and deduction are used.
- *Empiricism:* This implies that research is related basically to one or more aspects of a real situation and deals with concrete data that provide a basis for external validity of research results.

- *Replicability:* This principle allows research result to be verified by replicating the study and thereby building a sound basis for decision.
- *Economics:* Research should be completed within the allocated financial resources
- *Time Frame:* the research should be completed in the stipulated time frame.

1.7 PROBLEMS IN RESEARCH

Conducting research is not an easy task. A number of favourable conditions are to be met for the smooth and effective conduct of research. This is because research is not an individual task; various agencies are involved in a particular research programme. The smooth conduct of research in various fields is often obstructed by various elements. They include:

- *Inadequate training:* Majority of the researchers are not equipped with the knowledge of scientific methodology. It is essential that all the researchers should have knowledge of different scientific methods used in research.
- *Lack of confidence:* Improper understanding of the research problem as well as the scientific methods lowers the confidence level of the researcher.
- *Lack of interaction:* The interaction between researchers and various universities are important for effective completion of a research programme. Unfortunately majority of the researchers and universities are reluctant to communicate the information and share the resources.
- *Repetition:* This results from improper interaction among the scientific community. Overlapping of research studies occurs that results in wastage of money and time.
- *Deficiency in fund:* For conducting quality research adequate funding is required. Many of our research institutions do not have adequate support from the funding agencies.
- *Lack of coordination:* Proper coordination of the researchers, institutions, funding agencies, etc., is needed for effective research. The lack of coordination among these elements degrades the quality of research.

As far as the developing countries are concerned these problems are more intense. However, after the Second World War, in developing countries, there has been intense programmes concerned with training and developing research activities. In many countries the research was primarily concentrated on agriculture to provide the basic need of their people. Many programmes, nationally and internationally were launched under national, bilateral or UN agencies. Some countries concentrated on a variety of industrial research programmes. As a result of such serious approaches, a great advance has been achieved by many countries. However, there are a number of problems and limitations that have been facing by the developing countries all these years. A few of the major problems that decelerate the research activities in developing countries are discussed below [12].

Human resource management: At the initial stage one of the major problems that persisted in the developing countries was the unavailability of sufficiently educated people. Therefore, it became the primary duty of the government to provide education for its people and equip them for a conduct of scientific research. In recent days this scenario has changed with the improved facilities for training at graduate level. However, many developing countries do not have adequate training programmes at graduate and postgraduate level. This has led to the problems of reassimilation on return home. A few got the opportunity to study and train abroad through government sponsorships, scholarships or at their own expense. Many such people found their career in the developed countries because many developing countries failed to provide sufficiently paid employment and research facilities when they return to their own country. In order to solve this problem the improvement of the status and facilities in the developing countries is essential. Some countries opted the policy that they should send only those scientists who already have jobs and an established commitment to an institution in the country, abroad for a training on a contract basis so that they should return to serve their country.

Unavailability of sufficient funding: Another problem that persists in the developing countries is that the unavailability of sufficient funding to carry out research of world standard. It became the responsibility of scientists to convince the government about the significance of the research programme that was proposed by them and how it would help the development of the nation.

In most of the developing countries the research is application oriented. This is, of course, desirable to develop the facilities available in such countries. It is true that a research project with practical orientation will get funded more easily than a project in basic science. Also the government is careful in sanctioning projects that have long- term applications that could help the development of the country. Another common observation is that in developing countries there is not a broad spectrum of equal competence in all branches of science. A particular branch may be developed and there may be outstanding individuals in such areas. Such individuals and groups should be identified and should be provided with sufficient resources to carry out their research for the development of the country.

Poor research facilities: Scientific research is expensive. The cost of the equipment and other raw materials and the expense for their maintenance increases day by day. The interdisciplinary nature of scientific research made the situation still complicated because of the demand of wide range of sophisticated equipment. In developing countries many research institutions do not have sufficient facilities because the demand for resources far exceeds the funds available. Therefore, a serious and concentrated effort is needed for accelerating the scientific research programmes. It is desirable to identify certain centres of excellence to establish specialized facilities and equipment and to utilize them most efficiently. For example, as physics has become important for solving biological problems a medical research centre should have a well-developed physical research laboratory.

Lack of proper communication: Scientists in every research area should be up-to-date in their knowledge. They should always be in contact with the mainstream ideas. In many of the developing countries this is a problem. Very often scientists do not get opportunities for exchange of knowledge and other developments. They are not intellectually stimulated. This problem can be overcome; to a certain extent, through attending conferences and seminars of international standards. Such international meetings can renew contacts and the scientists can observe the developments that occurred all around the world. Some countries provide fellowships for short-term and long-term visits at well-established laboratories in the developed countries. Many international agencies such as International Atomic Energy Agency, Food and Agricultural Organization and various UN agencies, etc., are awarding a number of international fellowships and can be utilized by scientists in the developing countries.

1.8 RESEARCH IN ACADEMIC FIELDS

The term *academic discipline* is usually used to denote a wide range of subjects taught in universities and colleges. In the Indian context, the majority of the research works are taking place in educational institutes such as IITs, IISERs, universities, and colleges. In such institutes, the research works are conducted in phases with other programmes. In earlier stages, the academic disciplines are broadly classified as science, social science, humanities, and commerce. Recently, it has been realized that such classification is more or less meaningless and in many research areas, more than one disciples need to work in harmony. Therefore, most of the institutes promote interdisciplinary research more than research in pure subjects. For example, 'Econo-Physics' is one of the most recently developed research area, where the principles of physics are applied in the field of economics. In a similar way, linguistics and psychology come together in the field of 'psycholinguistics'.

Research in the academic fields is very important since it provides a vast opportunity for students from various disciplines to start a research career after their postgraduation. In Indian universities and other academic research institutes, admission for a research programme is through an entrance test conducted by the institute. However, a student who qualifiesfor JRF or NET is usually exempted from such a test. As India is a fast developing country, and central and state governments are promoting research in various disciplines at a large scale, there are a good number of opportunities for researchers in India.

1.8.1 Undergraduate Research Programme

The Council on Undergraduate Research (CUR) defines "undergraduate research" (UR) as "an inquiry or investigation conducted by an undergraduate student that makes an original intellectual or creative contribution to the discipline." CUR is the leading body that oversees and affiliate institutes to perform undergraduate research on campuses. The concept was introduced in 1969 by Margaret McVicar, who was the Dean of Undergraduate Education at the Massachusetts Institute of Technology (MIT), under the name of Undergraduate Research Opportunities Programme (UROP). In this programme, an undergraduate student in any discipline assists a faculty, researcher, graduate student, and/or other undergraduates in research in areas of similar interests. Undergraduate students support collaborative research by either pursuing their own research ideas or joining established research projects.

The most common ways of incorporating undergraduate research experience in institutes are:

- Undergraduate Research Experiences (URE), and
- Course-based Undergraduate Research Experiences (CURE)

In URE, a handful of students are selected to assist a mentor in their research in a laboratory, working over a few semesters. The students are selected through competitive tests and a perusal of their academic records. Meanwhile, CURE is a class-based programme presided over by a researcher or a graduate student and consists of lectures, grades, and assignments. The CURE are firmly integrated into undergraduate courses to enhance student learning. If students are systematically taught competent research at the undergraduate level, they will get interested in the subject and might become more inclined to take up research-intensive academic programmes and careers in the future. With this aim, research activities are introduced in the UG curriculum in Indian universities.

1.8.2 Improving Academic Research

In 2020, the Association of Indian Universities, New Delhi, published a book titled "Reimagining Indian Universities" [13]. In this book, Dr. Talat Ahmad, the Vice-Chancellor of the University of Kashmir, Srinagar, suggested ten remedial measures to improve the quality and outcome of research activities in Indian universities. These suggestions address many problems faced by Indian universities and other affiliated colleges that undertake research activities. A brief discussion of various remedial measures to improve research activities in universities and academic institutions in India is given below.

Funding for Research: Proper funding is the backbone of any research programme. It is very important to explore possibilities of funding for the research programme at regional, national and international levels through various sources such as public institutions, private industries, Non-Governmental Organizations (NGOs) as well as through opportunities for public-private partnership. There are many national-level funding agencies like the Department of Biotechnology (DBT), Department of Science and Technology (DST), University Grants Commission (UGC), Indian Council of Agricultural Research (ICAR), Science and Engineering Research Board (SERB), Indian Council of Medical Research (ICMR), Council of Scientific and Industrial Research (CSIR), etc. Also, funding opportunities from the industry, academia, and NGOs need to be exploited to their fullest. In addition to these, international organizations such as World Health Organization (WHO), World Trade Organization (WTO), World Bank, United Nations Organization (UNO), etc., also provide research grants for research in relevant areas.

Infrastructure for Research: This is one of the major problems faced by many academic institutes when undertaking research activities. This is more dominant in the field of science and applied science research were instruments that costs crores of rupees are to be used. Availability of proper funding can solve the problem to some extent. It is essential to develop our research infrastructure for enhancing our research and improving our research quality drastically. In addition to the availability of a sufficient equipment, trained manpower is also required. Further, access to literature through databases like Scopus, Science Direct, Pubmed, Medline, Toxline, etc., must also be provided to undertake systematic literature reviews, meta-analyses, and other updates.

Manpower for Research: Qualified and trained manpower is an essential part of any research programme. For developing such manpower courses related to research, the methodology needs to be incorporated into the UG and PG curriculum at college as well as university levels so as to develop a research culture and harness a research temperament among our students and scholars. Even at the school level, the curriculum should be modified to promote creativity, critical, analytical, and innovative thinking among our students. 'Atal Tinkering Lab' is one such innovative idea introduced by the central government of India, for students in classes 6 to 9 to make them aware of hands-on experiences and research. Motivated students can undertake research activities with great passion.

Policies for Research: To promote research quality in tune with fast-changing global trends, we need comprehensive, goal-oriented, and focused research policies for enhancing the quality of research. Keeping this in mind, the National Education Policy (NEP) 2019, encourages the development of research universities and autonomous colleges for carrying out intensive research on thrust areas identified by the experts.

Incentives for Research: Quality of research must receive priority over quantity of research under all circumstances and the peers and experts in the field must evaluate the quality of research based on their experience and by using journal metrics like impact factor and citation index. Further, there is a lot of debate going on whether research output should be used for providing financial incentives to teachers, awarding fellowships, and giving recognition to them in their academics or it must be reserved only for the quality of new knowledge added to the existing body of literature. Beyond any doubt, it is essential that good research needs to be encouraged through appropriate incentives, making its use in appointments and promotions of teachers.

Collaborative Research: Recently, many research policies promote collaborative research where the integration of various disciplines, research facilities, and manpower work together to achieve a particular research objective. Depending upon the balance of influence between different participating disciplines, collaborative interdisciplinary research can be further classified into:

- cross-disciplinary research, and
- multidisciplinary research.

The main objective behind the collaborative research is to dissolve all boundaries between various disciplines, and knowledge needs to be liberated from all such confines and made freely available to all those who seek it.

Contextual, Need-based Research: Society needs fast solutions to many problems. Therefore, research must address such problems to satisfy the needs of society. In many developed nations, universities organize research fairs and establish research shops that allow community members to visit them and register their problems, based on which research projects are prepared and research is undertaken to find solutions to their actual problems. We also need to start such practices in our country by the research institutes, universities, and other academic institutions. Also, the research outputs must be communicated to society in time.

Translational Research: One of the major drawbacks of research activities in India is that the result is not getting translated to society properly. Merely carrying out research is of no use unless the results generated out of scientific inquiry are translated into a policy and the policy is practised and implemented to resolve day-to-day issues of the society. Therefore, it is very important to have greater coordination between researchers and policy-makers for incorporating research findings and recommendations into our administrative policies.

Integrity in Research: A researcher must maintain integrity in research. Therefore, young researchers need to be educated adequately about related issues like plagiarism, duplication, gift and ghost authorships, salami publications, outsourced publishing, etc. It is hard to achieve quality in research unless awareness about the importance of integrity, transparency, and accountability is promoted and incorporated as an inherent component of the research. Recently, under the pressure of API scores set by the UGC norms for career advancement, many teachers publish their articles in predatory journals and this should be countered and controlled effectively.

Ethics in Research: Ethics in research is very important for maintaining quality and high standards of research. Research ethics must address issues like confidentiality, privacy, non-malfeasance, beneficence, autonomy, and justice without which no credible and substantive research is possible. Awareness about these components needs to be created from the very beginning of the research. Every institution must constitute an Institutional Ethics Committee and it should be vigilant, proactive about any such malpractices in research. A detailed discussion of research ethics and the related topic is given in chapter 9.

1.9 SCIENTIFIC ATTITUDE

Attitude can be regarded as a feeling or a pattern of thoughts. Scientific attitude is a way of looking into things or way of approaching a problem. It is necessary to become a successful scientist. Scientific attitude helps one to think reasonably, logically and clearly without any prejudice. Scientific attitudes, according to Gauld and Aukins (2000), are a composite of a number of mental habits or tendencies to react consistently in certain ways to a novel or problematic situation.

According to Bhaskara Rao (1989) the most useful scientific attitudes are open-mindedness, critical-mindedness, respect for evidence, suspended judgement, intellectual honesty, willingness to change opinion, search for truth, curiosity, rational thinking, etc.

Majority of the articles on scientific attitude pointed out the following scientific attitudes.

- *Belief:* A scientist believes that everything that happens in this world has a cause or reason. He/she should not believe in superstitions.
- *Curiosity:* A scientist is curious about what happens around. That is he/she shows interest and pays particular attention to objects or events. He/she asks questions and seeks answers.
- *Objectivity:* A scientist is objective if he does not allow his personal feelings and biases to influence the recording of observations, interpretation of data, and formulation of conclusions.
- *Critical-mindedness:* A scientist draws conclusions based on evidence. When in doubt, he/she questions the accuracy of a statement in relation to the evidence presented.
- Open-mindedness: A scientist listens to and respects the ideas of others. He/she doesn't reject any knowledge that conflicts with his idea. He/she

accepts criticism and changes his mind if reliable evidence contradicts his beliefs.

- *Inventiveness:* The main objective of research is the creation of knowledge. A scientist is expected to generate new and original ideas.
- *Risk-taking:* A scientist expresses his opinions and tries new ideas even at the risk of failure or criticism. This can be regarded as another phase of open-mindedness.
- *Intellectual honesty:* A scientist gives a truthful report of observations. He/she does not withhold important information just to please himself/herself or others. He should acknowledge the source of information.
- *Humility:* A scientist is a human being and it is natural to commit errors in observations or in recording the observations. A scientist is humble when he admits that he is not free from committing errors. He/she recognizes that there may be better ideas and realizes that there are individuals whom he/she may have to consult to arrive at correct observations and conclusions.
- *Responsibility:* Society as well as the scientific community expects great contributions from a scientist. A scientist should actively participate in a task and also should dutifully perform tasks assigned to him/her.

It is the duty of every individual to realize that scientific attitude is an important aspect of a personality of someone who wants to be successful in the field of science and in life as a whole. The scientific attitude encourages truthfulness; it also inspires humility, together with open-mindedness. Scientific attitude can be carried forward in all aspects of our life–personal, social, economic, political and even religious. In many of situations science offers a more realistic and practical solution than religion. Therefore, it is important to inculcate the scientific attitude among the individual right from the beginning of their education because children are found to be more curious to know about various things they observe around themselves in their daily life.

Realizing the importance of science Jawaharlal Nehru said:

"It is Science alone that can solve the problems of hunger and poverty, in-sanitation and illiteracy, of superstition and deadening customs and tradition, of vast resources running waste, of a rich country inhabited by the starving people. Who indeed could afford to ignore Science today? At every turn we have to seek its aid. The future belongs to Science and to those who make Friends with Science."

The main objective of science teaching is the development of scientific attitude among the students. Therefore, the curriculum should be properly designed and the teachers should be properly trained. Through developing appropriate classroom atmosphere, making use of scientific literature, effective use of laboratory time, inclusion of science programmes as co-curricular activities and proper planning of exercises the impact of science teaching can be improved. An effective method of developing scientific attitude is the use of discovery approach to teaching, especially for science, because we usually understand most by doing experiments. Many of the attributes of scientific temper are also the universal human values which are important for the well-being of an individual as also the society. Inculcation of these values can itself be done through education by using scientific methodologies [14].

1.10 SCIENTIFIC TEMPER

Scientific temper refers to the mental attitude which is behind acquiring reliable and practical knowledge. As Subodh Mahanti (2013) observed the word scientific temper was first used in India by Jawaharlal Nehru. Nehru believed that the scientific temper should be the guiding principle in governing human actions. Through the word scientific temper a number of social reformers, scholars and scientists advocated the need to instil a spirit of scientific enquiry in the society [15].

Nehru first explained the concept of *scientific temper* in The Discovery of India, through the following points (Nehru, 1946:509-15):

- There is an element of inevitability about the applications of science and technology. However, mere applications of science and technology will not be a sufficient condition. What is needed is 'the scientific approach, the adventurous and yet critical temper of science, the search for truth and new knowledge, the refusal to accept anything without testing and trial, the capacity to change previous conclusions in the face of new evidence, the reliance on observed fact and not on preconceived theory, the hard discipline of the mind — all this is necessary, not merely for the application of science but for life itself and the solution of its many problems.'
- Scientific temper is the temper of a free man.
- Scientific approach should be an integral part of our social interactions, as expressed by the quote "The scientific approach and temper are, or should be, a way of life, a process of thinking, a method of acting and associating with life, a process of thinking, a method of acting and associating with our fellowmen."
- While we live in a scientific age, there is no evidence of scientific temper in the people or their leaders.
- Even scientists who practise science do not necessarily have scientific temper.

1.11 QUALITIES OF A GOOD RESEARCHER

Research is not a mechanical process. Active participation on the part of the researcher is necessary for the effective conduct of the research. Therefore, the qualities of the researcher have an upper hand in determining the success of a research programme. A research is undertaken by a researcher because of his curiosity or interest and complete dedication is required in the research work. A researcher should have good knowledge of the scientific methods. Also he should be conscious of the possible errors in observing and recording the data. The qualities of a good researcher include: creativity, efficiency, scientific in approach, activeness, resourcefulness, honesty and economics. Yogesh Kumar Singh (2006) elucidates the qualities of a researcher observing the following points [16].

- A researcher should have full understanding of the functions and activities of the research work.
- A researcher should have the reflective thinking about various dimensions of the research.
- A researcher should be sensitive towards research. A sensitive person can perceive the problem.
- A researcher should be creative and imaginative. These abilities are essential for formulating the action hypotheses for the problem.
- A researcher should have the knowledge and training of action research.
- A researcher should have insightful into the area of interest.
- A researcher should have the scientific attitude for studying and observing things.
- A researcher should have objective thinking.
- Researcher's behaviour should be democratic.
- The most important characteristics are the patience and pursuance.
- A researcher should have knowledge and skill of measuring instruments and elementary statistics.
- A researcher should have open mind to discuss the problems with colleagues and experts of the field and to get valuable suggestions.
- A researcher should be economical in designing the project from time, energy and money point of view.

1.12 SCIENTIFIC COMMUNITY

A community which studies and practises science is referred to as scientific community. It doesn't have a sociological label rather an intellectual gathering. Recently, scientific community is regarded as the basic units within which science is socially and technically organized. Most of the contemporary researchers operate with some notion of such communities. Karin D. Knorr-Cetina describes scientific community as [17]:

"In general, these communities are seen as small social systems with fuzzy borders and internal mechanisms of integration; these systems are frequently circumscribed by a specialty area represented in the scientific literature".

The concept of scientific community can be seen frequently in scientific literature and articles. Scientific community is a broad concept as Thomas Schott (1991) explains:

"The scientists within a country form a national scientific community, a community within the world scientific community. They enter the national scientific community through the relatively similar scientific education which they have undergone and their acquisition of the shared culture of science at an elementary level. They perform their research in the framework of national institutional arrangements for research such as universities with similar patterns, the same national associations and journals, supported by the same national foundations and the same bodies which set the national science policies; thus, they perform their research within a common institutional and intellectual setting. All this prepares them for participation in the more differentiated and specialized traditions of scientific knowledge and research, and for personal connections which are more intense with colleagues who are nearby working on similar subjects and problems, and also with more remote colleagues" [18].

The scientific community in a country is responsible for the scientific activities performed by various agencies and institutions. It signifies the production and advancement of scientific knowledge. The basic indicator for the existence of a scientific community; as Jacques Gaillard, V.V.Krishna and Ronald Waast observed in the *Scientific Communities in the Developing Dorld;* is the steady state production of knowledge in special fields or sub-fields of science. The constitution of local, national disciplines, creation of university chairs, systems of national recognition and rewards, higher specialized educational structures for creating neophytes, full-time specialized research structures, and areas and networks of scientific research and communication with corresponding professional societies and journals signify the existence of local, national communities [19].

A national scientific community cannot be characterized simply by considering the scientific output produced by the members of that community within the boundaries of the nation. This is because a national scientific community is not limited to the members present in that country. A scientist can contribute to his national scientific community while working abroad, as he may be in a better working atmosphere, in terms of facilities; than available in his country. Some problems may need international attention and cooperation. Local or national scientific communities alone cannot address such problems. In such situations coordination of various national scientific communities all over the world and a collective contribution is required.

The idea of scientific community attracted the attention of the world after the publication of *The Structure of Scientific Revolution* by Thomas Kuhn (1962). According to Kuhn scientific community is strongly associated with a paradigm. A group of scientists working on a particular paradigm constitutes a scientific community. (Paradigm in a simple sense represents a theoretical framework based on which a scientist can do his research. This will be discussed in detail in the next chapter). Straun Jacobs (2006) explains Kuhn's observation; members of a scientific community of a particular scientific specialty have, as Kuhn sees it, been similarly educated and have undergone much the same professional initiation through higher degree research. They have studied the same specialist literature, drawing similar conclusions from it, and have learnt a common language. The typical Kuhnian scientific community is characterized by homogeneity, consensus and – scientific transactions and communication occurring between, but seldom beyond, scientists of the same community – well defined boundaries [20]. Researchers of the community utilize the resources of the paradigm and apply their inherent abilities to arrive at a solution to the problem.

In the present era scientific communities have various responsibilities in promoting scientific attitudes and research activities. Wensing et al. (2012) points out the necessity for the formation of a scientific society (even though the article is about healthcare, it can be well applied to other fields of science also) with the following objectives and responsibilities [21]:

- To provide an international forum for scientific exchange, which may take the form of traditional conferences, focused working groups, and online discussion networks. This would contribute to the quality of the science and also help individuals to develop an identity as a researcher in this field. It may also be possible to establish explicit links to dedicated scientific journals in the field.
- To announce relevant existing training and development opportunities in the field and possibly also facilitate such training programmes, particularly for young researchers. A journal, website, or online mailing group would be needed for this purpose. Providing information on relevant conferences, funding opportunities and job vacancies could be further services provided by the scientific society.
- Help researchers (in any field) to disseminate their research findings more effectively to stake holders and raise awareness about the determinants of improvements in various spheres of scientific research, management, and policy.

- Promote and lobby for our science across the world, particularly in countries where it is currently underdeveloped, so that it reaches a critical mass, which is needed to survive and flourish. In particular, a society could help develop core competency for scientists, lobby for career paths in academic and scientific structures for researchers, and lobby for sustained research funding to support implementation science.

1.13 NON-SCIENCE AND PSEUDOSCIENCE

The "fields of study" can be generally classified into science, non-science and pseudoscience. In previous sections we have seen the meaning and the importance of science in detail. Non-science is again a system in which knowledge is gained through research. Non-science refers to inquiry, academic work or disciplines that do not involve the process of empirical verification to generate knowledge. Non-sciences also are ways in which humans try to interact with and understand the universe. Non-sciences can easily be identified through their nature; they usually involve creative processes, the acceptance of subjective knowledge, belief in establishing their epistemological basis. Mathematics, logic, philosophy, metaphysics, religion, music, and art are all non-sciences. It is to be noted that there may be some scientific aspects used in these human endeavours, but they are not a part of science or of scientific thinking. Jeffrey A Lee (2000) says "mathematics and logic are closely related to science but are not necessarily based on real-world phenomena; they seek to explain the outcome of situations, but the situations do not have to exist in reality" [22]. Also Lee classifies religion, ethics, morality, etc., as belief fields which are based on faith in what is right and what is wrong and they are not necessarily based on reality.

Non-scientific theories attempt to explain non-empirical concepts or values. As theories they are not inferior to scientific theories and therefore must be evaluated under different criteria. Non-scientific theories are sometimes called conceptual theories and are tested by their utility, logical consistency and consistency with our knowledge and beliefs. A scientist should recognize the difference between scientific and non-scientific theories, and appropriate methods should be selected for the investigation and evaluation of each.

Pseudoscience: Pseudoscience is sometimes difficult to identify; but can be imagined as a discipline that pretends to be scientific. Pseudoscientific practices are often presented as scientific, without supporting evidence, and employ non-scientific methods. It cannot be reliably tested or verified.

According to Mario Bunge "pseudoscience is a body of belief and practices but seldom a field of active inquiry; it is tradition-bound dogmatic rather than forward looking and exploratory" [23]. Carl Sagan (1997) says, "...hypotheses are often framed precisely so they are invulnerable to any experiment that offers a prospect of disproof, so even in principle they cannot be invalidated. Practitioners are defensive and wary. Skeptical scrutiny is opposed." [24].

Pseudoscience is more often characterized by vague, exaggerated or unfalsifiable claims. (Karl Popper's idea of falsification will be discussed in the next chapter). Therefore, the theories of pseudoscience cannot be scientifically tested.

Pseudoscience can be identified based on the following criteria:

- Use of vague, exaggerated or untestable claims.
- Over-reliance on confirmation rather than refutation.
- Lack of openness to testing by other experts.
- Use of misleading language.
- Absence of progress.
- Personalization of issues.

A number of pseudoscientific practices exist worldwide; astrology, homoeopathy, acupuncture, graphology, palmistry, telepathy, etc., are examples. Popper placed Marxism and Freudian psychology in the domain of pseudoscience.

Barry L Beyerstein (1995) identifies the following characteristics of pseudoscience [25].

- *Isolation:* Pseudosciences are isolated from the mainstream research. They do not value or promote the link between data and theory from other applicable areas of inquiry.
- *Non-falsifiability:* Most of the theories in pseudoscience are non-falsifiable because they are not stated in a testable way.
- *Misuse of data:* Pseudoscientists usually misuse the available scientific data. Beyerstein cites the example of the advocates of bogus science; they usually start with a position that the opponents will readily concede. They then add contentious and unsupported arguments until the shred of truth with which they began can no longer supports the edifice that has been constructed upon it.
- *Pseudosciences are not cumulative and self-correcting:* Pseudosciences show little evidence of progress. They are stagnant and exhibit little of the intellectual excitement. Pseudoscientists rely on "sacred texts", the ideas of which, as they think, are not to be questioned or modified.
- *Special pleading:* Pseudoscience wants both the prestige of an established field and release from the strict rules of evidence in proving a theory.

Pseudosciences usually mislead the humanity. In a scientific world it is essential to create awareness about pseudosciences among people. For this a thorough understanding of science and scientific methods is required. Such an understanding demands a thoughtful skepticism. According to *Webster's*

Collegiate Dictionary "skepticism is the method of suspending judgement pending investigation; it includes inquiry, questioning, criticism and doubt". A skeptic demands reasonable evidence and logical justification before the approval of any claim. Further, he is always open to more definitive evidence and modifies his beliefs accordingly.

REVIEW QUESTIONS

1. What is science and what are its characteristics?
2. How the science and technology are related?
3. What are the characteristics of science?
4. What are the components of technology?
5. Write a note on techno-science.
6. Explain the meaning of research.
7. Why the collaboration of various groups is important in research?
8. What are the main activities of collaborating groups in research?
9. Explain the characteristics of research.
10. Give the characteristics of research extracted from various definitions.
11. Explain the characteristics of the basic and applied research.
12. Discuss the classification of research based on the objectives.
13. Explain the various importance of research activities.
14. What is the significance of the research?
15. Give the principles of quality research work.
16. Discuss the main problems in research work.
17. Give an account of the research in academic fields.
18. Write a note on the undergraduate research programmes.
19. Explain how the research in academic fields can be improved.
20. What are cross-disciplinary research and multidisciplinary research?
21. What is a scientific attitude? Give an account of various scientific attitudes.
22. Explain scientific temper.
23. How the scientific attitude and scientific temper are related?
24. What are the qualities of a good researcher?
25. What is a scientific community?

26. Explain the role of the scientific community in spreading the scientific attitude.
27. What are the responsibilities of a scientific community?
28. What are non-science and pseudoscience? How can we distinguish them?
29. How can we identify pseudoscience?
30. What are the characteristics of pseudoscience?

REFERENCES

[1] Michael P. Clough and Jerrid W. Kruse, "Characteristics of Science: Understanding Scientists and their Work", www.storybehindthescience.org

[2] Indo-US Science and Technology Forum, http://www.inodusstf.org/

[3] Leon Olive, *History and Philosophy of Science and Technology – Vol. III: An introduction to Ethics.*

[4] Javier Echeverria, *The Technoscientific Revolution*, 2003.

[5] Wayne Goddard and Stuart, *Research Methodology: An Introduction*, Juta & Co. Ltd (2006), p. 1.

[6] Indo-US Science and Technology Forum, http://www.indousstf.org/

[7] Yogesh Kumar Singh, *Fundamentals of Research Methodology and Statistics*, New Age International (2006), p. 4.

[8] C R Kothari, *Research Methodology-: Methods and Technique*, New Age International (2004), p. 5.

[9] S.Rajasekar, P. Philominathan, V. Chinnathampi, Research Methodology, (2006), p. 2.

[10] *Research Methodology- A self-Learning Manual*, p. 29, http://www.iilm.edu/iilm-online/Research%20Methodology%20Self-Learning%20Manual.pdf

[11] C.R Kothari (2004), p. 7.

[12] P. B. Vose and A. Cervellini, "Problems of Scientific Research in Developing Countries", IAEA BULLETIN, Vol. 25, No. 2, pp. 37-40.

[13] Pankaj Mittal and Sistla Rama Devi Pani "Reimagining Indian Universities" Association of Indian Universities (2020).

[14] P.L. Dhar, *Developing Scientific Temper*, p. 16.

[15] Subodh Mahanthi, "A Perspective on Scientific Temper in India", *Journal of Scientific Temper*, Vol. 1, January 2013, pp. 46-62.

[16] Yogesh Kumar Singh (2006), p. 12.

[17] Karin D. Knorr-Cetina, "Scientific Communities or Transepistemic Arenas of Research: A Critique of Quasi-Economic Models of Science" *Social Studies of Science* 12 (1982), pp. 101-130.

[18] Thomas Schott, (1991). 'The World Scientific Community: Glob ability and Globalization'. Minerva, 29, 442.

[19] Jacques Gaillard, V.V. Krishna and Ronald Waast, *Scientific Community in the Developing World*, Sage Publications (1997), p. 19.

[20] Struan Jacob, "Models of scientific community: Charles Sanders Peirce to Thomas Kuhn", *Interdisciplinary Science Reviews*, 2006, Vol. 31, No. 2, pp. 163-173.

[21] Wensing et al. "Does the World Need a Scientific Society for Research on How to Improve Healthcare?" *Implementation Science* 2012, 7:10

[22] Jeffrey. A.Lee, *The Scientific Endeavor- Methodology and Perspectives of Science*, Pearson India (2000), p. 2.

[23] Mario Bunge, 1984, *What is Pseudoscience*, The Skeptical Inquirer, 9, pp. 36-46.

[24] Carl Sagan, The *Demon-Haunted World: Science as a Candle in the Dark*, Headline Book Publishing (1997), p. 25.

[25] Barry. L. Beyerstein, *Distinguishing Science from Pseudoscience*," pp. 30-31.

Suggested Readings

1. James E. McClellan III and Harold Dorn, *Science and Technology in World History*, The Johns Hopkins University Press, Baltimore (2006).
2. Bryan Bunch and Alexander Hellemans, *The History of Science and Technology*, Houghton Miffun Company, New York (2004).
3. Sergio Sismondo, *An Introduction to Science and Technology Studies*, Wiley-Blackwell (2010).
4. I.R.Walker, *Reliability in Scientific Research,* Cambridge University Press (2011).
5. Finlay MacRitchie, *Scientific Research as a Career*", CRC Press, New York (2011).
6. Ad Maas and Hans Hooijmaijers, *Scientific Research in World War II: What Scientists Did in the War,* Routledge (Tylor and Francis Group), London (2009).

Chapter 2

Philosophy of Science

All men by nature desire knowledge.

—**Aristotle**

Man, being the servant and interpreter of Nature, can do and understand so much and so much only as he has observed in fact or in thought of the course of nature. Beyond this he neither knows anything nor can do anything.

—**Francis Bacon**

2.1 INTRODUCTION

This chapter is intended to provide a brief introduction to the philosophy of science. The question "what is philosophy of science" cannot be answered in a straightforward way. Right from the beginning of the scientific investigations various philosophers like Aristotle, Francis Bacon, Bertrand Russell, Karl Popper and Thomas Kuhn put forward various philosophical theories. A scientist adopts various methods for solving a problem. The main function of philosophy of science, as one could see, is to analyze the scientific methods. Philosophy of science presents a "view" and argues that the scientific theories should be based on these views. In a sense philosophy of science always questions the scientists. As John Losee points, "philosophy of science is a discipline in which the concepts and theories of the sciences are analyzed and clarified" [1]. The philosophy of science analyzes the characteristics of scientific inquiry, scientific procedure, conditions satisfied by the scientific explanations and cognitive status of scientific laws and principles [2]. Thus, philosophy of science is thinking over science and scientific methodology and it is different from doing science.

Philosophy of science is broadly classified into two:

- Epistemology, and
- Metaphysics.

Epistemology deals with the scientific claims and metaphysics deals with the philosophical features of the world described by the science. That is, epistemology is the reality and metaphysics refers to the possibility. Epistemology deals with the problem that occurs when practising science. James Ladyman defines epistemology as [3]:

"the branch of philosophy that inquires into knowledge and justification"

Even though philosophy is different from science, which is based on real knowledge gained through solid observation and experience; it is believed that philosophy is required to serve as an intellectual foundation for science. Many questions like "what is space?" can be treated philosophically as well as scientifically. The studiy of different philosophies of science are important as it provides a deeper insight into the assumptions of scientific practice.

2.2 SCIENTIFIC STATEMENT

Science is the study of nature and natural phenomena. A scientist always explains the facts about the real-world phenomena. Such a statement that attempts to describe a real world phenomenon is called a *scientific statement.*

EXAMPLES

Velocity is the rate of change of displacement.

Sun rises in the east.

Violet has the least wavelength in the visible spectrum.

Sun rays contains infrared radiation.

If a body is heated, it will expand.

For a statement to qualify as a scientific statement, it should be:

- Testable, and
- Reproducible.

The statement about some qualities of a person or a thing cannot be considered scientific, because such a statement cannot be tested. Of course, you can test the kindness of a person, but each time you will end up with different answers and such a test is beyond science. In other words such qualities cannot be proved scientifically. The infrared component in the sunrays can always be tested and verified.

The reproducibility of the scientific statement implies that the result obtained by one scientist should be the same as another scientist would get under similar experimental conditions. If the other scientist cannot get the same result then the original result is not scientific and the statement needs a modification. Sticking on the reproducibility principle creates difficulties in life sciences and atmospheric science; for the experimental condition itself is not reproducible in many occasions.

Further, the scientific statements can be in quantitative and non-quantitative form. In our example the statement '*velocity is the rate of change of displacement*' is a quantitative statement. On the other hand the statement '*if a body is heated it will expand*' is a non-quantitative statement.

It is to be noted that there is no absolute scientific truth. Therefore, all scientific statements are not equally valid or equally useful. Rather, the probability of the statement is important. All statements are not equally probable. Nobody doubts the probability of the statement '*sun rises in the east*', but the statement '*ripe apples are red*' has less probability. Such statements are often regarded as *statistical statements*. Also, some statements are more reliable and some others are more useful. For example, the statement '*sun rises in the east*' is more reliable whereas the statement '*use of CF lamp reduces the power consumption*' is more useful. A scientific statement when attains universal acceptance it becomes a *fact* and a generalization of facts leads to a *law*.

2.3 THE CLASSICAL APPROACH

The origin of philosophy of science can be traced back to Greece during the period of Aristotle and his contemporaries. As a philosopher Aristotle relied on logical reasoning for arriving at a conclusion. This is called *classical approach* which is the first systematic treatment of the principles of correct reasoning. In this approach the validity of the conclusion depends upon the strength of the argument. There is nothing to surprise that Aristotle did not use any experimentation because he was a philosopher and he had confidence in his philosophy. But as Lee (2000) pointed out this could work very well in mathematics and logic, but science is different [4]. The reality may be different from our perception. Aristotle says at the beginning of his *Physics* "the best way to reach information about the '*science of nature*' is to advance from universals to particulars, because universals are easier for us to grasp with the help of our senses. The tool used by Aristotle for this was the *deductive reasoning* in the form of *syllogisms*. Using syllogisms one could infer new truths from established ones.

In syllogism one examines two statements, say, a major statement and a minor statement and says whether or not a third statement (conclusion) follows from the first two statements.

EXAMPLE

Statement 1: A is true for all B.

Statement 2: B is true for all C.

Inference: A is true for all C.

This deduction is meaningful and its validity does not require any proof. But the question rests in the validity of the statements (premises). How can we find true premises which are the foundational principles of given science? Aristotle himself addressed this problem. He realized that the knowledge already present in our mind is not sufficient to recognize the premises in many of the situations. Instead, the foundational premises must be acquired through experiences derived from perceptions and memories of perceptions. After repeatedly assembled various perceptions into memories and then, processing these memories, one can comprehend the first principle.

Consider the following example.

Statement 1: All balls in the box are red.

Statement 2: I took two balls from the box.

Inference: Therefore balls in my hands are red.

The inference is necessarily true since the argument is sound. But the deductive reasoning does not always a valid inference. As already mentioned, the validity of the conclusion depends on the correctness of the premises. It is clear in the following example.

Statement 1: Aristotle advocates deductive reasoning.

Statement 2: Aristotle was a philosopher.

Inference: Therefore a philosopher advocate deductive reasoning.

The statements are true but the inference is not necessarily true.

Aristotle's philosophy and methodology worked well, especially for developing a taxonomical framework for biology. Many scientists like Archimedes and Euclid adopted Aristotle's method for proving their laws. Euclidian geometry can be regarded as a paradigm of deductive reasoning in science. Euclid and his followers formed numerous theorems on the properties of geometrical figures starting from a small number of premises. This methodology was mainly focused in developing the basic concepts of science but it became inadequate after some time, especially in physics and chemistry. This limitation was eventual and it does not degrade the methodology which ruled the scientific world for centuries.

In spite of its ability to discover universals by generalization Aristotle was not in favour of induction because as he believed it was not sufficient in identifying the cause.

2.4 EMPIRICISM

In the previous section we have seen the classical method based on deductive reasoning. This method was purely based on logic. During the course of development of science this became inadequate and the methodology broke out of Aristotelian framework. This happened during Renaissance in the western world and science became oriented towards the quantitative verification of hypotheses. New methodology called *empiricism* emerged up. According to empiricism reason alone is not sufficient in drawing conclusion; but all ideas must be tested through experimentation by making observations and measurements and subsequent analysis. Therefore, empiricism relies on experimentation for acquiring the scientific knowledge. In words of Ladyman:

"Empiricists believe that knowledge can only be obtained through the use of the senses to find out about the world and not by the use of pure thought or reason; in other words, the way to arrive at justified beliefs about the world is to obtain evidence by making observations or gathering data" [5].

The two approaches in empiricism are:

- Induction, and
- Deduction.

2.4.1 Induction

Induction is a *specific-to-general* approach. It is a type of pattern recognition. The researcher collects information of situations in which a particular phenomenon is involved and after analyzing the collected information he derives a conclusion regarding the phenomenon. Inductive logic was proposed by Francis Bacon (1561-1626) in place of Aristotelian deductive method. In his book *Novum Organum* (New Tool) published in 1620 he gave a detailed description of this method. Bacon provided a central role for experience and experiments in scientific research. In Bacons' words:

"Another form of induction must be devised than has hitherto been employed, and it must be used for proving and discovering not first principles (as they are called) only, but also the lesser axioms, and the middle, and indeed all. For the induction which proceeds by simple enumeration is childish" [6].

The best example one can suggest for the inductive type of reasoning is the formulations of Newton's law of gravitation which says *everybody in this universe exerts a gravitational force of attraction on every other body*. Newton observed the movement of planets around the sun, satellites like moon around the planets and a variety of motion on and near the surface of the earth. Definitely, he did not observe the entire universe; despite Newton inferred from his observation that the principle held true for all bodies.

As it is widely accepted induction requires two assumptions [7]:

- Representative sampling, and
- Uniformity of nature.

Representative sampling: In induction the generalization of a specific observation is carried out. Therefore it is important to consider the samples or events that are representative of the phenomena under consideration. Randomization and objective sampling are the ways to obtain a representative sample.

Uniformity of nature: This is one important aspect behind the inference. If a particular phenomenon exists in a number of events, the same phenomenon will be present in all such events.

To be an effective scientist one must be skeptical and his mind should be open to any sort of knowledge and should not jump into conclusions without proper investigation. Bacon called the things that could get in the way of right inductive reasoning as the *Idols of the Mind. James* Ladyman (2002) in his book *Understanding Philosophy of Science* explains each of these with suitable examples. They are [8]:

- *Idols of the Tribe:* This represents the tendency of all human beings to perceive more order and regularity in nature than there is in reality. For example, we assume all heavenly bodies move in perfect circles, and to see things in terms of our preconceptions and ignore what doesn't fit in with them.
- *Idols of the Cave:* These are individual weaknesses in reasoning due to particular personalities and likes and dislikes. For example, someone may be either conservative or radical in temperament and this may prejudice them in their view of some subject matter.
- *Idols of the Marketplace:* These are the confusions engendered by our received language and terminology, which may be inappropriate in a particular situation. For example, we may be led into error by our using the same word for the metal lead and for that part of a pencil that makes a mark on paper.
- *Idols of the Theatre:* These are the philosophical systems that incorporate mistaken methods, such as Aristotle's, for acquiring knowledge.

Bacon's inductive method starts with sensible experience, as Malherbe (1996) explains, "immediately drawn from things, as they appear". These *data* are then generalized to low-level axioms (commonly accepted principles) based on the tables of induction. According to Bacon there are three types of tables of induction [9].

The table of essence and presence: This enumerates all situations in which the nature under consideration is present.

The table of deviation or absence of degrees: This lists all situations which are as similar as possible to the first table, but where the nature under consideration is absent.

The table of counter-instances: This suggests experiments in order to search possible counter-examples.

By applying these tables to the observed event one can form low-level axiom. From a generalization of these axioms a law of nature can be derived. Thus, Bacon's method is founded on *observation and induction.* Many theories were developed through induction. The best example that one can cite is the theory of evolution developed by Charles Darwin through a dedicated observation and analysis of various species and then by generalization.

The simplest form of induction is called *enumerative induction.* Here we observe a large number of instances of a particular phenomenon which has some characteristics and conclude that the phenomenon always has that characteristic. This happens in the case of many medicines, they give relief; but how it works may not be known.

2.4.2 Hume's Problem of Induction

Bacon's method revolutionized the scientific world. Many scientists like Galileo and Newton worked on Bacon's framework and devised a number of theorems and laws. The wide acceptance gained by Bacon's method did not mean that it is absolutely correct. Nobody was worried about this until David Hume (1711 – 1776) pointed out the problem of induction in his work *An Enquiry Concerning Human Understanding*. Two simple questions were asked [10].

1. What justifies the faith we place in induction?
2. How should we go about persuading someone, who refuses to reason inductively, that they are wrong?

Hume presented a simple answer to these questions. According to Hume the use of induction cannot be rationally justified. We use induction all the time in everyday life and in science, but Hume insisted this was a matter of brute animal habit. If anybody challenged to provide a good reason for using induction, we cannot give a satisfactory answer.

To explain his arguments Hume attributed a "*uniformity of nature*" characteristic to the inductive inferences. By uniformity of nature Hume means the inference is valid also for the unexamined objects or instances. Hume argued that we cannot strictly prove the truth of *uniformity of nature* but we might hope to find a good empirical evidence for its truth. Samir Okasha (2002) puts this in another way: "it is an established fact that nature has behaved largely up to now; but we cannot appeal to this fact to argue that the nature will continue to be uniform. Because this assumes that what has happened in the past is a reliable guide to what will happen in the future: which is the *uniformity of nature* assumption. The argument for *uniformity of nature* on empirical grounds, we end up reasoning in a circle" [11]. An elaborate discussion on Hume's problem can be found in *Understanding Philosophy of Science*"by James Ladyman (2002) and also in *Philosophy of Science – An Encyclopedia*, by Sahotra Sarkar and Jessica Pfeifer (2005).

2.4.3 Deduction

Imagine that four 3″ × 5″ cards are on the table. You can see that each card has a single letter or number on its top: one has the letter 'A', one has 'B', one has the number '4', and one has the number '7'. You may assume that each card contains a single letter on one side and a single numeral on the other side.

What cards is it necessary to turn over, to evaluate the validity of this rule: 'If a card has an A on one side, then it has a 4 on the other side'?

This was a problem put forward by Wason (1996, cited in Richard D Jarrard, p.71). This is a good example where the deductive reasoning can be employed to obtain the answer. This seems to be a very simple question and many answered in a minute or two; A and 4. But the correct answer is A and 7. The justification is as follows:

- Certainly card B is not useful, because it cannot support and prove or disprove the rule regardless of the number written on the other side.
- In case of card 4, even if it has an *A* on the other side, it supports but neither proves nor disproves the rule.
- But the flipping of card 7 tests the rule. If the other side is *A* then the rule will be disproved.

This answer can be analyzed philosophically and scientifically. Philosophers interpret the answer *A* and *4* as a conformation bias. The chooser of the card 4 is expecting a result that confirms the rule, rather than choosing the card 7 and possibly disproving the hypothesis. Scientists may justify choice of the card *4* as a search for patterns where they are most likely to be found. As Richard D Jarrard (2001) pointed out this particular problem illustrates [12]:

- Confirmation bias is present in science, but to some extent it is a normal consequence of our pervasive search for patterns.
- Many people's 'deductive thinking' may actually be inductive pattern recognition of a familiar deductive form.

A description of deductive logic is given in section 2.3. In deductive logic, each statement in the argument is either true or false. For the conclusion to be true, the following two conditions must be satisfied:

1. The premises must be true.
2. The form of the argument must be valid.

A valid deductive argument is one in which the conclusion is necessarily true if the premises are true. The validity or invalidity of a deductive argument is totally independent of the correctness of the premises; *it depends only on the form of the argument* [13]. Consider the following classical examples:

EXAMPLE 1

Statement 1: All dogs are cats.

Statement 2: All cats are animals.

Conclusion: Therefore, all dogs are animals.

In this example the form is valid but one premise is false. Therefore, even though the conclusion happens to be correct; the argument is incorrect.

EXAMPLE 2

Statement 1: All dogs are mammals.

Statement 2: All mammals are animals.

Conclusion: Therefore, all dogs are animals.

Here the form is valid and the premises are true. So the argument is correct and the conclusion must be true.

EXAMPLE 3

Statement 1: All dogs are mammals.

Statement 2: All cats are mammals.

Conclusion: Therefore, all dogs are cats.

In this case the premises are true but the form of the argument is not valid. Therefore, it does not yield a correct conclusion.

From these examples it is clear that in deductive reasoning one should consider the correctness of the premises and form of the argument. The evaluation of premises is subjective based on local expertise and that of argument form is objective [14].

Deduction and induction constitute the two branches of scientific logic. In general, deduction is developing a theory from the general to the particular, whereas induction is moving from the particular to the general. It is believed that deduction is more efficient than induction, as it probably does not mislead the scientist. In fact ,induction is found to be more productive than deduction. René Descartes argued that science should be confined to the deduction whereas Francis Bacon advocated induction, as majority of scientific discoveries were empirical. Isaac Newton rejected Descartes and was in favour of Bacon's empiricism. But in the course of rapid scientific progress it was realized that both deduction and induction are necessary aspects of science; both deduction and induction have their own place in the methodology of science. The question of superiority is a fallacy.

The difference between the deductive and inductive arguments can be put in a single sentence. In a deductive argument, the conclusion follows necessarily from the premises whereas in an inductive argument, the conclusion follows probably from the premises (Richard D Jarrard, 2001). As a result deductive arguments are judged as valid or invalid: in a valid deductive argument, *if the premises are true, then the conclusion must be true.* Inductive arguments are judged as strong or weak according to the likelihood that true premises imply a correct conclusion.

2.5 POSITIVISM

Positivism is the most widely accepted scientific method. Positivism is a philosophy of science originated with David Hume in the early modern era which developed into *logical positivism* in 1930s. The term *"positivism"* was used for the first time by Saint-Simon to designate the method of the sciences and was taken over by Auguste Comte for his philosophy. According to Comte positivism referred to the last "scientific stage" in the development of "mind" which evolves through three stages: *theological*, *metaphysical*, and *positive*. In theological stage mind believes in religious superstition, which is the most primitive stage of the mind. In metaphysical stage such superstitious are replaced by impersonal forces such as gravity and are still considered mysterious. In positivism metaphysical forces are replaced by observational concepts. Comte suggests three defining characteristics of positivism.

1. Metaphysical and fictitious ideas should not be a part of science.
2. We should seek causes in Hume's sense, that is, as "invariable relations of succession and resemblance". (Comte was very much influenced by Hume and Emmanuel Kant through their empiricist approach.)
3. The explanation of facts is simply the establishment of a connection between single phenomena and some general facts.

There are several alternatives to positivism like falsification and realism. As Levin (1988) says positivists believe that reality is stable and can be observed and described from an objective viewpoint. They argued that the phenomena should be isolated and emphasis is given to the repeatability of the observations. Hirschheim (1985) observed that: "positivism has a long and rich historical tradition. It is so embedded in our society that knowledge claims not grounded in positivist thought are simply dismissed as 'ascientific' and therefore invalid".

This observation was supported by many people and, in general, the empirical studies were positivist in approach. As mentioned, positivism relies on observation and experimentation. In the words of Anne B Ryan "positivist researchers believe that they can reach a full understanding based on experiment and observation. Concepts and knowledge are held to be the product of straightforward experience, interpreted through rational deduction. Positivism seeks to reduce everything to abstract and universal principles, and tends to fragment human experience rather than treat it as a complex whole" [15]. Thus, positivism provides a framework to explore reality as a concrete, given entity which can be understood objectively.

Positivist approach holds certain beliefs:

- *Prediction and control:* There are general patterns of cause-and-effect that can be used as a basis for predicting and controlling natural phenomena. The goal is to discover these patterns.

- *Empirical verification:* We can rely on our perceptions of the world to provide us with accurate data.
- *Research is value-free:* A strict methodological protocol is followed; research will be free of subjective bias and objectivity will be achieved.

The various steps involved in positivism can be summarized as follows.

First, the scientist identifies a problem or phenomena for study. As a second step a hypotheses is set up. A hypothesis represents a possible explanation of the phenomenon. Usually, hypothesis is based on the observation of the present situation, past experience of the scientist or based on some established knowledge. The next task is to test the correctness of the hypotheses. This is done by experiments through careful collection of data and analysis. Various methods can be used for the collection and analyzing data. The experiments can be performed under controlled conditions such as in a laboratory or outside the laboratory (field). If the hypothesis fails the test, then it is to be modified in view of the observations and is to be tested again. If the hypothesis passes the test, the scientist can go ahead and it can again be tested under new conditions for a refinement. It is advised that the scientist should keep on testing the hypothesis so that it becomes more and more accurate and closer to the reality.

Within positivism the knowledge is treated as [16]:

- Only methodologies by which knowledge is arrived at are important. These methodologies must be objective, empirical and scientific.
- Only certain topics are worthy of enquiry, namely those that exist in the public world.
- The relationship between the self and knowledge has been largely denied – knowledge is regarded as separate from the person who constructs it.
- Mathematics, science and technical knowledge are given high status, because they are regarded as objective; separate from the person and the private world.
- Knowledge is construed as being something discovered, not produced by human beings.

Positivism is founded on an epistemological base which has some limitations. Positivism faces many challenges from various fields like anthropology, ethnography and critical psychology and from developments in qualitative research. Kuhn (1962) opinioned that positivist visions of science do not always reflect the actual practice of doing science. Nevertheless positivism is still a dominant model for research.

During the twentieth century positivism got its new birth as logical positivism. It was primarily concerned with theory of knowledge (epistemology), mathematical logic, and philosophy of science. As Robert Klee observes: 'the logical positivists had a peculiar "spirit," an overriding set of intellectual and philosophical attitudes, preferences, and commitments' [17]. Positivists depend strongly on mathematical logic because they understood mathematical logic has a universally applicability in understanding difficult-to-understand concepts, problems and issues. Klee continues: "...mathematical logic contains enough power and precision to allow us to capture and express in a useful and understandable way the most important causal and explanatory relationships that occur out in the real world, that is, in the domain of investigation in which a given scientist works" [18]. In logical positivism, a statement can be true only if it is a self-evident analytical, deductive truth as is found in mathematics and formal logic or if the statement describes reality precisely [19]. Logical positivists practised a healthy skepticism and specific methodology that was employed in the analysis of philosophical problems.

2.6 FALSIFICATIONISM

"In so far as a scientific statement speaks about reality, it must be falsifiable; and in so far as it is not falsifiable, it does not speak about reality".

—**Karl Popper**

Falsification is a modification of positivist approach and it was introduced by Karl Popper who had a considerable influence in the philosophy of science during twentieth century. It is developed on the idea that "no matter how many times a theory passes attest, scientists can never be sure that it will pass all future tests" [20].

Initially Popper had a study on Marxism of Carl Marx and psychoanalysis of Sigmund Freud. Popper considered these to be pseudosciences. Popper made important remark on these as: "it was easy to think of both these theories as very successful sciences if one assumed that scientific knowledge proceeds, and is justified, by the accumulation of positive instances of theories and laws" [21]. Popper noticed that, when the theory is so general it is too easy to accumulate positive instances to support a theory. For example, consider the statement "all metals expand on heating". One can give a number of examples, but it is hard to find a disproving situation. In other words the idea of conformation is fundamental to the scientific method. By this Popper means if we are thorough in a particular scientific method it is very easy to find confirming situations.

Popper was always impressed by novel predictions that were risky of being turned out to be false. A number of such predictions can be seen in science. Some are given below.

Einstein's prediction of gravitational lensing based on general theory of relativity.

- Newton's prediction on the return of Halley's comet during 1758.
- Mendeleeff's prediction of the existence of the unknown elements of gallium and selenium derived from the structure of the Periodic Table of the elements.

- Prediction on the existence of elementary particles based on Sakata model.

According to Popper presenting a novel and risky prediction is common in scientific theories and it made science intellectually acceptable. Popper argued that a conformation (positive instance) actually counts for anything if it is a potential falsifier of the theory. Thus, according to Popper it is not conformation but falsification is the heart of scientific method.

Falsification allows a scientist to formulate a hypothesis and then to show that it is not correct. That is, the scientist attempts to find an instance that could disprove the hypothesis. Falsification has the advantage that if the scientist finds one negative instance that disproves the hypothesis, it can be rejected or modified. The statement "all swans are white" cannot be proved if one sees millions of swans that are white; but the finding of one black swan could disprove the theory. For a positivist the hypothesis is to be tested continuously until it passes all possible tests. Thus, according to Popper the scientific method should be like this [22]:

- Scientific theories are proposed hypothetically;
- Propositions are deduced from these theories; and
- The propositions are then tested, that is, every effort is made to prove them false.

Therefore, it is the falsifiability that determines whether a statement is scientific or not. A scientific statement should be falsifiable. This leads to the idea that a theory can never be definitely accepted. In words of Popper "every scientific statement must remain tentative forever" [23]. Crotty put this in another way: "on Popper's accounting, what is put forward as scientific truth turns out to be, not something shown to be true, but simply something that scientists have so far been unable to prove false. This turns scientific truths into merely provisional statements" [24].

Well established scientific theories like Einstein's theory of gravitation clearly satisfied the criterion of falsifiability. On the other hand pseudosciences like Astrology could never pass the test. Popper made interesting observation on astrology: "by making their interpretations and prophesies sufficiently vague they were able to explain away anything that might have been a refutation of the theory had the theory and the prophesies been more precise. In order to escape falsification they destroyed the testability of their theory" [25].

Popper through his theory of falsification made an attempt for the demarcation of science and pseudosciences. Popper concluded that the criterion of falsifiability is a solution to the problem of demarcation.

Even though Popper's idea of falsification attracted the scientific world, there are so many criticisms against it. James Ladyman gives a detailed explanation of the limitations of falsification. They include [26]:

- *Some legitimate parts of science seem not to be falsifiable:* Probabilistic statements such as about the half-life period of a radioactive element, existential statements like statement about the existence of a black hole, unfalsifiable scientific principles like the conservation energy law, second law of thermodynamics and Newton's theory of gravitation and hypothesis of natural selection are beyond falsification.
- *Falsificationism is not itself falsifiable:* This was admitted by Popper himself; his theory is not a scientific theory. Instead, it is a philosophical or logical theory of the scientific method.
- *The notion of degree of falsifiability is problematic:* There cannot be an absolute measure of falsifiability, but only relative one. This is because for a universal generalization there can be infinite set of falsifiers. For example, Einstein's theory of gravitation is supposed to be more falsifiable than Newton's theory of gravitation.
- *Scientists sometimes ignore falsification:* It is often observed that scientists are never ready to abandon theory that they cherish. Scientists thought up modifications or extra assumptions to save it. This happened in the prediction of the orbit of Uranus. Instead of regarding their theory as falsified, most scientists of the time assumed that one of the basic parameters like knowledge about mass, position and motion was wrong, and some other scientists proposed the existence of another planet to accommodate the data.

Popper started with the question "When should a theory be ranked as scientific?" or "Is there a criterion for the scientific character or status of a theory?", and ended up with the idea of falsification. From a philosophical point of view Popper's idea is more relevant but for a scientist positivism works well; scientists look for verification rather than falsification.

Both positivism and falsification tests the hypothesis through experiments. If the hypothesis passes the test the scientist becomes confident on the validity of the hypothesis and he can go for further tests. On the other hand if the hypothesis fails a test then a modification of the hypothesis is required. For example, the prediction of the orbits of the mercury based on Newtonian theory was a failure; but could predict correctly based on Einstein's theory of relativity. When a hypothesis passes almost all convincible tests it becomes a law. It is to be noted that there is no specific criteria for a hypothesis to be regarded as a law; it becomes a law when almost all scientists familiar with it agree upon it.

2.6.1 The Duhem Problem

Popper's idea of falsification got wide acceptance among the scientific community as it seemed to be logically consistent. A single instance of failure was sufficient to reject a hypothesis. In 1954 Pierre Duhem pointed out a major limitation of this method. To quote Duhem's words:

"*The physicist can never submit an isolated hypothesis to the control of experiment, but only a whole group of hypotheses. When experiment is in disagreement with his predictions, it teaches him that one at least of the hypotheses that constitute this group is wrong and must be modified. But experiment does not show him the one that must be changed*".

(P Duhem, 1954, *The Aim and Structure of Physical Theory*, p. 185)

The idea of this statement is that it is impossible to test a scientific hypothesis in isolation, because an empirical test of the hypothesis requires background assumptions about background conditions, the reliability of measurements, the initial conditions of a system, etc. These background assumptions are called auxiliary assumptions or auxiliary hypotheses.

Duhem problem poses two problems before the philosophers of science.

1. There is no reason to suppose the fault lie with the hypothesis under test.
2. Which of the background conditions used to derive a predicted consequence should be rejected or disconfirmed when experiment disagrees with that prediction.

According to Duhem it is impossible to isolate a single hypothesis; but a hypothesis and its background assumptions as a *bundle* can be tested against the empirical world and be falsified if it fails the test. In order to test the hypotheses under consideration we assume that these background assumptions are correct. This is essential because a hypotheses by itself is incapable of making predictions. Thus, if there is a failure in the test we can validly infer that the conjunction (of the test hypothesis and auxiliary hypothesis) is false. But we cannot tell whether it is the test hypothesis or the auxiliary hypothesis that is responsible for the test failure. An American philosopher Willard van Orman Quine (1961) put forward the argument that "it would be quite reasonable to reject a law of logic, or change the meaning of our terms, if it was more convenient than rejecting a particular theory (James Ladyman, p. 79). This means under a condition of a falsifying evidence we should reject the hypothesis rather than rejecting a theory. To put this in other words when we have rational reasons to accept the background assumptions such as scientific theories as true, we will have rational reasons to conclude that the hypothesis tested is probably wrong under a falsifying evidence. This is a kind of Bayesian approach (named after Thomas Bayes, an English Mathematitcian, 1702-1761). In 1979 Jon Dorling presented a Bayesian analysis of Duhem problem where *Dorling* showed that "if a Bayesian personalist reconstruction is adopted then a natural and instructive resolution of Duhem's problem is automatically available to us" [27].

Duhem problem can be easily understood in a wider sense. That is, in any scientific experiment to test a hypothesis the entire physical world is involved. The experimental design, methods of data collection or observation, interpretation of the outcomes, etc., involve principles of optics, thermodynamics, electromagnetism and so on. If the hypothesis is falsified under an experimental test, then it is easy to conclude that the hypothesis might be wrong. However, from a logical point of view one can argue that any of the other principles might be faulted. In other words "no principled way exists to localize the bearing of evidence" [28]. This statement is often called Duhem-Quine thesis.

2.7 SCIENTIFIC SKEPTICISM

Skepticism is identified by its important characteristics that it continuously questions the beliefs and conclusions. This is a positive attitude so that scientists can question the accuracy and reliability of the knowledge claimed by others and the knowledge they already have. The word skepticism originated from the Greek word 'Skeptesthai', meaning 'to examine'. In this etymological sense, one can say that scientists must be true skeptics and vigilant enough to examine the evidence, arguments, and reasons for their beliefs. This means researchers should not surrender to self-deception and deception of themselves by fellow scientists, students, or others. Research with a skeptic attitude, always questions the reliability of knowledge claims by subjecting them to a systematic investigation. Skeptics have challenged the adequacy or reliability of any claims by asking what principles they are based upon or what they establish. They have questioned whether some such claims are, as alleged, indubitable or necessarily true, and they have challenged the purported rational grounds of accepted assumptions. In ancient times a chief form of skepticism was medical skepticism, which questioned whether one could know with certainty either the causes or cures of diseases.

It is interesting to notice that the science itself is a structured kind of skepticism. A researcher very often uses the idea of falsification to verify the experimental results or claims. Further, a scientific skeptic need not always be a researcher who conducts experiments but can be an ordinary person who accepts only those claims that are possibly true based on falsifiable hypotheses and testing. One of the major characteristics of a scientific skeptic is that he/she always approaches beliefs carefully, and they are ready to change their attitudes when new evidence or valid reasons are present before them. However, it is not easy to defend many conventional and long-accepted notions. For example, there are claims and practices prevalent in the society that scientists must see as unlikely to be true on scientific grounds. One such example is the health claims through the use of certain foods. However, with the passage of time and the acquisition of new knowledge, older knowledge and theories will be replaced by new ones, like geocentric theory is replaced by heliocentric theory. In one sense we can admit that the strong opposition against introducing extremely new ideas is characteristic of scientific skepticism.

Philosophical skepticism is based on an interesting argument that *we do not know propositions that we ordinarily think we do know*. This has a slight difference from scientific skepticism. In scientific skepticism, we usually question a preposition or a research outcome based on quantitative analysis. A scientific skeptic may say that we do not know that there is an even number of stars in the milky way. A philosophical skeptic thinks philosophically in a more general way to say that we do not know that the milky way exists in the universe. Most of us readily agree with the first statement; the vast majority of us does not even believe the statement that there is an even number of stars in the milky way and even those who believe they do not know it; that is, the proposition or the statement is not justified. It is widely accepted that knowledge requires belief and justification.

2.8 PARADIGM AND PARADIGM SHIFT (SCIENTIFIC REVOLUTION)

"In science, as in the playing card experiment, novelty emerges only with difficulty, manifested by resistance, against a background provided by expectation".

—Thomas Kuhn (1962, p. 76)

The principles paradigm and paradigm shift was introduced by Thomas Kuhn through the book *The structure of Scientific Revolution* published in 1962. In this book Kuhn presented a historical study of the development of science. Before Kuhn science was considered a rational endeavour in which progress and knowledge are achieved through the steady, painstaking accumulation of experimental data, accredited facts and new discoveries [29]. This was the *normal science* according to Kuhn. He used the word *paradigm* to represent *a conceptual model or intellectual framework within which scientists conduct their research and experimentation*. The exercise of scholarly activities is subject to the prevailing paradigm. All scientists of a particular field consider the problems within the framework set by the paradigm. Kuhn describes the two essential characteristics of paradigms [30]:

1. Their achievement was sufficiently unprecedented to attract an enduring group of adherents away from competing modes of scientific activity.
2. They were sufficiently open-ended to leave all sorts of problems for the redefined group of practitioners to resolve.

Examples of such paradigm includes: Ptolemaic astronomy, Copernican astronomy, Aristotelian dynamics, Newtonian dynamics, corpuscular optics, wave optics, quantum mechanics, etc.

The way of normal science is to seek and create a model that will account for as many observations as possible within a paradigm. The observation of instances that does not fit within the paradigm and a courageous attempt on the part of the scientist to move out of the paradigm to seek a solution indicates the beginning of a change. The history of science shows that a paradigm is subjected for a change at any time.

The paradigm shift or Scientific Revolution is usually said to have occurred after the middle-sixteenth century during the Renaissance. The Renaissance inspired a spirit of curiosity in the minds of people. Scholars began to question many beliefs that had been accepted for hundreds of years. The new ways of thinking started the new real science. Scientists began to replace old concepts with new theories, which were proved through the scientific method of observation and experimentation. A new approach to science has emerged. The Scientific Revolution was a new way of thinking about the natural world and it was based upon careful observation and a courageous and curious mind to question accepted beliefs.

It is to be noted that unlike a political revolution scientific revolution developed gradually over a period of time. It began in the sixteenth century and is still continuing. The most important changes in seventeenth-century science took place in the fields of astronomy, physics, chemistry, and biology. Copernicus, Keppler and Galileo redefined the concept of the universe. Isaac Newton proved that gravity acts on every object in the universe. William Harvey discovered how blood circulates. Revolutionary thinkers like Francis Bacon and Rene Descartes came up with their scientific methods. Kuhn argued that true breakthroughs arise in a totally different way: when the discovery of anomalies leads scientists to question the paradigm, and this in turn leads to a scientific revolution that he termed paradigm shift.

New model of universe–an example of scientific revolution

The consequences of new rational thinking were first reflected in astronomy. Scientists started questioning the widely accepted geocentric theory of Ptolemy. Nicolas Copernicus (1473 – 1543) came up with the heliocentric theory after studying the movements of planets for about 25 years. He reasoned that the earth and other planets are revolving around the sun. Copernicus did not publish his findings *On the Revolutions of the Heavenly Bodies*, until 1543, the last year of his life because he was aware that his theory contradicts the religious belief and will be rejected.

Based on the foundations laid by Copernicus, Danish Astronomer Tycho Brahe (1546–1601) carefully recorded the movements of planets for many years.

After the death of Tycho Brahe, Johannes Kepler (1571 – 1630) continued his work. After analyzing the data provided by Brahe, Kepler identified that some law governs the planetary motion. Kepler formulated three laws governing the orbital motion of planets. These laws showed that the Copernicus's basic ideas were true. These laws indeed, say that all planets are revolving around the sun in an elliptical orbit with specific time period.

Galileo Galilei (1564–1642) was a contemporary of Kepler, and was convinced by Copernicus's heliocentric model of the planetary system. He was the first to use the telescope to study the heavenly bodies. Using telescope he discovered the mountains and craters on the moon, and the four moons of the Jupiter. He observed that the planet Venus went through phases just as the Moon does. An important point to be mentioned is that Galileo is regarded as the *father of modern scientific method* because he was the first to apply scientific methods systematically. He was the first experimental scientist.

The discussion will be incomplete unless we say a few words about Isaac Newton (1642–1727). Isaac Newton is the most important contributor to the development of modern science. Newton's great discovery was that the same force determines the motions of the planets, the pendulum, and all matter on the earth and in space; the *force of gravity*. Newton formulated the universal law of gravitation. It says "every object in the universe attracts every other object and the strength of attraction depends on the mass of the objects and the distance between them". Newton published his discoveries in *Mathematical Principles of Natural Philosophy* in 1687. A nice description of the development of science all around the world is presented in *An Introduction to the History and Philosophy of Science*" by *R.V.G.Menon* (Dorling Kindersley India Pvt. Ltd, 2010).

Scientific revolution had its impact on every sphere of science. Medicine and chemistry also witnessed major discoveries during this period, for example, circulation of blood, vaccine for small pox, discovery of oxygen, development of Boyle's law, etc.,. A large number of scientific instruments were also developed. Discovery of microscope, X-ray, Laser etc. had great impact on the scientific studies. In fact, scientific revolution is an ongoing process.

Many people identify the following three internal causes for scientific revolution [31]:

- The research into motion conducted by natural philosophers in the fourteenth century;
- The scientific investigations conducted by Renaissance humanists; and
- The collapse of the dominant conceptual frameworks, or paradigms, that had governed scientific inquiry and research for centuries.

The third point is very important and needs a little bit elaboration. As already mentioned Kuhn regarded the scientific revolution as a paradigm shift. A scientific revolution occurs when the old paradigm collapses and a new paradigm replaces it. In astronomy the antiquity and the middle-age people worked on the paradigm of Ptolemaic model. Later on during the sixteenth century this became inadequate in providing satisfactory explanations for the material universe and Copernicus came with the new paradigm of heliocentric model within which Kepler, Galileo, and Newton all worked. Some more examples are listed below.

- Newtonian mechanics
- Huygens's wave theory of light

- Einstein's theory of relativity
- Maxwell's electromagnetic theory
- Bohr atom model
- Dynamical theory of heat
- Kinetic theory of gases
- Quantum mechanics
- Nanotechnology

These are paradigms evolved during the course of development of science by replacing the paradigm existed prior to them. A paradigm shift or scientific revolution is actually a change in world view. The new paradigm will not be just extension of the old paradigm. Rather, it brings out a change the way in which then terminology is defined, how scientists view the subject, and what questions are regarded as valid. Scientists will make attempts to encompass and explain all unexplained phenomenon within the new framework.

Science is often regarded as objective in majority of the scientific methods; scientists are not influenced by prejudices. They approach the problem with an open mind and present the results as such. However, scientists are human being; not machines. This means we cannot eliminate the subjectivity that arises in their thoughts. Very often they are guided by some preconceived ideas.

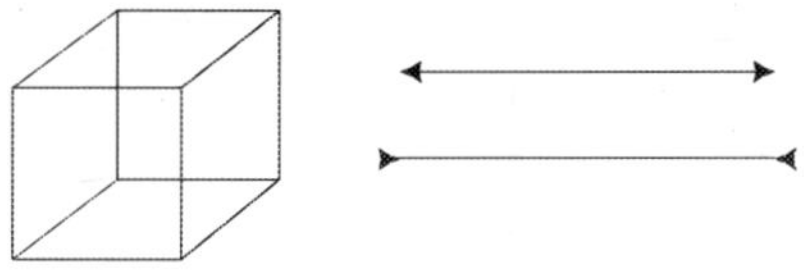

(a) (b)

Fig. 2.1

Our perceptions of the real world actually involve the effect of such ideas. As Lee pointed out "many psychologists believe that what we see is influenced by what we expect to be there and what we expect is influenced by past experience" [32]. This indicates that there is possibility to ignore observations that are insignificant according to the theory that he believes. As Hanson (1958) stated "Seeing is not only the having of a visual experience; it is also the way in which the visual experience is had ". The observations made on the same event by two observers will be different. There are some classical examples that show our perception depends on our previous experience and concepts. Consider the cube given in the Figure (a). It is often confused that whether the top right square is facing us at the front of the cube or the bottom left square is facing us as the front. Such a change in experience is called '*Gestalt shift*'.

Similarly, the bottom line in the Figure (b) appears to be longer than the top line (Müller–Lyer illusion). We can measure the length of the lines and see that they are the same. Even then we perceive one to be longer than the other.

Thus, what we observe is not neutral with respect to the theories and beliefs. In other words, the content of observation (data) is a function of the theories that the observer believes. Since the observations are theory-laden, science also is often regarded as *theory-laden*. James Ladyman makes wonderful observation regarding paradigm in this context. "…all observations are contaminated by background theories then the merits of each paradigm cannot be compared by subjecting them to experimental test because the proponents of the competing paradigms will not necessarily agree about what is observed ". Kuhn compares the paradigm shift to a Gestalt shift.

The theory-ladenness of observation, supports the so called *incommensurability* (which has a literary meaning *lack of common measure*) of two competing paradigm. In Kuhn's words "…the choice between competing paradigms proves to be a choice between incompatible modes of community life" (Kuhn, 1962, p. 94). The concepts ontology, etc., of different paradigm will be different. The incommensurability manifests in two ways:

- Meaning incommensurability, and
- Eeference incommensurability [33].

Meaning incommensurability refers to the fact that theories within different paradigms are incommensurable. This means that the terms and concepts in the scientific theories in different paradigms are not mutually intertranslatable. The *reference incommensurability* means that a particular term in a paradigm', for example, '*atom*' may refer to a different thing in the second paradigm. Thus, the term incommensurability provides the idea that *different theories correspond to the different worlds of different theories, and the proponents of competing paradigms inhabit different worlds; for example, the world of Einstein is literally a different world from that of Newton* [34]. This observation has important consequences, for example, nobody can say Ptolemy was wrong and Copernicus is correct because Ptolemy's earth was a different object from Copernicus's. This is the beginning point of the idea of social constructivism of reality.

2.9 INTERPRETIVISM OR ANTI-POSITIVISM

"Interpretivism holds the belief that truth is a construct invented or influenced by the observer and that reality is relative and not separate from the individual who observes it".

—**Sean**

Interpretivism denies what positivism asserts. Interpretivists argue that the reality can be fully understood only through the subjective interpretation of the reality and through an intervention in it. The interpretivist philosophy demands the study of phenomena in their natural environment. They acknowledge that scientists cannot avoid affecting those phenomena they study. Therefore, as opposed to positivism where objectivity has a key role, interpretivism has a subjective element. They are not ignorant of the fact that there may be many interpretations of reality and consider all these interpretations as a part of scientific knowledge. They argue that data have to be interpreted; it does not present explanations by itself. Interpretation is supposed to be the ability to extract meaning from observation and this is different from the process of following scientific method. As the followers if interpretivism argues:

"Subjectivity at its best has a link to creativity".

Creativity represents new ways of thinking and understanding. The creativity can work out more effectively in subjective interpretations. Many times achievements are made by stepping out of the rules of scientific methods and relying on inspiration. Anti-positivists like Wilhelm Dilthey (1833-1911) and Max Weber (1864-1920) pointed out the positivist failure to "appreciate the fundamental experience of life, and instead favour physical and mental regularities, neglecting the meaningful experience that was really the defining characteristic of human phenomena". Interpretivism relies on interviews and subjective observation.

Interpretivism is developed based on the idea that there are fundamental differences between the natural world and the social world and hence the logic and methods of science cannot be applied as such to the society. Scientific methods aim to generate knowledge based on cause-effect relationship whereas social sciences focus on the subjective experience. By its very nature interpretivism is a qualitative research paradigm in which exploration and insight into the subjective experience are valued. People like Hammersley, Mackenzie & Knipe (2006) agree on the fact that researchers should recognize that all participants involved, including the researcher, bring their own unique interpretations of the world or construction of the situation to the research and the researcher needs to be open to the attitudes and values of the participants or, more actively, suspend prior cultural assumptions. Therefore, the researcher himself becomes a part of interaction with the subject of research and this allows him to communicate with the cultural background of a society. The inferences then drawn by the researcher will be more realistic because his understanding is from personal experience rather than from a manipulation of collected data.

2.10 POST-POSITIVISM

"The world is not what I think, but what I live through. I am open to the world, I have no doubt that I am in communication with it, but I do not possess it; it is inexhaustible".

—**Merleau Ponty**

The basic philosophical assumptions that define research paradigms according to Guba and Lincoln (1994); can be summarized from the responses to three fundamental and sequential questions [35]:

1. The ontological question: What is the form and nature of reality and what can be known about it?
2. The epistemological question: What is the nature of the relationship between the knower or would-be knower and what can be known?
3. The methodological question: How can the inquirer go about finding out whatever he or she believes can be known?

The various perspectives of research paradigms like positivism, interpretivism, post-positivism, etc., are discussed in response to these questions.

Post-positivism emerged as a critique of empiricism. Positivism argues that the reality can be discovered by conducting scientific research. The experiments can be repeated to prove the validity of the result. Post-positivism is considered an empirical, explanatory approach that maintains belief in observables (Gortner, 1999). Post-positivism has emerged from the recognition that [36]:

- True reality is not apprehensible,
- Objective and subjective realities are not mutually exclusive,
- Findings cannot be proven to be true, and
- Inquiry is not value free.

According to Popper (1937) post-positivists argue that one can never be sure that the next research study will not be the one that shows your theory is wrong. This simply means that one can never confirm that a theory is absolutely correct. This is the essence of falsification. When a theory is falsified the researcher should go for a modification of the theory. Nowadays, falsification is not practiced in its complete meaning. Post-positivists employ a modified version of falsification. The occurrence of a negative instance is sometimes attributed to instrumentation, misinterpretation of the data, misapplication of the theory, poor sampling, etc. Therefore, from a one-time failure one cannot conclude that the theory is wrong.

It is quite interesting to notice the arguments of post-positivists. In positivism theory is derived from careful observations. The essence is being the assumption that the data are independent of the theory. Post-positivists are against this concept. According to them any collection of data is based on theory. They share the view that a single study is not sufficient to find out the truth, and it requires a number of attempts; each time attaining a closer approach to the truth. The post-positivist approach is widely employed in social sciences research. As Anne B Ryan observes: "post-positivist research principles emphasize meaning and the creation of new knowledge, and are able to support committed social movements, that is, movements that aspire to change the world and contribute towards social justice" [37]. Further Anne B Ryan points out the following characteristics of post-positivist approach.

- Research is broad rather than specialized–lots of different things qualify as research.
- Theory and practice cannot be kept separate. We cannot afford to ignore theory for the sake of 'just the facts' [38].
- The researcher's motivations for and commitment to research are central and crucial to the enterprise (Schratz and Walker, 1995: 1, 2, as cited in Anne B Ryan p. 13).
- The idea that research is concerned only with correct techniques for collecting and categorizing information is now inadequate (Schratz and Walker, 1995: 3, as cited in Anne B Ryan p. 13).

Post-positivism places the researcher in a key position and provides full freedom for conducting the research as opposed to positivism where strict scientific methods are to be followed. Post-positivism provides an opportunity to investigate the epistemology followed by the researcher himself. "The post-positivist stance asserts the value of values, passion and politics in research. Research in this mode requires the ability to see the whole picture, to take a distanced view or an overview. But this kind of objectivity is different from 'just the facts', devoid of context – it does not mean judging from nowhere" (Eagleton, 2003: 135, cited in Anne B Ryan, p.18).

It is to be noted that post-positivism does not set an aim like *'finding the absolute truth'*. This is due to the fact that they had the realization that the world and experiences are very complex and such a search will prove meaningless. They only acknowledge the reaching of a valuable conclusion as there are no universal solutions to the problems. As Frances E Racher and Stevan Robinson (2002) puts the goal of research is explanation, prediction, and control and involves making generalizations and cause-effect linkages. Knowledge may be gleaned through a variety of quantitative and qualitative research methods that may complement each other and move knowledge closer to the truth, which can never be fully verified [39].

2.11 RELATIVISM

"Relativism is tantamount to the denial of the principle of non-contradiction for if man is the measure of all things, then different people would assign the value true or false to the same assertion rendering it both true and false. Such a move, however, contravenes the principle of non-contradiction, the most certain of all basic principles and a presupposition of all thought and speech".

—**Aristotle**

Consider the statement "Roger Bacon was born in the year 1214". Can anybody say whether this statement is true or false? Nobody for sure can tell the answer. This statement is either true or false – it is true relative to the Julian calendar. This is what is meant by relativism; there is no absolute truth. Our perception of the world is relative.

Relativism can be better understood in contrast to its opposite concept, i.e., *absolutism*, because absolutism is far more easy to define and to understand. Absolutism says that "there exists one description of the world, true for all viewpoints, that determines whether any statement is true or false" [40].

Frank (1952) defined relativism as "*the negation of absolutism*". Frank arrived at this definition from his question about the concept 'truth'. Does the concept refer to something whose nature is absolute or something whose nature is relative? [41]. His answer was that the concept of truth in science does not refer to an absolute property. Relativists argue that most truths are merely relatively true. By the very nature of science it could never attain the absolute reality. Because science lives by its simplifications, idealizations, limitations of scope, compromises, selectivity, conjectures and approximations [42]. Relativists also insist that since truth is relative, what is taken as true may also be false.

The sociology of scientific knowledge became a matter of discussion in 1970s after the publication of *The Structure of Scientific Revolution* by Thomas Kuhn. Relativism emphasizes the role of social factors in conducting the scientific research and in decision making. As Montaigne (1533-1592) pointed out: "with changes in our bodily and emotional conditions one and the same judgment may appear true to us on one occasion and false on another; therefore no absolute truths on such matters exist" [43]. In other words relativism questions the objectivity of truth. According to relativists all concepts such as rationality, truth, reality, right, etc., must be understood as relative to a specific conceptual scheme, theoretical framework, paradigm, form of life, society, or culture [44].

Robert Klee (1997, p.147) rightly observed that 'all scientific knowledge is relative to a paradigm under which it qualifies as knowledge'. This conclusion was made based on Kuhn's observations that the paradigms are incommensurable. Therefore, different paradigms construct different natural universes. As a result the conclusions that one may reach should necessarily be relative. These decisions are based on the evidence independently of their own personal interests and goals. The interests and the aim set by a researcher would definitely be influenced by a number of social factors. The paradigm itself can be based on some social interest. For example, the Ptolemaic concept of the universe was well appreciated by the church and they even refused Copernicus's observations. A scientist adopts a paradigm because of the social, political and ideological preconceptions prevailing in the community to which he belongs. As Lee pointed out in its extreme version relativism states that science is purely social (Lee, p. 21).

The relativism can be viewed, in social context; as *ethical relativism* and *cultural relativism*.

Ethical relativism is a view against the ethical universalization. This means there are no universal moral judgments. The judgments can be relative to either individual or culture. Ruth Benedict (1887-1948) views morality as dependent on the varying histories and environments of different cultures. It is a generally accepted fact that the concept of ethical universalization is very weak whereas the ethical relativism is hard to defend. To prove it false one has to present a moral judgment that has universal acceptance.

Further, the morality is sometimes related to our culture. Cultural relativism includes many things. The attitude, belief, taste, etc., of people are greatly influenced by the culture. The actions are counted as right or wrong accordingly whether they obey the cultural norms. As Maria Baghramian writes "the doctrine of cultural relativism, inspired by the work of social anthropologists, where it is argued that there can be no such thing as a culturally neutral criterion for adjudicating between conflicting claims arising from different cultural contexts, has become one of the best known forms of relativism and has shaped not only the theoretical framework of the social sciences but also the ethical and political outlooks of many non-specialists" [45].

Conceptual relativism, on the other hand, is form of relativism where ontology is related to conceptual schemes or scientific paradigms. Relativism is again viewed as *ontic*, *cognitive* and *aesthetic* in various contexts.

2.12 REDUCTIONISM AND HOLISM

The idea of *reductionism* was first introduced by Rene Descartes in 1637 in his book *Discourse on the Method of Rightly Conducting One's Reason and of Seeking Truth in the Sciences*. Reductionism means a bottom-up approach to understanding complex natural phenomena by 'reducing' them to their fundamental parts and studying their interactions under controlled conditions. Scientific reductionism is most often used to describe the notion that everything in the universe can be broken into smaller and smaller parts and that the whole can be understood in this way. This includes many things such as objects, theories, phenomena, and explanations. The reduction can be applied successfully in science disciplines such as physics, chemistry, zoology, botany, etc. Reductionism seeks the explanation of the whole by looking at the relationships between the component parts that can be experimentally tested, thus eliminating the requirement of assumptions on extra forces such as consciousness, vital force, and others. However, a major problem with this approach is that scientists often fail to revert to see how their findings fit in with totality.

Holism is just the opposite of reductionism, where the principle of reduction is rejected and focuses on the totality or consolidated effect of the subject. The word 'holism' came from Greek, and has a meaning equivalent to totality. Holism is based on the principle that 'the whole is different from the sum of its parts'.

The supporters of holism argue that the properties of a given system such as physical, biological, chemical, social, and economical cannot be explained based on its component parts alone. For example, common salt is a combination of two elements, sodium and chlorine. Sodium is a solid substance and chlorine is a gas; both have entirely different properties. However, the common salt (sodium chloride) has properties entirely different from those of sodium and chlorine. In the late 1990s, the term holistic is modified to *wholistic* for better clarification of the concept.

It is to be noted that both the reductionism and wholism is an essential part of scientific research. As mentioned in the example, the properties of the sodium chloride molecule cannot be described by the properties of its components at the atomic level. However, when we deal with living organisms such as human beings, we need reductionism. By observing the whole body, one cannot understand the functions of internal organs. Therefore, reductionism and wholism are supplementary to each other, just like the interdisciplinary nature of various subjects.

2.13 SCIENTIFIC REALISM

"Nothing is divine but what is agreeable to reason".

—**Immanuel Kant**

In modern science many phenomena were explained by postulating entities that we do not observe. Many entities like electrons, radiation, dark matter, electric and magnetic fields, etc., fall under this category. We are unable to observe them directly and this raises the question whether they really exist and if so how? This question divided the philosophers of science into two categories: *realists* and *anti-realists* or *instrumentalists*. Scientific realism is subscribed to the idea that theoretical entities or scientific objects do exist. Instrumentalists do not agree with this; for them theoretical postulates are just instruments for generating predictions [46].

Realist philosopher Christopher Norris (2004, cited in T Jayaraman, 2007) describes scientific realism as: "scientific realism, broadly speaking, accepts the existence of objective reality as a fundamental premise. This objective reality exists independent of our theories and descriptions, and beliefs and thoughts concerning the same. These theories and descriptions acquire the status of truths or falsehoods depending on how they stand with respect to that objective reality, rather than on whether they coincide with systems of beliefs, some favoured paradigms and the like. Among this limitless set of truths, there are some that we know now, some that we don't know now and will find out later and some that may lie beyond the furthest reach of our knowledge-seeking powers" [47].

Stathis Psillos (2000) describes the three dimensions of scientific realism as:

- The metaphysical thesis,

- The semantic thesis, and
- The epistemic thesis.

The *metaphysical thesis* implies that the world has a definite and mind independent structure. This implies that if the unobservable natural kind s posited by theories exist at all , they exist independently of our ability to be in a position to know, verify, recognize, etc., that they do [48]. It distinguishes the scientific realism from all sort of anti-realism. The *semantic thesis* implies that scientific theories should be taken at phase values because they are true descriptions of the observable and unobservable. The theories can be turned out to be true or false. If the theory is true the unobservable they suggest represent the real world. The *epistemic thesis* implies that established and predictively successful scientific theories are approximately true of the world. It is meant to distinguish scientific realism from agnostic or skeptical versions of empiricism. It emphasizes that science can deliver both theoretical and observational truth at an equal extent [49].

[**Note:** The rejection of any of the above commitments results in anti-realism. So we have a variety of anti-realisms in practice. For example, skeptics deny metaphysical thesis, social constructivists deny semantic thesis and reductive empiricists deny epistemic thesis.]

According to scientific realism scientific theories are approximations of universal truths about reality. Lee explains this in another way: "realism is based on the idea that there is a real world which is separate from our perception of it and that the methods of science allow us to get closer to understanding that reality, even the unobservable parts" (Lee, p.21). Scientific realism holds that scientific theories go beyond data to posit the existence of non-observable entities, such as quarks, mental representations, and social cognition; which actually exist (Thagard, 2002, cited in John T Cacioppo et.al., 2004). The attempt is to describe the world as it really is. Realists argue that the product of scientific research is the knowledge which is independent of theory or methodology [50].

Realists justify their argument by saying that if a theory provides the best explanation of evidence then the theory must be true. For example, the evidence for the existence of the electron came through the various results available about the behaviour of cathode rays – their deflection, their power of penetration, the fact that they can do 'work'. Electrons were the best explanation of these results.

Another justification of scientific realism comes through the 'no-miracles (or ultimate) argument'. To make it clear Pierre Cruse presents a very nice example:

"Suppose that a friend of yours comes up with an implausible conspiracy theory, for example, that the government has rigged the result of the national lottery, so that a prominent member will win it at the next draw. Then suppose that a prominent government member does in fact win the next lottery draw. In that case it is possible that your friend's theory is false, and that the government member won just by chance. But for this to be the case your friend's prediction would have to have been incredibly lucky. Given her successful prediction, a more reasonable conclusion seems to be that her theory was actually true" [51].

The same is true for the scientific theories. If one makes predictions based on a scientific theory and if the events that the theory predicts genuinely occur, then it is reasonable to conclude that the theory is true. This is the base for the no-miracles argument of scientific realism. The obvious and manifest success of science is not through a series of miracles. Throughout the history many scientific theories have made novel predictions and were true. The same was expected by scientific realism.

Now consider the theories that were regarded as failures, for example, Aristotle's theory of crystalline spheres and Newton's theory. From a realist's view, Newton's theory was a limiting case of Einstein's theory of relativity. So a realist claims that Newton's theory is approximately true or close to the truth. Thus, as Pierre Cruse opinioned: "instead of saying that successful theories are completely true, the realist can claim that they are generally approximately true, and that this, rather than their actual truth, explains their success" [52]. A false theory may be able to make some predictions but it can never come up with a *novel* prediction. If a false theory could make a number of correct *novel* predictions, then it is a miracle. This leads to the conclusion that only theories that could make successful novel predictions are approximately true.

2.14 MODERNISM AND POSTMODERNISM

Modernism is a philosophical movement that, along with cultural trends and changes, arose from enormous transformations in Western society during the late 19th and early 20th centuries. Modernism can be considered a trend of thought that affirms the power of human beings to create, improve, deconstruct and reshape their environment, with the aid of scientific knowledge, technology, and practical experimentation, and is thus in its essence both progressive and optimistic. Modernism usually refers to neo-classical and enlightenment assumptions such as concerning the role of reason, rationality, scientific reasoning, etc., which enhance our understanding of the human condition. It believes in the principle that reason and science provide an accurate, objective, reliable foundation of "knowledge". Reason transcends and exists independently of our existential, historical, cultural contexts; it is universal and "true". Modernism became a great social movement since it was believed that the truth, we found would help to make a new world.

The modern period has been a period of significant development in the fields of science, technology, politics, and warfare. It has also been an age of discovery and globalization: it is during this time that the European powers and later their colonies, began their political, economic, and cultural colonization of the rest of the world. By the late 19th and early 20th century, modernist art, politics, science, and culture had come to dominate not only Western Europe and North America, but almost every civilized area on the globe, including movements, thought of as opposed to the West and globalization. The modern era is closely associated with the development of individualism, capitalism, urbanization, and a belief in the positive possibilities of technological and political progress.

Postmodernism is one of the intellectual currents that had the greatest prominence in and effect on various areas in the contemporary era and emerged in the seventeenth to eighteenth centuries. Postmodernist thinkers were at first concerned mostly with literature but soon extended their criticism to all forms of culture, including beliefs about morals, gender, and science. In the present world, it is seriously discussed in scientific circles, various political, social, educational, and many other areas. The postmodern is characterized by questioning these sorts of knowledge and belief in progress. The fundamental differences between modernism and postmodernism are that, while modernism emphasizes realities and the discovery of realities, postmodernism emphasizes the instability of everything and the creation of realities. Therefore, the term 'postmodernism' is applied to several movements in the above-mentioned disciplines that reacted against modern movements. "Modernism believes in certainty, necessity, and meta-narrative, while, in postmodernist view, under no circumstances should any mention of these categories be made. Particularly in the area of values, moralities, politics, and education no mention should be made of universal and constant theoretical foundations. Values are relative things that differ from culture to culture. Therefore, the postmodernism movement is actually the era after modernism and a type of transition from it" [53].

The origin of postmodernism was introduced by the French philosopher Jean Francois Lyotard in 1979 in the book *The Postmodern Condition: A Report on Knowledge* which was originally published in French and translated to English in 1984. Lyotard characterized the postmodern era as one that has lost faith in all grand, totalizing "metanarratives" — the abstract ideas in terms of which thinkers since the time of the Enlightenment have attempted to construct comprehensive explanations of historical experience. According to Lyotard "scientific knowledge does not represent the totality of knowledge; it has always existed in addition to, and in competition and conflict with, another kind of knowledge" [54] Moreover, once scientific knowledge is understood in terms of other kinds of knowledge, that is, not as the basis on which all other forms of knowledge are constructed, it is possible for Lyotard to claim that "the game of science is thus put on par with the others".

A few characteristics of postmodernism are given below.

- It is characterized by questioning all sorts of knowledge and belief in progress.

- Postmodernism insists that there are many truths as opposed to a single pure truth.
- In postmodernism, all social institutions fall under scrutiny, as do all philosophical assumptions.
- There is a questioning of all ideologies rather than accepting certain truths as natural.
- Postmodernism questions the idea of presence, or immediate experience as being reliable and real.
- It believes that there are no universal, objective means of judging any given concept as "true".
- For most postmodern people, the concept of absolute truth does not exist.
- The truth may exist independent of human consciousness but there are no objective means of nailing it down. All postmodern conclusions lead from this assumption.

In science also, one can find the elements of postmodern thinking. The term "postmodern science" was first used by Stephen Toulmin in 1981 [55]. Scientific authorities are not only questioned on the quality and honesty of their experts but also attacked, by postmodernism, on the scientific method and its universality. "Postmodernists often blame modern science for all the environmental problems of the world. They campaign for traditional knowledge and local practices, most of which are not led by rational and scientific criteria. Postmodernists assert that modern science is just one culture-bound way of looking at nature. According to them, the content of all knowledge is socially constructed; even the 'facts' and 'laws' of modern science are western constructions reflecting dominant interests and cultural biases of western societies" [56]. In a postmodernists world, truth becomes relative, and often knowledge is replaced by interpretation. The book *The Tao of Physics: An Exploration of the Parallels between Modern Physics and Eastern Mysticism*, Fritjof Capra, can be considered a landmark in the whole history of postmodernism. In this book, Capra makes an assertion that both physics and metaphysics are leading to the same system of knowledge. Capra, who was a critic of Descartes' reductionism, exhorted the society to abandon conventional linear thoughts and the mechanistic view of Descartes.

However, it is interesting to note that, postmodernism has not much to do in science since science and scientific knowledge are based on the scientific method and postmodernism is more or less a cultural movement. In 2004, Nanda pointed out that, "while there could be different styles in art, literature, mythology, architecture, and the like, science cannot have different styles, and the criteria of scientific inquiry are applied uniformly to all, cutting across national and cultural differences" [57]. Toulmin (1981), observed that by the twentieth century, the main features of postmodernism have been transformed to:

- the separation between scientific thought and technological practice;

- the philosophical foundations that underlie scientific theories;
- the presumed connection between objectivity and value neutrality; and
- the professionalization of scientific activities.

In a simple sense, modernism has allowed the possibility of science and has even encouraged its ascendancy. Postmodernism, on the other hand, has many components that deny the possibility of modernist science.

REVIEW QUESTIONS

1. Explain the classification of the philosophy of science.
2. What are epistemology and metaphysics?
3. What is a scientific statement? Explain with an example.
4. What do you mean by the reproducibility of a scientific statement?
5. Explain the classical approach of the philosophy of science.
6. Write a detailed note on empiricism.
7. Explain the inductive approach and the deductive approach.
8. Explain the assumptions of the inductive approach.
9. Discus Hume's problem of induction.
10. Explain deduction with a suitable example.
11. Explain the features of positivism.
12. What are the beliefs of positivism?
13. Discuss the various steps involved in positivism.
14. In positivism, explain how knowledge is treated.
15. What is logical positivism?
16. Discuss falsificationism.
17. Discuss the criticism against falsificationism.
18. Discuss the Duhem problem.
19. Write a note on scientific skepticism.
20. With a suitable example explain the paradigm and paradigm shift.
21. Explain the scientific revolution with a suitable example.
22. Discuss the features of interpretivism.
23. Discuss the features of post-positivism.
24. Explain relativism. What are its characteristics?
25. Write a note on reductionism and holism.
26. Discuss scientific realism.
27. What is modernism in the philosophy of science?
28. Distinguish between modernism and postmodernism.

29. What are the characteristics of postmodernism?
30. What are the twentieth-century features of postmodernism?

REFERENCES

[1] John Losse, *A Historical Introduction to the Philosophy of Science*, Oxford University Press, 1972, p. 1.

[2] Ibid., p. 2.

[3] James Ladyman, *Understanding Philosophy of Science*, Routledge, New York, 2002, p. 5.

[4] Jeffrey A Lee, *The Scientific Endeavor: Methodology and Perspectives of Science*, Dorling Kindersley Pvt. Ltd, 2000, p. 13.

[5] James Ladyman (2002), p. 21

[6] Francis Bacon, *Novum Organum*, (1620), p. 32.

[7] Richard D Jarrard, "*Scientific Methods*", Online Book, 2001, p. 42.

[8] James Ladyman (2002), pp. 22-23.

[9] Steffen Ducheyne, "*Bacon's Idea and Newton's Practice of Induction", Philosophica* 76 (2005) pp. 115-128.

[10] Samir Okasha, *Philosophy of Science: A Very Short Introduction*, Oxford University Press, 2002, p. 24.

[11] Ibid., p. 25.

[12] Richard D Jarrard (2001), p. 72.

[13] Ibid., p. 75.

[14] Ibid.,

[15] Anne B Ryan, Post Positivist Approach to Research, http://eprints.nuim.ie/874/1/post-positivist_approaches_to_research.pdf

[16] Ibid., p. 15.

[17] Robert Klee, *Introduction to the Philosophy of Science: Cutting Nature at Its Seams*, Oxford University Press, 1997, pp. 28-29.

[18] Ibid., p. 30.

[19] John T Cossioppo, Guun R. Semin and Gary G. Berntson, "Realism, Instrumentalism, and Scientific Symbiosis", *American Psychologist*, 2004, Vol. 59, No. 4, 214 –223.

[20] Jeffrey A Lee (2000), p. 16.

[21] James Ladyman (2002), p. 66.

[22] Michael J Crotty, "The Foundations of Social Research: Meaning and Perspective in the Research Process" Sage Publications, 1998.

[23] Karl Popper, *The Logic of Scientific Discovery*, Routledge, New York, 1992, p. 280.

[24] Crotty, p. 33.

[25] Karl Popper, *Conjectures and Refutations*, London: Routledge and Keagan Paul, 1963, pp. 33-39.

[26] James Ladyman (2002), pp. 81-86.

[27] Dorling, Jon. "Bayesian Personalism, the Methodology of Research Programmes, and Duhem's Problem" *Studies in History and Philosophy of Science* 10 (1979): 177-87.

[28] Merrilee H. Salmon et al., *Introduction to the Philosophy of Science*, 1992, Hackett Publishing Company, Cambridge.

[29] Joseph L Goldstein, "Paradigm Shifts in Science: Insights from the Arts", *Nature Medicine*, Vol. 18, No. 10, October 2012, pp. iii – vii.

[30] Thomas S Kuhn, "The Structure of Scientific Revolutions", *International Encyclopedia of Unified Science,* 1966, Vol. 2, No. 2, p. 10..

[31] Brain P Levack (2007), "The West: Encounters and Transformations"; Pearson Education, Chapter 17, p. 11.

[32] Jeffery A Lee (2000), p. 19.

[33] James Ladyman (2002), pp. 117-118.

[34] Ibid., p. 118.

[35] Frances E Racher and Steven Robinson, "Are Phenomenology and Post-positivism Strange Bedfellows?", *Western Journal of Nursing Research* (2002) 25(5), 464-481.

[36] Ibid.

[37] Anne B Ryan, p. 13.

[38] Ibid.

[39] Frances E Racher and Steven Robinson (2002), 25(5), 464-481.

[40] Louis Caruana, "Science and Ethics: tracing parallels and contrasts between Science, Relativism and Utilitarianism", *Revista Portuguesa de Filosofia* 62 (2006), pp. 119-36.

[41] Steven D Hales " A companion to Relativism", Wiley-Blackwell, 2009, p. 3.

[42] Ibid., p. 4.

[43] Maria Baghramian, "*A Brief History of Relativism*" Columbia University Press, 2010, p. 8.

[44] Richard J Bernstein, *Beyond Objectivism and Relativism: Science, Hermeneutics, and Praxis*, University of Pennsylvania Press, 1983, p. 8.

[45] Maria Baghramian p. 1.

[46] Pierre Cruse, "*On Scientific Realism*", Richmond Journal of Philosophy 3 (Spring 2003), http://w ww.richmond-philosophy.net/rjp/back_issues/rjp3_cruse.pdf

[47] T Jayaraman, "Scientific Realism for the Contemporary Materialist", *The Marxist*, Vol. XXIII, 1, January to March, 2007.

[48] Stathis Psillos, "The Present State of Scientific Realism Debate", *Brit. J. Phil. Sci*, 51(2000), 705-728.

[49] Ibid.

[50] John T Cossioppo, Guun R. Semin and Gary G. Berntson, "Realism, Instrumentalism, and Scientific Symbiosis", *American Psychologist*, 2004, Vol. 59, No. 4, 214 –223.

[51] Pierre Cruse, "On Scientific Realism", *Richmond Journal of Philosophy* 3 (Spring 2003)

[52] Ibid.

[53] Nooshin Forghani et al., "*A Critical Examination of Postmodernism Based on Religious and Moral Values Education*", International Education Studies; Vol. 8, No. 9; (2015)

[54] Jean Francois Lyotard "*The Postmodern Condition: A Report on Knowledge*". p. 7, (1984)

[55] Toulmin, Stephen. "The emergence of post-modem science. In: *The Great Ideas Today*", 68-114. (1981) Chicago: Encyclopedia Britannica.

[56] C George Thomas "*Research Methodology and Scientific Writing*", Springer (2021).

[57] Nanda, M. "*Prophets Facing Backward: Postmodernism, Science, and Hindu Nationalism*". Permanent Black, New Delhi, (2004)

Suggested Readings

Robert Nola and Howard Sankey, *Theories of Scientific Method*, Acumen Publishing Limited, UK, 2007.

1. Rudolf Carnap, *Philosophical Foundations of Physics*, Basic Books Inc. New York, 1966.
2. Sahotra Sarkar and Jessika Pfeifer, *The Philosophy of Science: An Encyclopedia*, Routledge, 2006.
3. David Papineau, *The Philosophy of Science*, Oxford University Press, 1996.
4. Wilbur Applebaum, *The Scientific Revolution and the Foundations of Modern Science*, Greenwood Press, London, 2005
5. Marcus Hellyer, *The Scientific Revolution*, Blackwell Publishing, 2003.
6. Walter Roy Laird and Sophie Roux, *Mechanics and Natural Philosophy Before Scientific Revolution*, Springer, 2008.
7. Colin Howson, *Hume's Problem: Induction and the Justification of Belief*, Clarendon Press, Oxford, 2000.
8. Carveth Read, *Logic: Deduction and Induction*, Dodo Press, 1914.
9. Wilbur Applebaum, *Encyclopedia of the Scientific Revolution: From Copernicus to Newton*, Garland Publishing Inc. 2000.
10. Larry Laudan, *Beyond Positivism and Relativism: Theory, Method and Evidence*, Westview Press, 1996.
11. Morris R Cohen and Ernest Nagel, "*An Introduction to Logic and Scientific Method*", Allied Publishers Limited, 1998

12. Barry Gower, *Scientific Method: A Historical and Philosophical Introduction*, Routledge, 1997.
13. Hall A R, *Scientific Revolution-1500-1800: The formation of Modern Scientific Attitude*, Logmans Green and Co., 1954.
14. Friedrich waismann, *Causality and Logical Positivism: Vienna Circle Institute Year Book,* Springer, 2011
15. Toby E Huff, *Intellectual Curiosity and the Scientific Revolution: A Global Perspective*, Cambridge University Press, 2011.

Chapter 3

Getting Started with Research

Experience without theory is blind, but theory without experience is mere intellectual play.

—**Immanuel Kant**

... what goes into the word dies: what goes into the work lives.

—**Friedrich Waismann**

3.1 INTRODUCTION

Research is the investigation of a particular topic through a series of well-organized processes. The major goals of research include establishing facts, analyzing information, reaching *new* conclusions, etc. Attempting to replicate previously published (and thus not new) results with the aim of confirming them is also research. A systematic-critical review and compilation of previous results in a certain area can also raise knowledge levels, and can therefore also be regarded as research [1]. As we have seen in the first chapter research can be categorized based on various criteria. In all types of research the ultimate aim is to produce something novel and the result is often evaluated in terms of its utility, of course basic research need not have immediate practical applications. It is important to assure high quality in research programmemes in order to produce knowledge that is applicable outside of the research setting.

In recent years researches in various fields are gaining an unprecedented focus and attention. All over the world government as well as various industries are investing huge sums for research activities. The effectiveness of a research programmeme depends on a number of factors right from the identification of a relevant problem. The probability of success of a research project is greatly enhanced when the "beginning" is correctly defined. The term *beginning* indicates the planning of research, a precise statement of goals and justification. The resources necessary to conduct the study is to be determined by the researcher. This is important because, due to deficiency of such resources the researcher may fail to complete research on time and in budget.

Undertaking research is a challenging and rewarding experience. It provides an opportunity for an in-depth study of the selected topic. Good research produces results that are appreciated by peers, whose methodologies can be replicated and knowledge that can be applied to real-world situations. Considerable time may be needed to develop an objective that is clear and specific and to plan and design the research. It is to be noted that scientific research does not include routine activities carried on as an incident to commercial or industrial operations, as, for example, the ordinary testing or inspection of materials or products or the designing or construction of equipment, buildings, etc.

The success of a research programmeme inevitably depends on the interest of the researcher as well as on the planning and execution. A desire to know and understand serves as a motivation for research. It is a commonly accepted fact that nothing contributes to the quality of your work more than your commitment to it. A researcher should always keep in mind:

"Your readers expect you to do more than just mound up and report data; they expect you to report it in a way that continues the ongoing conversation between writers and readers that creates a community of researchers. To do that, you must select from all the data you find just those data that support an answer to a question that solves a problem your readers think needs solving"

—**(Brenda Booth et al., 2002).**

Research contributes in many ways to the development of the individual and the society. Besides providing new knowledge, research educates the individual and the society, also provides training in critical thinking. Scientific research and knowledge are of value not only as an instrument, i.e., a means of achieving something else by which we set store. The knowledge gained is also of value in its own right, regardless of how it might conceivably be used [2].

3.2 MOTIVATION IN RESEARCH

Motivation is an important requirement for good scientific research. It is observed that many of the researchers lack sufficient motivation and after one or two years of their research work, they feel bored, irritated, unsatisfied, or even hate the research work. The immediate motivation for joining a research programme may be any of the following.

- Desire to get a research degree along with its consequential benefits.
- Desire to face the challenge in solving the unsolved problems, i.e., concern over practical problems initiate research.
- Desire to get the intellectual joy of doing some creative work.
- Desire to be of service to society.
- Desire to get respectability.

The majority of the researchers stop their research after acquiring a research degree. Among the above-stated motivations, the desire to get the intellectual joy of doing some creative work is the most important one as it is essential for a researcher to carry out the research work for a long period of time. Often, failure may arise in research and can overcome such situations only if the researcher is motivated by long-term goals and not just by the desire for a short-term achievement. Many more factors such as directives of the government, employment conditions, curiosity about new things, desire to understand causal relationships, social thinking, and awakening, etc., may motivate people to perform research activities.

With the aim of making the students aware of the research activities, UGC in India and other such regulatory bodies and universities all over the world included a research project in undergraduate and postgraduate curriculum. However, its aim is not fulfilled completely due to many reasons such as lack of proper motivation, unavailability of research facilities, time constraints involved in the semester system, etc. The study of Natalia V. Kozlova et al. [3] showed this factor very clearly. In their words, "most of the participants starting their research projects as part of their university curriculum were not aware of the role of research work in their professional development. They did not understand how their research work could contribute to their future professional career. Most of the respondents were oriented towards external factors such as tutors' or other people's respect and tutors' requirements as well as formal academic achievements (e.g., good marks and success in exams). Moreover, the results revealed that a majority of the study participants had positive attitudes towards students' research work at university but only less than a third were interested in their own research projects". Therefore, it is essential to give proper motivation and guidance for the undergraduate and postgraduate students about the research activities and about the role of research activities in their professional development.

3.3 INTUITION IN RESEARCH

The fundamental goal of research can be the acquisition of knowledge. Knowledge can refer to both theoretical and practical understanding of a subject, which includes facts, descriptions, information, and skills acquired through experience or learnt through books or other means. Knowledge becomes information when it is communicated through any media such as journals, books, web media, etc.

3.3.1 Methods of Acquiring Knowledge

Kerlinger [4] identified the four methods for acquiring knowledge. A brief description is given below.

- *Method of Tenacity:* From childhood onwards, we receive certain information from our parents and elders and we believe and accept those things to be true. This is mainly due to our social structure. The majority of such beliefs are well rooted and not easy to remove. Therefore, it is very difficult to change such beliefs even if there are some conflicting pieces of evidence. Many of our beliefs about superstitions originate like this and society is not ready to change such beliefs even in this modern era.
- *Method of Authority:* In the olden days, kings or other authorities such as priests or other leaders were the decision makers. Nobody was allowed to question the authority. The information passed by such authorities was taken for granted by society and they believed and lived their lives based on such information and knowledge. Most of the religious beliefs written in holy books also come under this category. In a simple sense, the authority is considered infallible. Compared to the method of tenacity, the method of authority is acceptable if it is corroborated by evidence and reasoning. Definitely, this is not possible in science. The authority of the king or politicians is accepted, we may have to end up with the fate of Galileo.
- *Method of Intuition:* The method of intuition or a priori method of acquiring knowledge is considered to be superior to the two already mentioned. The method of intuition is called a priori method as reasoning is done from what is 'prior' or 'before'. In the method of intuition, the propositions accepted are self-evident. An important characteristic of this method is that, in intuitive propositions, the reason is considered the criterion of truth rather than experience. This method mainly draws reason from cause to effect or from observed fact to another fact or principle not observed. Even though there is no scientific evidence for the conclusion, it may be accepted sometimes. The difficulty of this method is that if two or more reputed individuals arrive at different conclusions based on their intuitions, then one cannot accept any of those conclusions.
- *Method of Science:* The scientific method is the superior method to acquire knowledge systematically by formulating a hypothesis, testing and verifying the hypothesis through observations and other experimental activities. It helps for finding the answer systematically. Among the four methods of acquiring knowledge, the scientific method is the most reliable. More details about the scientific method and its characteristics will be discussed in the coming chapters.

In research, intuition plays an important role in many circumstances. Sometimes, new ideas can be originated quite unexpectedly from observation, some incident, or while doing an experiment in the laboratory. Through such incidents, a scientist may get a sudden stimulation or idea to solve a particular problem or to explain a certain observed phenomenon. A well-known example is the story of Archimedes who got a method to find the specific gravity suddenly when he entered a bathtub.

3.3.2 Favourable Conditions for Intuition

Intuitive ideas are very useful in research. Jerrald [5] pointed out the following favourable conditions for intuition.

- *Define the Problems Correctly:* A researcher needs to have clarity in identifying the problem. If the problem is clear, then, the chances of success through intuition are better. Describing the problem to other competent individuals may also help in solving the problem because it forces the researcher to define the problem in simple terms.
- *Accumulate All the Needed Facts:* It is important to collect all the relevant information related to the problem. Then, the collected information is to be analyzed with a critical mind. Sometimes, a good literature search may prompt insights, because it compels us to go through the data and its interpretation in a logical sequence.
- *Desire a Solution:* It is experienced that a researcher may have a personal stake in a problem, which may help or hinder intuition. This can be affected in two ways. In one way, an interest in solving the problem can promote concentration and imagination. The interest pulls the researcher to pursue a task and help to overcome difficulties and disappointments. In a second way, an intense desire for a solution can become destructive, if it leads to disturbing worry and anxiety.
- *Discuss with Others:* Discussing the ideas with others may help in intuitive ideas. As they have different perspectives, the researcher may benefit from those perspectives directly or may be able to combine the existing knowledge with theirs. When carried out in a relaxed and friendly manner, these discussions are most likely to encourage insight.
- *Temporary Abandonment of the Problem:* Another option is to simply relax and abandon the work temporarily whenever you encounter anomalous data in your experiment or observation. These can create a favourable environment for the insight to happen.

Effective practice of these principles will help the researcher to make use of his intuition effectively in making conclusions about the research problem. However, it is very important that the conclusion must be supported by sufficient scientific evidence that could be verified.

3.3.3 Serendipity

Serendipity is an unconventional way to intuition. Sometimes, researchers make some wonderful discoveries or inventions simply by accident or luck. The term serendipity is used to denote the chances of such accidental discoveries. There are some good examples of such discoveries in science.

- In 1828, Friedrich Woehler was trying to produce ammonium cyanate by mixing potassium cyanate and ammonium chloride. In the process, he obtained urea, which is the first organic chemical compound synthesized.
- In 1895, Roentgen was conducting some experiments on the effects of cathode rays at the University of Wurzburg. He noted that certain rays were emitted when a current was passed through a covered discharge tube. This was indicated by the illuminations on a barium-covered screen kept nearby. Through further experiments, he confirmed that some were produced by the impact of cathode rays on objects. Since the content of the rays was not known, he called them the X-rays.

3.4 PLANNING AND DESIGNING RESEARCH

Planning and designing a research form an important part of the research activity. It requires some interest, experience and expertise. For a beginner, it not an easy task. Care should be taken that the topic involves original research and critical thinking. It is common to suggest the following steps to plan a research:

- Find a topic specific enough to let you master a reasonable amount of information on it.
- Ask questions about that topic until you find some that catch your interest.
- Determine what kind of evidence that your readers will expect in support of your answers.
- Determine whether you can find sources that have those data.

According to Justin D Congdon and Arthur E Dunham (1999) research plan consists of two general areas: (i) research concepts and contexts and (ii) research logistics. They have presented an outline of each.

An outline of the logistics section of a research plan [6]:

1. Identify what information needs to be collected and how it will be collected.
   (a) Develop a logistical and quality control plan for data collection, handling, and storage procedures, including who will be responsible for each procedure.
   (b) Design data sheets that prompt the person collecting data for each measurement. Data sheets should also be made with data recording, entry into computer files, editing and data analyses in mind.
   (c) Prior to actual data collection, carefully "walk through" as many of the techniques and procedures as feasible to detect problems with protocols and equipment.
   (d) Evaluate incoming data for recurring errors; frequent review may reveal unsuspected patterns that, if identified and responded to quickly, provide opportunities to improve research protocol or direction.

2. Talk to other investigators currently working with similar types of studies about logistics, research protocol, quality control plans, and types of data analyses.
3. As soon as possible following data collection, the data should be entered into computer files and then reviewed to detect problems associated with data sheets, data recording and entry, and with computer files.
4. Order supplies as far in advance as reasonable and funds are available.
5. Plan for the unexpected (equipment failure, accidents, illness, unusual weather).

An outline of the concepts and context sections of a research plan [7]:

1. Literature survey
   (a) Become familiar with the literature to identify a research problem and to explore the areas that may impact the research plan.
   (b) Explore ways that enhance the integration of the study into broader areas and ways to collect, analyze, and present your data so that they are useful to others.
2. Carefully and clearly state the problem in the form of concise questions or as working or null hypotheses.
3. Make a list of possible scenarios related to your research questions and then rank the most probable ones.
4. Discuss all aspects of the research proposal with colleagues as you develop them. Investigators that work in the same or closely related fields are valuable sources of information.
5. Write a detailed research proposal.

Each of the points mentioned above are very important and needs detailed understanding. A good integration of the steps mentioned above is essential for the successful completion of a research programmeme. During planning informal discussion can be made with other researchers which may help in refining the ideas. In addition some logistical problems may already have been faced and solved by other researchers. This is then followed by the writing of detailed research proposal. In words of Justin D Congdon and Arthur E Dunham (1999), "the process of writing a detailed proposal will help identify problems with research concepts, questions, and logistics and will enhance integration of various aspects of the proposed research".

3.4.1 Research Design

Research design is an important part of a research programmed because it facilitates the smooth sailing of the various research operations, thereby making research as efficient as possible yielding maximal information with minimal expenditure of effort, time and money (Kothari, p. 32). The research design is to be done carefully as any error in the design will reflect in the entire project. The research can be viewed as a three-stage process: (i) identification of a relevant problem; (ii) planning and deigning research; and (iii) execution of research. In the previous section we have discussed the various considerations regarding the selection of a research problem. Research design is not just a plan of action. The purpose of research design is to ensure that the data collected through various methods must lead us to a solution to the defined problem. In other words a research design is the arrangement of conditions for collection and analysis of data in a manner that aims to combine relevance to the research purpose with economy in procedure [8]. Research design involves several decisions and assessments, including selecting a design and proposed data collection strategies [9]. The research design has the following aspects

1. A clear statement of a scientific problem.
2. Procedure and techniques to be used for gathering information.
3. Methods to be used for processing and analyzing data [10].

Proper planning is necessary for the successful execution of the research. Therefore at any stage of the planning the researcher can make a revisit to the first stage if he finds it necessary; and can make necessary modifications if facing any difficulty in finding necessary resources. A researcher should have a clear plan and design before he starts the execution of the research.

A good design is often characterized by adjectives like flexible, appropriate, efficient, economical and so on (Kothari, p. 33). Also the type of research design to be selected is determined by the problem under consideration. A particular design that proved efficient in solving a problem may not be that much useful in another situation. Therefore the research design demands a good understanding of the research problem. As Kothari observes a research design appropriate for a particular research problem, usually involves the consideration of the following factors [11]:

- The means of obtaining information;
- The availability and skills of the researcher and his staff, if any;
- The objective of the problem to be studied;
- The nature of the problem to be studied; and
- The availability of time and money for the research work.

Scientific research is usually focused around two activities: *data collection and analysis*, and *testing the hypothesis*. Data collection and analysis involves the identification of important variable and identification of the behaviour and the relationship between those variables. Hypothesis testing usually involves the possible explanation of the observed behaviour and relationships. The observed relationships can be of two types: *causal relationship* and *correlational relationship*. In causal relationship there is a direct influence of one variable on the other. In correlational relationship changes in one variable influence the other but detailed experimentation is required to confirm whether the variable actually influences the other. The researcher should have knowledge of different types of data collection and of their relevance and reliability in the present situation. Appropriate and reliable methods for data collection are to be adopted (detailed discussion is in chapter 5).

The adoption of a suitable methodology from a number of available methodologies is also an important fact that determines the progress and success of the research. The researcher should have sufficient knowledge about different methodologies. He should have the ability to understand the strengths and limitations of those methodologies so that he can select the most appropriate methodology for solving the problem defined by him. [For a detailed discussion of different types of research design refer to *Research Methodology: Methods and Techniques* by C R Kothari.]

3.4.2 Selecting a Research Topic

The first and most important challenge in a research programme is the selection of an appropriate topic for research. The manner in which the problem is defined is crucial to the success of the study. This first step is the recognition of a felt difficulty, an obstacle or a problem that strikes the mind of the researchers. The choice of a relevant topic is not an easy task especially for the beginners. This demands much time and deep thought. The personal experience of a researcher on the events and occurrences in the related discipline will help him in identifying the topic. The selection of the topic includes various criteria that may set by the interest of the researcher, the society, the government, etc. Therefore, the researcher must consider a number of factors related to the purpose, planning, design, execution and completion of the research. Broad classification of these factors is made by Van Dalen and Meyer (1966) as personal considerations and social considerations. They are listed below as cited by A Y Abdulkarrem in the book *Research Methodology in Education: An Interdisciplinary Approach*" edited by S A Jimoh [12].

Personal Considerations

1. The research problem should be in line with the researcher's goal, expectations and the expectations of others.
2. The researcher should be genuinely interested in the problem and should be free from strong biases.
3. The researcher should possess or should be able to acquire the necessary skills, abilities, competency, and requisite knowledge to study the problem
4. The researcher should have access to the subjects and facilities necessary to carry out the study.
5. The researcher should possess the time and money required to complete the research work.

6. The researcher should obtain the administrative support, guidance and cooperation for the conduct of the study.

Social Considerations

1. Will the solution of the problem appreciably advance knowledge in the field?
2. Will the findings be of practical value to people?
3. Will the findings have a wide coverage and for long period of time?
4. Does the topic need to be extended beyond its present limit?
5. Is the topic sufficiently delimited to permit an exhaustive treatment and yet is sufficiently significant to warrant its investigation?
6. Will the conclusions of the study be of doubtful value because the tools and techniques available for use are not adequate and sufficiently reliable?
7. Will the study lead to the development of other one?

The considerations presented above are not a complete list, depending upon the locality, society, government and the interests of various funding agencies more considerations may be added.

One major problem associated with the selection of a topic is that the topic may be too broad. This places the researcher in an ocean of published papers. In order to avoid this refining or narrowing of the research topic is essential. The broadness of the topic may be due to one or more of the following reasons:

- It deals with a complex issue studied over an entire century or more, for example: "The development of science from 1600 to 1700" instead of selecting a small period of one decade.
- It deals with the complete works of a prolific author, for example, *Life and Works of Einstein* instead of selecting a particular work
- It focuses on general rather than on specific problems, forexample, "Factors affecting the quality of thin film" instead of stating "the effect of chamber pressure on the quality of thin films".
- It asks no question or presents no hypothesis; example: "The black holes" rather than "how we can detect a black hole?"

There is no step-by-step procedure to narrow a topic. It is integrated with the literature survey and the knowledge of the researcher. The researchers have to take effort to acquire sufficient background knowledge in the relevant area for defining a well refined and sharp research problem. The various stages involved in selecting a topic for research can be summarized as follows.

- Brainstorm for ideas
- Read general background information
- Focus in on a manageable topic
- Define your topic as a focused research question

3.4.3 Criteria for Good Research Problem

There are no universal criteria for the evaluation of a research problem. According to Kerlinger (1973) good research problems must meet three criteria. First, the research problem should describe the relationship between two or more variables. Second, the research problem should take the form of a question. Third, the research problem must be capable of being tested empirically (i.e., with data derived from direct observation and experimentation).

As per the guidelines issued by the University of Amsterdam (2008) a good research problem satisfies the following criteria:

- *Question mark:* The research problem can be in a declarative or in a question form. The University recommends formulating the research problem as a question. This gives something to hold on to during the rest of your thesis because it is simple: there is a question and in the text you look for an answer.
- *Possibility to respond:* Some questions are impossible to answer in a scientific way, for example, 'how beautiful is the colour yellow'. We don't have the scientifically justified instruments to answer this question. It must also be possible to answer the question in a practical way so it must be researchable, meaning you have to be able to collect evidence that will answer the question.
- *Relevance in connection to the research goals:* The research problem and the research goals are closely connected to each other. If an answer to the problem doesn't meet the goal of the research, one of them should be adapted. For example, it's wrong to connect a describing problem to an advising goal.
- *Attainability:* The problem must be one that can be solved during the amount of time you have. So it can't be too broad (ex: 'How can we have world peace?'). But it also can't be too narrow (ex: How does my neighbour think about Americans?').
- *Open question:* The research problem should be an open question. That means it cannot be answered by "yes" or "no". But also with open questions you should watch out for the possibility of a shallow answer.
- *Unmistakability:* Your research problem must be clear and there has to be only one way to interpret it. For example, the question: 'What do Vietnamese people think about the West?' is unmistakable because it is not clear what is meant by 'the West', it can be a lot of things.
- *Punctuality:* The problem must be clearly specified. For example: Don't write 'How can prejudices against Americans be combated?' if you mean: 'How can prejudices that live among Vietnamese students for American businessmen be combated?'

- *Brevity:* Although your research problem should be as punctual and specific as possible, not all fencings must be placed in your research problem. It must be a brief and fluent sentence. You can specify your terms with definitions in a commentary.

It is essential to use simple words in defining the technical terms and scientific words used in the statement of the problem. Also the basic assumptions related to the research problem should be made clear.

3.4.4 Sources of Research Ideas

Ideas for research can be obtained through various sources. Kenneth S Bordens and Bruce B Abbot (2011) categorized these sources into three: 'experience, theory and applied issues' [13].

Experience: Experience and observation of things and events that goes on around is a rich source of research ideas. Such observations generate a number of questions in our mind. These observations can be unsystematic or systematic. For example, reading of newspaper or watching an event on television is an unsystematic observation. On the other hand reading a journal paper for the preparation of a class is a systematic observation. The depth of questions that arouse in our mind in these situations will be different. Compared to unsystematic observation, systematic observation is a powerful source of research ideas. Systematic observation is planned and is carried out for a period of time. For example, the observation of a group of children for some time will provide some idea for developing a hypothesis about their behaviour.

Theory: Theory can be regarded as a set of assumptions about the causes of behaviour and rules that specify how those causes act (Kenneth S Bordens and Bruce B Abbot, 2011, p. 60). Theories are also rich sources of research ideas. Theories can lead to the development of research questions in two ways [14]:

1. A theory allows you to predict the behaviour expected under new combinations of variables.
2. A theory can generate research ideas when two or more alternative theories account for the same initial observations.

Applied Issues: Many of the research ideas originate from an attempt to solve the problems faced by the humanity. This constitutes the applied research. Majority of research programmes falls under this category. It is easy to identify the problems in the society because the consequences of such problems can be felt directly. Population explosion, energy crisis, drinking water, medicine, etc., are some fields where the issues are plenty and scope for research is wide.

3.4.5 Principles of Good Research

Research programmes in different areas are different in terms of problems, procedure and outcomes. However, all types of research shares a common objective of knowledge generation. Therefore, we can find some common principles and characteristics among these research programmes. Tamsin White (2006) identified the following points as the principles of good research [15].

- There is a clear statement of research aims, which defines the research question.
- There is an information sheet for participants, which sets out clearly what the research is about, what it will involve and consent is obtained in writing on a consent form prior to research beginning.
- The methodology is appropriate to the research question. So, if the research is into people's perceptions, a more qualitative, unstructured interview may be appropriate. If the research aims to identify the scale of a problem or need, a more quantitative, randomized, statistical sample survey may be more appropriate. Good research can often use a combination of methodologies, which complement one another.
- The research should be carried out in an unbiased fashion. As far as possible the researcher should not influence the results of the research in any way. If this is likely, it needs to be addressed explicitly and systematically.
- From the beginning, the research should have appropriate and sufficient resources in terms of people, time, transport, money, etc., allocated to it.
    - ♦ The people conducting the research should be trained in research and research methods and this training should provide:
    - ♦ Knowledge around appropriate information gathering techniques.An understanding of research issues.An understanding of the research area. An understanding of the issues around dealing with vulnerable social care clients and housing clients, especially regarding risk, privacy and sensitivity and the possible need for support.
    - ♦ Those involved in designing, conducting, analyzing and supervising the research should have full understanding of the subject area.
- In some instances, it helps if the researcher has experience of working in the area. However, this can also be a negative factor, as sometimes research benefits from the fresh eyes and ears of an outsider, which may lead to less bias.
- If applicable, the information generated from the research will inform the policy making process.

All research should be ethical and not harmful in any way to the participants.

3.5 CRITERIA OF GOOD RESEARCH

Research methodologies were evolved over centuries through the contributions of many philosophers and scientists. Among the various methodologies we cannot point out an ideal one. Majority of these methodologies rely on observations and logical mathematics. As we have seen in the previous chapter each methodology leads the researcher towards the truth. But remember, the absolute reality or absolute truth is still a myth. All methodology has its own limitations; extension and refinement of these methodologies enlarge the scope of research and improves the results. The quality of the research is determined by the fact that 'how closely the result approximates the truth'. James Harold Fox (1958) suggested a few criteria of a good research programme. They are [16]:

1. The purpose of the research, or the problem involved, should be clearly defined and sharply delineated in terms as unambiguous as possible.
2. The research procedures used should be described in sufficient detail to permit another researcher to repeat the research.
3. The procedural design of the research should be carefully planned to yield results that are as objective as possible.
4. The researcher should report, with complete frankness, flaws in the procedural design and estimate their effect upon the findings.
5. Analysis of the data should be sufficiently adequate to reveal its significance; and the methods of analysis used should be appropriate.
6. Conclusions should be confined to those justified by the data of the research and limited to those for which the data provides an adequate basis.
7. Greater confidence in the research is warranted if the researcher is experienced, has a good reputation in research, and is a person of integrity.

These seven criteria do not form a complete list. These are intended to provide some assistance for the researcher while formulating the research problem and for the subsequent planning and execution. It can be seen that the principles of research and the criteria of good research share some common views and it is obvious. Further, the criteria serve some guidelines for evaluating the worth of a research project. However, such as assessment based on these criteria is subjective in nature and there is no method available to evaluate a research project objectively.

3.6 GUIDELINES FOR RESEARCH SKILLS AND AWARENESS

For the successful completion of a research programme the researcher must possess some skills and he/she must have awareness about all spheres of the research process. In order to access these skills and awareness the researcher must be able to provide a positive response to the following questions (issued by the UCD Graduate School, Dublin).

- Have I good knowledge of advances and developments in my field?
- Can I demonstrate knowledge of research in related fields and disciplines?
- Do I comprehend and can I effectively employ appropriate research methodologies?

- Can I critically analyze and synthesize new and complex information from diverse sources?
- Can I formulate and apply solutions to research problems and effectively interpret research results?
- Can I exercise critical judgement and apply critical thinking to create new ways of understanding?
- Can I formulate and apply solutions to research problems and effectively interpret research results?
- Do I need to know health and safety procedures and their application in my research environment?
- Have I broad awareness and knowledge of key relevant funding sources and grant application procedures?
- Do I appreciate basic principles of project and time management?

3.7 VALIDITY OF RESEARCH

Validity is a term usually associated with a research programme and its related empirical investigations. It refers to the conceptual scientific soundness of a research study (Graziano & Raulin, 2004, cited in Geoffrey Marczyk et. al., 2005). In other words validity describes the extent to which a measure accurately represents the concept it claims to measure. Messick (1999) defines validity as "an overall evaluative judgement of the degree to which empirical evidence and theoretical rationales support the adequacy and appropriateness of interpretations and actions". The notion of validity implies that the research data have such a character that one can move legitimately from the level of the empirical variables to the level of theoretical concepts. Knowledge about the validity creates a certain level of consciousness in the mind of the researcher because the primary aim of any type of research is to produce valid conclusions. The researcher becomes vigilant about the accuracy and usefulness of the findings. Care will be taken to prevent or control any confounding variables. There are several measures of validity that determine the quality of a research work. The validity is broadly classified into two: *external validity* and *internal validity*.

External validity addresses the ability to apply with confidence the findings of the study to other people and other situations, and ensures that the 'conditions under which the study is carried out are representative of the situations and time to which the results are to apply' (Black 1999, cited in Roberts P et al., 2006). In other words, it relates to whether or not research findings can be generalized beyond the immediate study sample and setting [17]. External validity is otherwise known as the *generalizability*. In all forms of research design, the results and conclusions of the study are limited to the participants and conditions as defined by the contours of the study. A study has external validity when the results generalize to other populations, settings, and circumstances [18]. The external validity is more for the research studies in science discipline due to the very nature of science, whereas in other disciplines like sociology or economics external validity strongly depends on the sampling. A narrow sampling may not give a result that has good external validity.

There are some factors and characteristics that impose limitations on the generalizability of results and these are referred to as *threats to external validity*. Kazdin, (2003c, cited in Geoffrey Marczyk et. al., 2005) identified some of these threats and they include sample characteristics, stimulus characteristics and settings, reactivity of experimental arrangements, multiple-treatment interference, novelty effects, reactivity of assessment, test sensitization, and timing of measurement. A detailed explanation of all these factors is available in Geoffrey Marczyk et al.(2005). I would like to quote them here in brief [19]:

- *Sample characteristics:* This threat to external validity refers to a phenomenon whereby the results of a study apply only to a particular sample. Accordingly, it is unclear whether the results can be applied to other samples that vary on characteristics such as age, gender, education, and socioeconomic status (Kazdin, 2003c, cited in Geoffrey Marczyk et al.).
- *Stimulus Characteristics and Settings:* This threat to external validity refers to an environmental phenomenon in which particular features or conditions of the study limit the generalizability of the findings (Brunswik, 1955; Pedhazur & Schmelkin, 1991, cited in Geoffrey Marczyk et al.).
- *Reactivity of the Experimental Arrangements:* This threat to external validity refers to a potentially confounding variable that is a result of the influence produced by knowing that one is participating in a research study (Christensen, 1988, cited in Geoffrey Marczyk et al.).
- *Multiple-Treatment Interference:* This threat to external validity refers to research situations in which (1) participants are administered more than one experimental intervention (or independent variable) within the same study or (2) the same individuals participate in more than one study (Pedhazur & Schmelkin, 1991, cited in Geoffrey Marczyk et al.).
- *Novelty Effects*: This threat to external validity refers to the possibility that the effects of the independent variable may be due in part to the uniqueness or novelty of the stimulus or situation and not to the intervention itself. In other words the novelty of the intervention or situation acts as a confounding variable, and it is that novelty (and not the independent variable) that is the real explanation for the results (Geoffrey Marczyk et al., 2005).
- *Reactivity of Assessment:* This threat to external validity refers to a phenomenon whereby participants' awareness that their performance is being measured can alter their performance from what it would otherwise have been (Christensen, 1988; Kazdin, 2003c, cited in Geoffrey Marczyk et al.).

- *Pretest and Post-test Sensitization*: These related threats to external validity refer to the effects that pretesting and post-testing might have on the behaviour and responses of the participants in a study (Pedhazur & Schmelkin, 1991, cited in Geoffrey Marczyk et al.). It is generally observed that in many forms of research, participants are pretested to quantify the presence of some variable of interest and to provide a baseline of behaviour against which the effects of the experimental intervention (independent variable) can be evaluated.
- *Timing of Assessment and Measurement:* This threat to external validity is particularly common in longitudinal forms of research, and it refers to the question of whether the same results would have been obtained if measurement had occurred at a different point in time (Kazdin, 2003c, cited in Geoffrey Marczyk et al., 2005).

Internal validity addresses the reasons for the outcomes of the study, and helps to reduce other, often unanticipated, reasons for these outcomes [20]. Internal validity relates to the extent to which the design of a research study is a good test of the hypothesis or is appropriate for the research question (Carter and Porter 2000, cited in Roberts P et al., 2006). It is the ability of a research design to rule out or make implausible alternative explanations of the results, thus demonstrating that the independent variable was directly responsible for the effect on the dependent variable and, ultimately, for the results found in the study [21]. Internal validity sometimes helps the researcher to provide a *plausible revival hypothesis*. "A plausible revival hypothesis is an alternative interpretation of the researcher's hypothesis about the interaction of the independent and dependent variables that provides a reasonable explanation of the findings other than the researcher's original hypothesis" (Geoffrey Marczyk et al., 2005, p. 159).

In general, a research problem studies some cause-effect relationship and the researcher aims to establish the causation between the independent and dependent variables. In words of Pedhazur & Schmelkin (1991): "the researcher ultimately wants to know whether the observed effect or phenomenon is due to the manipulated independent variable or variables or to some uncontrolled or unknown extraneous variable or variables". These uncontrolled extraneous variables are the threats to the internal validity. Through a proper design of the research the validity can be increased and rival hypotheses and influences of extraneous variables can be minimized.

In addition to the extraneous variables, Geoffrey Marczyk et al.(2005, pp. 161-174) discussed a number of other threats to internal validity identified by many people like Cook & Campbell (1979) and Pedhazur & Schmelkin (1991). They include: history, maturation, instrumentation, testing, statistical regression, selection biases, attrition, diffusion or imitation of treatment, and special treatment or reactions of controls. A summary of the discussion is represented below [22].

- *History:* History as a threat to internal validity refers to events or incidents that take place during the course of the study that might have an unintended and uncontrolled-for impact on the study's final outcome or the dependent variable; (Kazdin, 2003c, cited in Geoffrey Marczyk et Al.).
- *Maturation:* Maturation refers to intrinsic changes (biological and psychological) within the participants that are usually related to the passage of time.
- *Instrumentation*: Instrumentation as a threat to internal validity refers to changes in the assessment of the independent variable, which are usually related to changes in the measuring instrument or measurement procedures over time (Christensen, 1988; Kazdin, 2003c, cited in Geoffrey Marczyk et Al.).
- *Testing:* This threat to internal validity refers to the effects that taking a test on one occasion may have on subsequent administrations of the same test (Kazdin, 2003c, cited in Geoffrey Marczyk et al.).
- *Statistical Regression:* This threat to internal validity refers to a statistical phenomenon whereby extremely high or low scores on a measure tend to revert toward the arithmetic mean or average of the distribution with repeated testing (Christensen, 1988 and Kazdin, 2003c, cited in Geoffrey Marczyk et al.).
- *Selection Biases:* This threat to internal validity refers to systematic differences in the assignment of participants to experimental conditions. Selection biases are prevalent in quasi-experimental research in which participants are assigned to experimental conditions or comparison groups in a nonrandom fashion (Christensen, 1988; Kazdin, 2003c, cited in Geoffrey Marczyk et al.)
- *Attrition:* This threat to internal validity refers to the differential and systematic loss of participants from experimental and control groups.
- *Diffusion or Imitation of treatment:* This threat to internal validity is common in various forms of medical and psychotherapy treatment effectiveness research. This reason for this is (i) unintended exposure of a control group to the actual or similar intervention (independent variable) intended only for the experimental condition or (ii) the experimental group does not receive the intended intervention at all (Kazdin, 2003c; Pedhazur & Schmelkin, 1991, cited in Geoffrey Marczyk et al.).
- *Special Treatment or Reactions of Controls:* These common threats to internal validity may be caused by the special, often compensatory, treatment or attention given to the control group. Even in the absence of special attention or treatment, controls may realize that they are in a "lesser" condition and react by competing or otherwise improving their performance.

Internal validity can be approached in three ways: *content validity*, *criterion-related validity*, and *construct validity*.

Content validity: "Content validity has to do with how adequately the content of a test samples the knowledge, skills, or behaviours that the test is intended to measure' (Kenneth S. Bordens and Bruce B. Abbott, 2011). *Face validity* is a component of content validity and is established when an individual reviewing the instrument concludes that it measures the characteristic or trait of interest. A more rigorous way to assess content validity is to ask recognized experts in the area to give their opinion on the validity of the tool.

Criterion-related validity: Criterion-related validity is assessed when one is interested in determining the relationship of scores on a test to a specific criterion. *Concurrent validity* and *predictive validity* are measures of criterion validity. "Concurrent validity uses an already existing and well-accepted measure against which the new measure can be compared and predictive validity measures the extent to which a tool can predict a future event of interest" (Alison Twycross and Linda Shields, 2004). "If the scores of a test and the criterion are collected at about the same time then the validity assessed is the concurrent validity. On the other hand comparing the scores of a test with the value of a criterion measure observed at a later time assess the predictive validity' (Kenneth S. Bordens and Bruce B. Abbott, 2005).

Construct validity: Construct validity is the degree to which an *instrument* measures the trait or theoretical construct that it is intended to measure. In other words construct validity tests the link between a measure and the underlying theory. In essence, construct validity asks the question of whether the theory supported by the findings provides the best available explanation of the results (Geoffrey Marczyk et al., 2005). "If a test has construct validity, you would expect to see a reasonable correlation with tests measuring related areas. Evidence of construct validity can be provided by comparing the results obtained with the results obtained using other tests", (Alison Twycross and Linda Shields, 2004). Cook and Campbell (1979, cited in Geoffrey Marczyk et al., 2005) make the following suggestions for improving construct validity:

- Provide a clear operational definition of the abstract concept or independent variable.
- Collect data to demonstrate that the empirical representation of the independent variable produces the expected outcome.
- Collect data to show that the empirical representation of the independent variable does not vary with measures of related but different conceptual variables.
- Conduct manipulation checks of the independent variable.

3.8 RELIABILITY IN RESEARCH

The term reliability represents the consistency of a measurement. According to Geoffrey Marczyk et al. (2005, p. 103), "it is concerned with the consistency or stability of the score obtained from a measure or assessment over time and across settings or conditions". If the measurement is reliable, then there is less chance that the obtained score is due to random factors and measurement error. "Reliability describes how far a particular test, procedure or tool, such as a questionnaire, will produce similar results in different circumstances, assuming nothing else has changed' (Roberts P et al., 2006).

There are three aspects of reliability: *equivalence, stability and internal consistency (homogeneity)*. An understanding of the distinction between these three is important in assessing the reliability of a research programme.

Equivalence: Equivalence refers to the amount of agreement between two or more instruments that are administered at nearly the same point in time. Equivalence is measured through a parallel forms procedure in which one administers alternative forms of the same measure to either the same group or different group of respondents. This administration of the various forms occurs at the same time or following some time delay. The higher the degree of correlation between the two forms, the more equivalent they are [23]. Equivalence will be important when the measurement process entails subjective judgments or ratings being made by more than one person.

Stability: Stability is said to occur when the same or similar scores are obtained with repeated testing with the same group of respondents. Stability represents the consistency of the score from one time to the next. Stability is assessed through *test-retest.*

Internal consistency (homogeneity): Internal consistency concerns the extent to which items on the test or instrument are measuring the same thing. The individual items in a test must be highly correlated with each other so that high reliability can be achieved for the entire test. The appeal of an internal consistency index of reliability is that it is estimated after only one test administration and therefore avoids the problems associated with testing over multiple time periods [24].

There are different methods to determine the reliability. They include: *test-retest reliability, split-half reliability, alternate form reliability* and *inter-rater reliability*. It is to be noted that the errors of measurement that affect reliability are random errors and errors of measurement that affect validity are systematic or constant errors.

Test-retest reliability: Test-retest reliability refers to the degree to which scores are consistent over time. It indicates score variation that occurs from testing session to re-testing session as a result of errors of measurement. Since the same test is used on each occasion, changes in scores on the test cannot be due to such factors as different wordings of the questions or nonequivalent items. Testing the academic achievement on two separate occasions, say six months apart is an example of this type of reliability. The time interval between tests should be considered with this form of reliability because test-retest correlations tend to decrease as the time interval increases.

Split-half reliability: Split-half reliability refers to the administration of a single test that is divided into two equal halves. This is appropriate when the test is very long. "This approach is much less susceptible to time-interval effects because all of the items are administered at the same time and then split into separate item pools afterward' (Geoffrey Marczyk et al., 2005). Since longer tests tend to be more reliable, and since split-half reliability represents the reliability of a test only half of the actual test, a correction formula must be applied to the coefficient. Statistical procedures such as Cronbach's alpha coefficient and the Kuder-Richardson formula can be used for this purpose. Both provide an estimate of the average of all split-half estimates of reliability.

Alternate form reliability: 'Alternate-form reliability is expressed as the correlation between different forms of the same measure where the items on each measure represent the same item content and construct. This approach requires two different forms of the same instrument, which are then administered at different times. The two forms must cover identical *content and have a similar difficulty level. The two test scores are then correlated' (Geoffrey Marczyk et al.,* 2005). The obtained coefficient is called the *coefficient of stability* or *coefficient of equivalence.* Problem associated with this is the difficulty of constructing two forms that are essentially equivalent.

Inter-rater reliability: "Inter-rater reliability is used to determine the agreement between different judges or raters when they are observing or evaluating the performance of others" (Geoffrey Marczyk et al., 2005). In simple words any research tool with inter-rater reliability should provide the same information if used by different people.

Leary (2004) in the book *Introduction to Behavioral Research Methods* suggests the following strategies for increasing reliability and minimizing measurement error:

1. The administration of the instrument or measurement strategy should be standardized—all measurements should occur in the most consistent manner possible. In other words, the administration of measurement strategies should be consistent across all the participants taking part in the study.
2. The researchers should make certain that the participants understand the instructions and content of the instrument or measurement strategy. If

the participants have difficulty understanding the purpose or directions of the measure, they might not answer in an accurate fashion, which may bias the data.

3. Every researcher involved in data collection should be thoroughly trained in the use of the measurement strategy. There should also be ample opportunity for practice before the study begins and repeated training over the course of the study to maintain consistency.
4. Every effort should be made to ensure that data are recorded, compiled, and analyzed accurately. Data entry should be closely monitored and audits should be conducted on a regular basis.

3.9 ARTEFACT AND BIAS

Research is a process that is sequential and systematic. The researcher aims at establishing a relationship between an independent variable and a dependent variable (variable of interest). During the course of research he may encounter with certain unintended variables that may influence the result of the study. Such variables are referred to as a potential confound, artefact, or source of bias. Confounds refer to variables that should have been held constant within a specific study but that were accidentally allowed to vary. Confounds are threats to internal validity. Artefacts, on the other hand, refer to variables that should have been systematically varied, either within or across studies, but that were accidentally held constant. Artefacts are thus threat to external validity. In order to minimize or to eliminate the sources of bias proper designing of the research is required. During the designing process possible sources of artefacts and biases should be identified and care must be taken to eliminate their effects on the result of the study. There are endless sources of artefacts and biases. Some examples are mentioned below.

- The heterogeneity of research participants alone can contribute innumerable sources. Research participants bring a wide variety of physical, psychological, and emotional traits into the research context' (Geoffrey Marczyk et. al., 2005, p. 68).
- Environmental factors can influence the result of a study.
- Issues related to measurements can introduce artefacts and bias.
- The use of poorly validated or unreliable measurement strategies can contribute to misleading results (Leary, 2004, cited in Geoffrey Marczyk et al., 2005, p. 68).

These different characteristics can directly affect the results of a study. Majority of the sources of artefacts and bias we have already seen as the threats to different types of validity. Researchers must be aware of these potential threats and control for them accordingly. Let us discuss some of them in brief.

Experimenter Bias: Many people pointed out that the researcher themselves are the first source of artefact and bias. The individual variations, both physiological and psychological, can very well influence the study made by a researcher. For example, some beliefs hold by the researcher will intentionally or unintentionally influence the way of study of the researcher and then the result. Experimenter bias manifests in different form (Geoffrey Marczyk et al., 2005, pp. 69-70).

- A researcher might interpret data in such a way that it supports his or her theoretical orientation or a particular theoretical paradigm.
- The researcher might be tempted to change the original research hypotheses to fit the actual data when it becomes apparent that the data do not support the original hypotheses.
- Sometimes researchers blatantly ignore findings that do not support their hypotheses.
- Subtle errors in data collection and recording and unintentional deviations from standardized procedures.

A number of strategies can be employed to minimize the experimenter bias. Careful control over the research procedures can reduce the experimenter bias to a greater extends. Geoffrey Marczyk et al. (2005, p. 72) suggests the following strategies for reducing the experimenter bias:

- Carefully control or standardize all experimental procedures.
- Provide training and education on the impact and control of experimenter effects to all the researchers involved in the study.
- Minimize dual or multiple roles within the study.
- When multiple researcher roles are necessary, provide appropriate checks and balances and quality control procedures, whenever possible.
- Automate procedures, whenever possible.
- Conduct data collection audits and ensure accuracy of data entry.
- Consider using a statistical consultant to ensure impartiality of results and choice of appropriate statistical analyses.
- Limit the knowledge that the researcher or researchers have regarding the nature of the hypotheses being tested, the experimental manipulation, and which participants are either receiving or not receiving the experimental manipulation.

Participant effects: The participants involved in a research process can be a source of artefact and bias. There are commonly referred to as the participant effects. "It refers to a variety of factors related to the unique motives, attitudes, and behaviors that participants bring to any research study" (Kruglanski, 1975; Orne, 1962, cited in Geoffrey Marczyk et al., 2005, p. 77). Participant effects are otherwise known as *demand characteristics*. "Demand characteristics are the tendencies of research participants to act differently than they normally might simply because they are taking part in a study. At their most severe, demand characteristics are changes in behavior that are based on assumptions about the underlying purpose of the study, which can introduce a significant confound into the study's findings"(Geoffrey Marczyk et al., 2005, p. 77). *Hawthorne effect* can be read in this context. It says that the level of motivation and behaviour of the participant changes only because the person is aware of his participation in the research study.

Participant effects can be controlled through double blind technique, deception and a process of inquiry. "In *double blind* technique neither the researcher nor the participant know which experimental or control condition the participants are assigned to" (Leary, 2004, cited in Geoffrey Marczyk et al., 2005, p. 75). This method requires the presence of a supervisor. In *deception* the participants are provided with misinformation about the true hypothesis. Therefore the study will not be affected by the behaviour of the participants. In the inquiry method "the researchers can simply ask the participants about any number of issues related to participant effects and the overall purpose and hypotheses of the study" (Geoffrey Marczyk et al., 2005, p. 80). The question can be related to the hypothesis, experimental procedure, motivation, the role of the participants, etc. The information provided by the participants can be used for a statistical analysis.

3.10 MANAGERIALISM AND SCIENTIFIC RESEARCH

Managerialism is an important factor that determines the quality of scientific research. In a blog John Quiggin gave a description of managerialism: "The main features of managerialism policy are incessant organizational restructuring, sharpening of incentives and expansion in the power and remuneration of senior managers, with a corresponding downgrading of the role of skilled workers and particularly professionals" [25]. This comment is to be viewed in the context of scientific research. Research process demands creativity and dedication from the researcher. A scientist who is working under an organization is supposed to fulfill the demands set by the organization. Definitely he feels lack of freedom in expressing his scientific discoveries openly. It is true that "some scientific research organizations reached a stage where they are being run efficiently but have no genuine motivated scientists left to do the research" [26].

"The prevention of scientists from expressing their opinions on areas of their expertise is an effect of the encroachment of management on science" (Finlay MacRitchie, 2011, p. 58). This is because the primary concern is not given to the scientists but to the administrators. Hugh Tyndale-Biscoe (Scientist CSIRO) said that "modern management practices were comparable to 1930s Soviet science minister Nicolai Bukharin's determination to 'neutralize' scientific opposition to government planning, and this was compromising scientific independence" (cited in *Finlay MacRitchie*, 2011, p. 59) . This degradation started with the incorporation business management principles in scientific research. Management theories cannot be applied as such in scientific research. The opposite premise is erroneous. Business management principles are vested on the aim of profit making with a minimum investment. Their performance is based on assessments and pre-calculations. The essence of the reason why scientific research cannot be run by business management principles is made clear through the words of Nobel Laureate, Szent-Gyorgi, "if I knew what I was doing next year it wouldn't be worth doing because it would mean that it was pedestrian stuff " (cited in Finlay MacRitchie, 2011, p. 61). Managerialism is a hindrance to imagination and creativity.

In some research organizations the performance criteria of scientists are formulated by the administrators. This acts as a rigid framework thereby reducing the freedom of scientists. Such guidelines act as criteria for issuing grant for conducting research work. They demand detailed projects, want applicants to tell exactly what they will *do* and *find* during the tenure of their grants. "When it comes to their performance reviews, they are locked into the tasks to be performed and the time lines and try to fit the milestones and conclusions to the expected outcomes in order to obtain a successful evaluation. Their creativity is thus stifled, and the research becomes pedestrian" (Finlay MacRitchie, 2011, p. 62).

The constraints imposed by the managerialism on the scientists results in the loss of confidence and job satisfaction and the steep increase in stress and health problems. Another fact is that there comes a chance for the opportunists who want to please the administrators by acting as lackeys. Double agents will also be in action. People who want to make some changes remain powerless. The mental stress developed will force them to resign and thus creating a great loss to the science and scientific community.

Another impact of managerialism in scientific research is that it opens up a chance for corrupt practices. Very often opportunists get protection. The rigid hierarchy of the system that demands a profitable output within the specified time period sometimes forces the researcher to manipulate the result. This is against the research ethics. Finlay MacRitchie (p. 66) raises important questions:

- What if scientists had been allowed to use their creative talents to change direction and follow insights that could have culminated in unexpected discoveries of major importance?
- What if long-term research for the public good by scientists who had achieved international prestige had been strongly supported?
- What if this could have been done in an atmosphere of collegiality, kindness, and humour?
- What then would have been the bottom line?

These questions are deep and thought provoking. Managements and their principles cannot answer these questions. The authority of managerialism will continue. Research scientists will be in a situation in which they feel oppressed. They are forced to cope with their situations until they find another chance that matches their expertise.

3.11 LEADERSHIP IN SCIENTIFIC RESEARCH

"Lead and inspire people. Don't try to manage and manipulate people. Inventories can be managed, but people must be lead."

—**Ross Perot**

Effective coordination of the participants is essential for the success of a research programme. The efficiency in coordination is reflected through the leadership qualities of the researcher. Every researcher is assumed to have some leadership qualities. It incorporates a maturity, conviction, and expertise that translates into a purpose and direction. Once formulated, the vision represents a direction and must be reinforced by consistent and reliable actions of the leader. Leadership is a skill usually defined as a tool, behaviour and capabilities that a person needs in order to be successful at motivating and directing others [27]. Hagen et al. (1998) defined leadership as "the process through which leaders influence the attitudes, behaviours and values of others" [28]. The term *leader* presents different image for different people and strictly it is different from authority. It is true that living in fear and insecure circumstances present in many organizations, people generally confuse positions of authority with leaders. One vivid fact is that leadership cannot be easily defined, but is possible to recognize when it is present and when it is not.

At one stage of his career a researcher has to lead a group of less experienced or some of his students. Such a situation demands integrity as an essential quality. The subordinates or the students should have the confidence that the advisor will always be honest and will always be supportive. "It is essential that there be a relationship of complete trust between advisor and student. Even if a student errs in behavior, the supervisor must never betray the trust" (Finlay MacRitchie, 2011, p. 72). It is the responsibility of the advisor to ensure that the students are provided with the necessary requirements for carrying out their work. Students should be provided with a variety of research problems. The advisor should also ensure that the students are getting their remuneration in time to meet their financial needs. Patience and discipline are required on the part of the advisor because solution to the problem assigned to the student may not come in an easier way and it may take some time.

It is widely accepted that essential attribute that a researcher must develop is self-motivation. "Self-motivation in research means developing a burning passion to tackle the unknown to the point that it is almost an obsession".

(Finlay MacRitchie, 2011, p. 74). There are some researchers who work out the problem assigned to their students without their knowledge, definitely with a positive attitude, not for defeating the students. This provides an opportunity for them to access the student's progress in a better way.

In order to lead a scientific research programme, a researcher must possess some qualities. Finlay MacRitchie (2011, p. 75) pointed out the following qualities of a researcher that makes him a successful leader.

- Broad vision and an understanding of science, including its philosophy and history.
- Recognition that the greatest asset in research is not the million dollar grant or instrument but the individual scientists, because it is their creativity that produces original ideas that can lead to advances in knowledge.
- Concern for the well-being of scientists in their charge, meaning that they should be treated fairly and equally without the cronyism and favouritism that is sometimes evident where the ruthless managerial style operates; a good scientific leader works to enhance the careers of subordinates, whereas poor leaders trample on their best interests.
- Minimization of hierarchy, but where it exists, the most gifted scientists should be given the leadership roles.

As a leader a researcher should support his subordinates and students. He should provide all requirements and should encourage others to use their creativity to pursue research.

Brew.A and Peseta.T (2004) suggested some criteria for good practice in research supervision. They are [29]:

- Interest in, and enthusiasm for, the supervision of research students and research projects;
- Appreciation of a range of good practice approaches to supervision and an understanding of what constitutes a productive research environment;
- Establishment, for and with students/researchers, of clear goals and expectations in the light of up-to-date knowledge of requirements from organizations and funding bodies;
- Productive and regular meetings held with students/researchers which provide them with sympathetic, responsive and effective academic, clinical, professional and personal support and guidance;
- Careful management of the supervisory process to achieve timely and successful completion of the research aims and written reports;
- Development of a partnership with students/researchers which takes account of the need to assist them to develop a range of generic attributes and skills and to introduce them to the research community;

- Open communication established with students/researchers with timely feedback, which is both supportive and challenging, given on progress;
- Utilization of a repertoire of supervisory strategies to take account of the differing and diverse needs of individual students/researchers including providing assistance for remote learners and those with diverse educational needs to achieve success in their research;
- Evidence of systematic evaluation of competency in supervisory skills and of critical reflection and engagement with salient and emerging issues in their own field, to improve supervisory practice;
- Use, by the supervisor, of the literature on the scholarship of supervision pedagogy, and of relevant policy issues in research education to enhance the research experience of students/researchers.

In short, leadership involves motivating and inspiring others, not by pushing them but by satisfying their need for achievement, sense of belonging, recognition, and self-esteem.

3.12 SOME RESEARCH METHODS

The method used for research basically depends on the discipline or subject of the research topic. The same method cannot be applied to researches in all disciplines. For example, the experimental method is most suited for research in the science discipline, but it is not suited for a research programme in literature. In this section, we will see some of the research methods employed in various disciplines in brief [30].

Natural Observation: Natural observation refers to the observation of nature and natural phenomena. The researcher observes and records the phenomenon and its surroundings. This kind of approach is usually employed in social sciences, health, and ecology. These are also called natural experiments or observational studies. Before the development of experimental techniques, observation dominated almost all scientific disciplines. For example, Charles Darvin recognized several patterns in nature through observation of nature. Observation is also successfully employed in astronomy, geology, oceanography, natural sciences, and meteorology. The major strength of natural observation is that it allows researchers to observe behaviour in a natural setting in which it normally occurs.

Historical Research: The research that involves the study of past events is mainly referred to as historical researches. Based on the hypothesis, a historical researcher tries to recreate past events and draws a conclusion from them. It is to be noted that it is very difficult to apply the scientific method in historical research. The main limitation is that the researcher tries to set the events under controlled conditions, but the events that occurred in the past need not occur at conditions, instead, it may include uncontrolled variables or parameters.

Ethnographic Research: This is different from historical research. In this type of research, a researcher studies the current events rather than past events. Ethnography encompasses any study of a group of people for describing their socio-cultural activities and patterns and it involves the collection of many descriptive data. The most common approach in ethnographic research is participant's observation where the researcher lives with the subjects being observed as active participant and records extensive field notes. The researcher involves in all the activities of the society and the subjects. Direct observation can also be used for data collection in this method without becoming a participant and watching the events with a dispassionate perspective.

Cross-Sectional Study: This type of study is also known as prevalence studies, one-shot studies, or status studies. It is usually employed in social sciences, health sciences, and ecology. "A cross-sectional study can be considered as a survey of the prevalence of a phenomenon, situation, problem, risk factors, or other characteristics by taking a cross section of a defined population at one particular time. Measurements are made at a single instance only" [30, p. 68]. It is to be ensured that the sample must be the representative of age, gender, and social class structure of the population being studied. Through a study of a population at a certain time, conclusions across a wide population can be drawn. Cross-sectional study method has a great advantage that hat subjects are not deliberately exposed or treated, and therefore, there are no ethical issues involved.

Longitudinal Study: "An observational study performed over a period of years to determine the pattern of change in a phenomenon, situation, problem, attitude, or other similar attributes in relation to time is a longitudinal study" [30, p. 69]. Therefore, this is different from the cross-sectional study. However, this can be regarded as a series of cross-sectional studies over a long period of time. This type is usually employed in social studies and medical studies. Here, the researcher contacts the respondents many times for a longer period at regular intervals to gather necessary information. There are no fixed intervals; the interval can vary depending upon the nature of the study. This has the main disadvantage that people may respond to the questions as they become aware of the answer that the researcher expects from them. This occurs as a result of frequent contact between the researcher and the respondents. This is referred to as a conditioning effect.

Cohort Study: "In a cohort study, a particular group having a common trait within a subgroup of population (the cohort) is observed over a period. The common traits can be age, marriage, disease, height, weight, education, etc" [30, p. 70]. The term 'cohort' represents a group or a sub-group. We can see a number of examples for a cohort. A group of people affected by a particular disease, people living in a polluted area, a group of animals living in a particular condition, etc., are examples. Cohort studies can be *retrospective* or *prospective.*

In a retrospective cohort study, phenomena that have already happened are considered, while in a prospective cohort study, the study starts from the present. Therefore, the retrospective cohort study is historical in nature and often uses data already collected for other purposes. "A retrospective study is comparatively less costly, and it is an effective method for studying health and environment related risks due to pollutants, toxins, and drugs". In a prospective cohort study, certain variables are observed over a period for their effects.

Case Study: A case study is one of the most commonly used research methods in social sciences and in natural sciences. The case may be a person, plant, group, process, disease, event, community, or any other similar unit. A case study is characterized by three features [30, p. 71]:

- it is focused on a selected case,
- the researcher has the desire for an in-depth understanding of an issue; and
- data can be collected in many different ways such as observations, interviews, and secondary data review.

The case study involves the combinations of information from different sources. Case studies are frequently used by agricultural and life scientists to study a crop, animal, or a single instance of some physical phenomenon.

Correlational Research: Correlational research is usually used to study the correlation between two variables. That is, this method examines whether changes in one or more variables are related to changes in other variables. The study involves how the variables are related, to how many degrees they are correlated, etc. If two variables are correlated, we may be able to use the relationship to predict the value of one variable over the other. Essentially, the method of correlation is a mathematical tool for summarizing and finding the relation between variables. Like correlation, regression and curve-fitting are also used by the researchers to predict the behaviour of variables.

3.13 MONITORING THE RESEARCH ACTIVITIES

Research is an activity that needs continuous supervision or monitoring. There are many factors on which the outcome of a research programme depends. Starting from the identification of the research problem, formulation of hypothesis, collection of data, experiments, data analysis and finally arriving at a conclusion; all phases of the research programme need continuous and proper monitoring. Proper monitoring helps the researcher to complete his research work in the anticipated period of time. The importance of monitoring in various aspects of research is discussed below.

- *Data Handling:* It is very important to record the data accurately and consistently. For many researchers missing data is a major problem.

Therefore, proper monitoring will help to overcome such problems. This will also help to avoid inaccurate and inappropriate data including out-of-range data. Further, in experimental research, the accuracy of data depends on the quality of the experimental setup. Preferred standards must be kept for all experimental arrangements. The missing data can be estimated using a formula based on the experimental design.

- *Proper Tabulation and Periodic Reporting:* Proper recording of data is very important. It should be in a sequential manner. This will help to avoid the problem of missing data. The periodic reporting of the research progress will also help the researcher to be active in research. Most of the funding agencies and universities demand the submission of such progress reports at least once in six months.
- *Ethical Commitment:* Ethical commitment can be viewed as scientific honesty. The ethical practices that a researcher must follow are discussed in the last chapter. From the starting level itself, the researcher must maintain scientific honesty. For example, the data collection and recording must be done accurately and honestly. Any manipulation of the recorded data like making a deliberate change to get a required result is a scientific fraud. Proper and continuous monitoring will help to avoid such fraud and malpractices.

Proper Utilization of Funds: Many of the institutes and companies, provide financial assistance to the researcher in the form of grants. They expect its proper utilization and timely completion of the research work. It is to be ensured that the available fund must be used for the purposes mentioned in the research proposal submitted initially. The expenses can be recorded and financial reports must be submitted to the sanctioning authority. Most of the funding agencies in India demand the audit of the fund utilization by a chartered accountant.

REVIEW QUESTIONS

1. What is the importance of motivation in research?
2. Mention some immediate motivations of a young researcher.
3. What is the importance of research activities at undergraduate and postgraduate levels?
4. Briefly explain various methods of acquisition of knowledge.
5. What are the favourable conditions for intuition?
6. What is serendipity? Give some examples.
7. Give the importance of planning in research.
8. Give an outline of the logistics section of a research plan.
9. Give the outline of the concepts and context sections of a research plan.

10. Explain the various aspects of a research design.
11. Explain the criteria for the selection of a good research topic.
12. What are the personal and social considerations involved in the selection of a research topic?
13. What are the causes of broadness of a research topic? How can we rectify it?
14. Explain the various criteria of a good research problem.
15. What are the sources of good research ideas?
16. Explain the principles of good research.
17. Explain the criteria of a good research.
18. Discuss the guidelines for the research skills and awareness.
19. Discuss the role of validity in research.
20. Explain internal and external validity.
21. Explain various threats to external validity.
22. Discuss internal validity in detail.
23. Explain various threats to internal validity.
24. Discuss content validity, criterion-related validity, and construct validity.
25. What is the role of reliability in research?
26. Discuss the various aspects of reliability.
27. What are the various methods to determine reliability and explain each of them?
28. Describe the strategies for increasing reliability and minimizing measurement error.
29. What are artifacts and bias?
30. What are the various sources of artifacts and bias?
31. What is experimenter bias? Explain.
32. Explain participant bias? How it can be controlled?
33. Discuss various artifacts and bias in research. Explain the methods to control them.
34. Explain the role and effect of managerialism in research.
35. What is the importance of leadership in research?
36. What are the qualities of a good researcher?
37. Explain the criteria for good practice in research supervision.
38. What is natural observation?
39. Explain the historical approach in research.
40. What is ethnographic research?

41. Give an account of cross-sectional study.
42. What is longitudinal study?
43. Explain the features of cohort study.
44. What is a case study method? What are its features?
45. Write a note on correlational research.
46. Explain various methods in research.
47. Discuss the importance of monitoring the research activities.

REFERENCES

[1] Göran Hermerén, *Good Research Practice*, The Swedish Research Council's expert group on ethics, 2011, p. 29.

[2] Bengt Gustafsson, Göran hermerén and Bo Petersson, *Good Research Practice – What is it?*"The Swedish Research Council's expert group on ethics, 2006, p. 13.

[3] Natalia V. Kozlova and Inna V. Atamanova, "*The development of undergraduates motivation for research work*" Procedia - Social and Behavioral Sciences, 93 (2013) 498-502.

[4] Kerlinger, F.N. "*Foundations of Behavioural Research*" (3rd Ed.). Holt, Rinehart and Winston,USA, 1986, p. 667.

[5] Jarrard, R. D. "*Scientific Methods*" 2001. http://www.emotionalcompetency.com/sci/sm_all.pdf.236p.

[6] Justin D. Congdon and Arthur E. Dunham, *Defining the Beginning: The Importance of Research Design*, IUCN/SSC Marine Turtle Specialist Group Publication No. 4, 1999, p. 2.

[7] Ibid., p. 1.

[8] Research Methodology: Self Learning Manual, p. 19. http://www.iilm.edu/iilm-online/Research%20Methodology%20Self-Learning%20Manual.pdf

[9] Leonard Bickman and Debra J. Rog, *Applied Research Design – A Practical Approach*, Sage Publications, 1998, p. 4.

[10] Research Methodology: Self Learning Manual, p. 20.

[11] C R Kothari, *Research Methodology: Methods and Techniques*, New Age International, 2004, p. 33.

[12] A Y Abdulkareem, "Selecting a Research Topic", *Research Methodology in Education –An Interdisciplinary Approach,*" Ed. S A Jimoh, p. 89.

[13] Kenneth S. Bordens and Bruce B. Abbott, *Research Design and Methods: A Process Approach*, Mc Graw Hill, 2011, p. 56.

[14] Ibid.

[15] Tamsin White, *Principles of Good Research & Research Proposal Guide*, Policy, Performance and Quality Assurance Unit, London Borough of Richmond upon Thames, 2006, p. 1.

[16] James Harold Fox, "Criteria of Good Research", *The Phi Delta Kappan*, Vol. 39, No. 6, "*What Research Says about Teaching and Learning*", 1958, pp. 284-286.

[17] Alison Twycross and Linda Shields, "*Validity and reliability – What's it all about?, Part 1 Validity in quantitative studies*", Paediatric Nursing, Vol. 16, No. 9. November 2004, p. 28.

[18] Geoffrey Marczyk, David DeMatteo and David Festinger, *Essentials of Research Design and Methodology*, John Wiley & Sons, Inc., 2005, p. 176.

[19] Ibid., pp. 178-188.

[20] Roberts P et al., "Reliability and Validity in Research", 2006, *Nursing Standard.* 20, 44, 41-45.

[21] Geoffrey Marczyk, David DeMatteo and David Festinger (2005), p. 158.

[22] Ibid., pp. 159-174.

[23] Michael J. Miller, *Reliability and Validity*, Western International University, RES 600: Graduate Research Methods, p. 1.

[24] Ibid., p. 2.

[25] Finlay MacRitchie, *Scientific Research as a Career*, CRC Press, Taylor & Francis Group, 2011, p. 57.

[26] Ibid., p. 58.

[27] *Leadership Skills*, MTD Training & Ventus Publishing ApS, 2010, p. 10.

[28] Hagen.A, Hassan.M and Amin. S, "Critical Strategic Leadership Components: An Empirical Investigation", *SAM Advanced Management Journal*, Vol. 63, 3, (1998), p. 39. Harvard Business Review on Leadership, Harvard Business School Publishing.

[29] Brew. A and Peseta. T "Changing Postgraduate Supervision Practice: A Programm to Encourage Learning through Reflection and Feedback", *Innovations in Education and Teaching International,* 2004, 41 (1), 5-22.

[30] C George Thomas, "*Research Methodology and Scientific Writing*" 2021, Springer.

Suggested Readings

1. Jerome L. Myers and Arnold D. Well, *Research Design and Statistical Analysis*, Lawrence Erlbaum Associates Publishers, London, 2003.
2. Anthony E. Kelly, Rutgers University and Richard A. Lesh, *Handbook of Research Design in Mathematics and Science Education*, Lawrence Erlbaum Associates Publishers, London, 2000.
3. Mark L Mitchell and Janina M Jolley, *Research Desigh Explained*, Wadsworth Cengage Learning, 2010.
4. Thomas P Ryan, *Modern Experimental Design*, John Wiley & Sons Inc. Publication, 2007.
5. Jerome L. Myers and Arnold D. Well, *Research Design and Statistical Analysi*s, 2003, Lawrence Erlbaum Associates, Inc. London.
6. Anthony E. Kelly et al.(ed), *Handbook of Research Design in Mathematics and Science Education*, 2000, Lawrence Erlbaum Associates, Inc. London.

Chapter 4

Scientific Methodology

"We can't solve problems by using the same kind of thinking we used when we created them."

—**Albert Einstein**

"Begin at the beginning and go on till you come to the end; then stop."

—**Lewis Carrol, *Alice in Wonderland***

4.1 INTRODUCTION

A methodology provides a framework for the generation of knowledge. 'Scientific methodology defines and differentiates scientific knowledge from other types of knowledge' [1]. In its simplest form, scientific methodology denotes the way in which science is approached. Scientific methodology deliberately excludes the subjective aspects and stresses the objectivity of scientific knowledge. The main aspect of scientific methodology is that all theories are subjected to improvements or alterations due to contextual changes. This forms the basis of the idea that there is no absolute truth; every successful theory is only a way towards it. "*Research Methodology: A Self-Learning Manual*" (p. 17) defines the scientific method as '*a method of investigation by which scientific or any other impartial systematic knowledge is acquired is called a scientific method. The scientific method is a universally applicable, systematic method of understanding a phenomenon and verifying the truth.*' The scientific methodology includes a systematic observation, collection, and interpretation of data. In the words of *Shaughnessy & Zechmeister*, (1997, cited in Geoffrey Marczyk et al., 2005. p. 4), 'specifically, rather than relying on mere casual observations and an informal approach to learn about the world, scientists attempt to gain new knowledge by making careful observations and using systematic, controlled, and methodical approaches.'

The development of the scientific method is usually attributed to the thirteenth century scientist and philosopher Roger Bacon (1214-1294). This was then developed by Galileo Galilei (1564-1642), who is known as the father of modern scientific methodology. Philosophers like Francis Bacon and

Rene Descartes also made important contributions in developing the scientific methodology. Scientific methodology constitutes the defining characteristic of scientific research and it distinguishes science from non-sciences. How science is being studied is different from the study of other disciplines. 'Science is thus more deliberate and dedicated than non-science in following the method' [2].

No doubt; the scientific method is the best method for the acquisition of knowledge. The scientific method is not much different from our day-to-day ways of observing and learning about the world. The difference lies in the fact that in the scientific method we seek supporting evidence and logical-mathematical proof for the observations. This leads to the creation of knowledge. From our daily observations, we are learning but we are not creating knowledge. A systemic knowledge of the physical or material world is gained through observations and experimentation. There are different methodological practices in science. For example, we have *astronomy*, the observational science. In astronomy, the success of a theory depends on how powerful and effective the observation is. Therefore special and accurate methods of observation must be employed. Another method is experimentation; this requires a careful design of the experimental setup and proper execution of the experiment for a successful outcome. For example, Thomson's method is more refined than Millikan's method for the determination of the specific charge of electrons. Then the former method provides a more accurate result than the latter.

The scientific method possesses some general characteristics. In *Research Methodology: A Self-Study Manual* (pp. 17-18) the following characteristics of the scientific method were identified:

- Scientific methods employ a systematic approach.
- Scientific methods are based on objectivity.
- Every conclusion viewed through a scientific method must be verifiable.
- Scientific laws are universally applicable and testable.
- The scientific conclusions are predictable.

These points will become clear in the following sections. Another important point is that science depends heavily on mathematics for the formulation of its theories. Without mathematics, one cannot imagine the existence of science. This demands expertise in mathematics in order to be a good scientist.

The scientific method consists of well-organized elements. Although there are some disagreements the commonly accepted elements that comprise the scientific method are (Geoffrey Marczyk et al., 2005. pp. 5-16):

- *Empirical approach*: In this approach, decisions are made based on data derived from direct observations and experimentation. It emphasizes direct systematic and careful observation.

- *Observation*: Observation includes the creation of awareness of the world around us and making careful measurements. Observations should be systematic and free from biases.
- *Questions*: From careful observations, we get the idea for our research. The next step is to transform that idea into an *answerable* question. The question can be answered through available scientific methods and procedures.
- *Hypotheses*: Through a hypothesis, a researcher tries to present a possible explanation of the identified problem. This is then tested by analyzing data and can be either supported or refuted.
- *Experiments*: An experiment is conducted to verify the hypothesis. The researcher has to design the experiments properly to conduct the study.
- *Analysis*: After gathering the required data, the next procedure is its analysis. Statistical techniques are required for this. The type of statistical technique depends on the nature of the question and the nature of the accumulated data.
- *Conclusion*: After the analysis of the data the researcher will draw a conclusion about the result. The conclusion should be supported by the data analysis. Researchers should not draw broad conclusions.
- *Replication*: Replication means conducting of the same experiment for the second time with another group of participants to verify that the same results are obtained. The replicability of a conclusion increases its reliability of it.

4.2 RULES AND PRINCIPLES OF SCIENTIFIC METHOD

The scientific methodology can be understood based on the rules and principles it follows. Generally, a rule or a principle provides a direction of action. The collection of such rules and principles constitutes the *theory of methodology*. 'Principles of the method have some important features, such as reliability, robustness, power and speed' [3] and such principles are common throughout the history of science. Those principles were proposed during the course of the development of the scientific method and are still influential. 'Rules, like values, need not be followed one at a time; two or more rules may be conjoined, telling us to perform the action that conforms to them all. But, as with values, it may not be possible to follow all the rules at the same time since they offer inconsistent advice' (Robert Nola and Howard Sankey, 2007 p. 58). For example, sometimes we can follow, "choose the simplest hypothesis" and "choose the most accurate hypothesis"; but in all cases, the most accurate hypothesis need not be the simplest one.

Robert Nola and Howard Sankey, (2007, pp. 59-60) mention some principles that can also be treated as values. They are called *methodological principles* or M-*Principles*. They are:

1) One ought to accept only those theories that explain all the successes of their predecessors and repeat none of their failures; otherwise, they are to be rejected.
2) One ought to avoid ad hoc theories.
3) Given some rival hypotheses and a large set of facts they all explain, then one ought to pick that hypothesis that best explains all the facts.
4) One ought to accept the simplest theory and reject those that are more complex.
5) One ought to avoid theories that postulate unobservables unless they have an operational basis.
6) When a new theory replaces its immediate predecessor but also contradicts it, then one ought to adopt as the new theory the theory that contains the older theory, approximately under certain assumptions of values of parameters of the new theory.
7) In conditions of experimentation on human subjects, one ought to prefer double-blind over single-blind experiments and never perform an unblinded experiment.

An alternative form of these principles is that *hypothetical ought-imperatives.* We ought to follow some rules in order to realize some value. Here *antecedent* is a rule telling what we ought to do, and the *consequent* is some value that is to be realized by following the rule. The principles can be summarised as follows, (Robert Nola and Howard Sankey, 2007, p. 60):

1) One ought to avoid ad hoc hypotheses (r) rather than accept ad hoc hypotheses (r*) if one values highly refutable theories (v).
2) One ought to accept those theories that explain all the observational successes of their predecessors (r) since this is more likely to lead to truth (v) than accepting those theories that do not explain all their predecessor's observational successes (r*).
3) Where there is a conflict between theory T and what we observe, then one ought to form new theory, T* by making the minimum adjustments to T (r), since new T* is more likely to evade falsification (v) than would any other theory in which the modifications are more than the minimum (r*).
4) One ought to accept the theory that makes novel true predictions (r) since this is more likely to lead to the truth (v) than accepting any rival theory that captures the same observational information but makes no novel predictions (r*).

These principles are not given much consideration in the works of many philosophers. Some philosophers like Popper and Lakatos were in favour of practising these principles. According to *Larry Laudan* (1996), hypothetical imperatives are the main constituents of methodologies rather than categorical imperatives.

Geoffrey Marczyk et al. (2005. pp. 270-272), suggest the following principles that would act as guidelines for scientific research. These are general principles and are based on the nature of the scientific method.

- *Keep your eyes open*: The researcher should have an open eye to see what is happening around. The formulation of a powerful research problem results from the careful observation of the world around.
- *Be an Empiricist*: Scientific method strongly relies on empiricism. Empiricism emphasizes direct and systematic observation. This feature distinguishes science from pseudo-sciences. Therefore a good researcher should be an empiricist.
- *Be creative*: Creativity is the characteristic of a scientific mind. Creativity should play a vital role in all stages of the research process. It is particularly important in generating research ideas and in experimental design.
- *Research budgets research*: This principle emphasizes the importance of following a logical progression when conducting research. Each research study should be the next logical step in the overall line of research. Therefore, it is important that research studies answer discrete questions that flow logically from prior research studies.
- *Adhere to ethical principles*: This principle is very important but is not overemphasized. Ethical guidelines (detailed description is in chapter 8) should be observed by all researchers. The violation of ethical principles may hurt the participants and the reputation of other researchers.
- *Have fun*: The researcher should enjoy the research. Research can be exciting, so take pride in being part of something that will advance science and potentially improve the way we all live.

Observation of these principles will help the researcher in designing and conducting the study in an effective manner.

4.3 HYPOTHESIS

The purpose of scientific inquiry is first to understand a phenomenon, then to explain and to predict, and, if possible, to control the occurrence of a phenomenon. The first step of a research programme is the selection of a relevant topic and the formulation of a research problem. The selection of the topic is based on a number of factors and criteria (refer to section 3.4.2). After the formulation of the research problem, the researcher suggests a possible solution for the problem. This possible solution of a research problem is called a *hypothesis*. It is a declarative statement that predicts an expected outcome.

The term hypothesis is derived from the ancient Greek, *hypotithenai* meaning "*to put under*" or "*to suppose*". "Hypotheses can be considered intelligent hunches, guesses or predictions that help the researcher to seek a solution or to answer a research question" [4]. It is a tentative prediction about the nature of the relationship between two or more variables. As Lee (2000, p. 26) pointed out "hypotheses vary tremendously in their precision and complexity depending on such factors as how much is known about the topic and style of the scientist."

A number of definitions of hypotheses are available in the literature. Sarantakos (1993) defines a hypothesis as "*a tentative explanation of the research problem, a possible outcome of the research, or an educated guess about the research outcome.*"

According to Macleod Clark J and Hockey L (1981), '*a hypothesis is a statement or explanation that is suggested by knowledge or observation but has not, yet, been proved or disproved.*' C. R. Kothari (1990) says: '*A hypothesis may be defined as a proposition or a set of propositions set forth as an explanation for the occurrence of some specified group of phenomena either asserted merely as a provisional conjecture to guide some investigation or accepted as highly probable in the light of established facts*'. As Richard D Jarrard (2001) opinioned 'hypotheses are the key, and a hypothesis formation is a creative act, not a compelling product of observations. After developing a hypothesis, the scientist temporarily assumes that it is correct and then determines its logical consequences. This inference may be deductive, a necessary consequence of the hypothesis, or inductive, a probable implication of the hypothesis. To be fruitful, this inference must generate a testable prediction of the hypothesis. An experiment is then undertaken to confirm or refute that prediction. The outcome affects whether the hypothesis is retained, modified, or refuted' [5].

A hypothesis has got various functions. Some of them are mentioned below:

- It is a temporary solution to a problem concerning some truth that enables an investigator to start his work.
- It offers a basis in establishing the specificity of what to study and may provide a possible solution to the problem.
- Each hypothesis may lead to formulating another hypothesis.
- A preliminary hypothesis may take the form of a final hypothesis.
- Every hypothesis provides the investigator with a definite statement which may be objectively tested and accepted or rejected.

The formulation of a hypothesis is very important. It serves as a link between theory and investigation. It provides a basis for selecting variables, samples, and the research methodology. 'It serves the important function of linking together the related facts and information and organizing them into the whole' (Good, 1963). It prevents unfocused research. As Young (1965) stated, 'the use of hypothesis prevents a blind search or research and saves the researchers from gathering of masses of data which may later prove irrelevant to the study'.

4.3.1 Sources of Hypothesis

The formation of a hypothesis depends on the knowledge and ability of the researcher. It is a tentative statement and is considered the possible solution to the research problem. Formulation of a hypothesis requires a careful and logical analysis of the problem. However, there is no unique way of forming the hypothesis; the researcher may depend on different sources for the formulation of the hypothesis. They include *induction by simple enumeration, method of agreement, analogy, concomitant variation, and scientific theory* [6]. A brief discussion of these sources is given below.

- *Induction by simple enumeration*: This can be made clear through an example. Suppose we observe that ice cubes float on water irrespective of their sizes. Then we can formulate a hypothesis that the density of ice is less than that of water.
- *Method of agreement*: Here we notice some common factors among a group of objects, situations, or events. This can lead to the development of a hypothesis. For example, a number of metals like copper, silver, and iron are found conducting electricity, then the hypothesis all metals conduct electricity can be formulated.
- *Analogy*: This is the case where similar situations may lead to similar results. If we observe that many atoms emit spectral lines when excited, the same may be expected from the molecules. We can form a hypothesis that molecules emit spectral lines when excited.
- *Concomitant variation*: Here one can form a hypothesis about two phenomena by correlating the relationship between them. For example, if we find that the addition of iron increases the speed of a chemical reaction, we can hypothesize that iron is a catalyst for that particular reaction.
- *Scientific theory*: Often the established scientific theories provide a platform for the formulation of a hypothesis. Hypotheses can be formed through deduction based on such scientific theories. The hypothesis thus formed could be an extension or corollary of those scientific theories. For example, the knowledge of the oxidation reaction of metal iron may lead to the prediction of better methods to prevent corrosion of iron.

These are the sources of hypothesis usually a researcher depends on. However, a hypothesis may even arise from dreams; it was said that the German chemist F. A. Kekule obtained the idea about the structure of benzene from the dream of a snake biting its own tail. Another possibility is '*eureka intuition*'; where a scientist makes a sudden connection between the problem and an observation (remember the famous *eureka* story of Archimedes). Newton's observation of a falling apple and the sudden insight about gravity is an example of eureka intuition. *Serendipity* is another source of hypothesis. 'In *serendipity,* an unexpected observation is made and this along with the wisdom of the observer, leads to a novel insight or important discovery', (Lee, 2000, p. 28). The discovery of natural radioactivity by Becquerel is an example. He observed that the photographic plate kept well wrapped was affected by some radiation. The subsequent studies led him to the realization that the radiations were emitted by a radium sample kept near. The discovery of penicillin by Alexander Fleming also resulted from such an unexpected observation of the breakdown of bacteria colonies in the culture dishes and its subsequent study.

We have seen that the idea for a hypothesis can be originated from almost everywhere. However, the formulation of a hypothesis is not an easy task. It requires a clear theoretical and logical background and knowledge of the scientific method. Complete and perfect knowledge of the principles and practices of hypothesis formulation has to be acquired through training, special programme, conferences, seminars, etc.

4.3.2 Types of Hypotheses

Hypotheses are generally classified into two:

- Null Hypothesis (H_0 or H_N)
- Alternate Hypothesis (H_1 or H_A)

A *null hypothesis* represents a theory that has been put forward, either because it is believed to be true or because it is to be used as a basis for argument, but has not been proved. '*A null hypothesis always predicts that there will be no relationship between the variables being studied whereas the alternate hypothesis always predicts that there will be a relationship between the variables being studied'* (Geoffrey Marczyk et al., 2005. p. 38). The *alternative hypothesis* is a statement of what a hypothesis test is set up to establish. An alternate hypothesis can be considered the opposite of the null hypothesis and is reached only if the null hypothesis is failed. In such cases, *alternate* is the required conclusion for the researcher. The null hypothesis is considered to be most important because the null hypothesis relates to the statement being tested, whereas the alternative hypothesis relates to the statement to be accepted when the null is rejected.

For a null hypothesis, we have only two choices: 'it can be either rejected or not rejected, but it can never be accepted' (Geoffrey Marczyk et al., 2005. p. 11). The conclusion '*do not reject*' the null hypothesis does not necessarily mean that the null hypothesis is true; it only suggests that there is not sufficient evidence against the null hypothesis in favour of the alternate hypothesis. Rejection of the null hypothesis suggests that the alternative hypothesis may be true.

Suresh Chandra and Mohit Kumar Sharma (2013) stated the following rules for deciding the statements as the null hypothesis and alternate hypothesis [7].

- The null hypothesis is the hypothesis that is tested.

- What we hope or expect to be able to conclude as a result of the test usually should be placed in the alternate hypothesis.
- The null hypothesis should contain the statement of equality. That is for the null hypothesis we would have either =, ≤ or ≥. Hence the hypothesis is tested only for three situations (i) not equal to, (ii) greater than, and (iii) smaller than.
- The null and alternate hypotheses are complementary to each other. That is both of them together exhaust all possibilities regarding the values that the hypothesized parameter can assume.

A further classification of hypotheses is as follows:

- *Explanatory hypothesis*: A hypothesis about the cause of a phenomenon is called an explanatory hypothesis. For example, the statement, 'increase in the amount of carbon dioxide causes global warming', is an explanatory hypothesis.
- *Descriptive hypothesis*: A hypothesis about the law which governs a phenomenon is called a descriptive hypothesis. The statement, 'colours observed in thin oil films are due to the interference of light', is a descriptive hypothesis.
- *Tentative hypothesis*: This type of hypothesis is formed when the phenomenon cannot be fully understood. Here we see that how far the hypothesis is successful in explaining the phenomenon. Famous examples are the corpuscular theory and wave theory of light. Both are able to explain some phenomena but are not final.
- *Representative fictitious hypothesis*: It is an assumption about a certain phenomenon. One cannot prove it directly. When it is proved to be correct it becomes a theory or a law. Newton's idea of gravitation was fiction but became a law when it was proved.

4.3.3 Characteristics of Good Hypotheses

A hypothesis is a clear statement of what is intended to be investigated. Usually, it is formulated on the basis of observations or extensions of scientific theories. 'It should be simple, specific and stated in advance' [8]. A hypothesis should not leave any ambiguity about the subjects, variables, and statistical tests. Kothari identified some important characteristics that must be possessed by a hypothesis. They include (Kothari, p. 185):

1) A hypothesis should be clear and precise. If the hypothesis is not clear and precise, the inferences drawn on its basis cannot be taken as reliable.
2) A hypothesis should be capable of being tested. In a swamp of untestable hypotheses, many times the research programmes have bogged down. Some prior studies may be done by researchers in order to make the hypothesis a testable one. A hypothesis "is testable if other deductions

can be made from it which, in turn, can be confirmed or disproved by observation."

3) A hypothesis should state the relationship between variables if it happens to be a relational hypothesis.
4) A hypothesis should be limited in scope and must be specific. A researcher must remember that narrower hypotheses are generally more testable and he should develop such hypotheses.
5) A hypothesis should be stated as far as possible in most simple terms so that the same is easily understandable by all concerned. But one must remember that simplicity of the hypothesis has nothing to do with its significance.
6) A hypothesis should be consistent with most known facts, i.e., it must be consistent with a substantial body of established facts. In other words, it should be one that judges accept as being the most likely.
7) A hypothesis should be amenable to testing within a reasonable time. One should not use even an excellent hypothesis if the same cannot be tested in a reasonable time for one cannot spend a lifetime collecting data to test it.
8) A hypothesis must explain the facts that gave rise to the need for explanation. This means that by using the hypothesis plus other known and accepted generalizations, one should be able to deduce the original problem condition. Thus, hypothesis must actually explain what it claims to explain; it should have an empirical reference.

According to Karl Popper, a hypothesis must be *falsifiable*. Popper considered falsifiability as the basic qualification for a hypothesis being considered as a good hypothesis. If it is not falsifiable, then scientific methods cannot be used to verify it.

The quality of a hypothesis can be evaluated based on the following guidelines (Lee, 2000, p. 29).

- A hypothesis should be logically consistent, the hypothesis should not contradict itself.
- A hypothesis should be consistent with most of the available knowledge on the subject.
- The hypothesis should offer an explanation of the phenomenon under consideration.
- It should have the potential of *adding to* the existing theory.
- It must suggest consequences that are empirically testable, i.e., it should make a prediction that can be checked against reality.

A hypothesis is the basis of any research programme. All research activities are oriented towards the verification of the hypothesis.

4.4 TESTING A HYPOTHESIS

Any useful hypothesis will enable predictions, by reasoning. It might predict the outcome of an experiment in a laboratory setting or the observation of a phenomenon in nature. Hypothesis testing or significance testing is a method for testing a hypothesis about a variable in a phenomenon, using data measured through careful observation. The goal of hypothesis testing is to establish a cause-effect relationship between an independent variable and a dependent variable. Through the analysis of the collected data using statistical approaches, a researcher can draw a conclusion and based on the conclusion he can either accept or reject the hypothesis. If a hypothesis was rejected he should go for an alternate hypothesis and then follow by its verification. Only those hypotheses that survive rigorous testing are incorporated into the framework of reliable knowledge.

The conformation or rejection of a hypothesis is not an easy task; for several reasons [9]:

- Many hypotheses have inherent ambiguities that prevent simple confirmation or falsification. An experiment may favour one interpretation of a hypothesis, but the door is left open for other interpretations.
- Most experiments, in spite of careful experimental design, have at least some inherent ambiguity.
- Most hypotheses and their tests have associated assumptions and concepts. Refuting evidence indicates the inadequacy of either the main hypothesis or corollaries and one may not know confidently which to reject. Typically, the 'hard core' of a theory is relatively invulnerable to attack, and we refute or modify the 'protective belt' of ancillary hypotheses, assumptions, and conditions [Lakatos, 1970].
- Instead of directly testing a hypothesis, we usually test deductive or inductive predictions derived from the hypothesis. This prediction may be wrong rather than the hypothesis.

Due to these difficulties, most of the experiments are now targeted at inductive prediction rather than proving or disproving the hypothesis.

In hypothesis testing, we try to answer the question: whether the null hypothesis is to be rejected or not to be rejected. However, hypothesis testing is not a single-step procedure. Different steps involved in hypothesis testing may be summarised as follows [10]:

- *Formulation of the statement*: This is the first stage in which a null hypothesis and an alternate hypothesis are stated. These statements should be very clear and should be in accordance with the nature of the problem under consideration. It is to be noted that an indication of the type of test to be used is given by the statement of the hypothesis itself (for details see *Research Methodology* by *Suresh Chandra* and *Mohit Kumar Sharma*).

- *Choice of significance level*: Usually, a null hypothesis is tested on a predetermined level of significance and it is to be specified before calculating the test parameter, called *test static*. It is also known as Type I error.
- *The decision for a law of distribution*: Once the level of significance is decided, the next task is to determine an appropriate sample distribution like a normal distribution.
- *Computation of statistical parameters for random sampling*: After selecting a random sample, the statistical parameters are calculated. They include arithmetic mean, variance, standard deviation, etc.
- *Computation of test parameter*: Test parameter like *test static* is calculated from the statistical parameters obtained in the previous step.
- *Making a decision*: This is the final step where a comparison of the calculated *test static* is made with the value obtained corresponding to the significance level. From such a comparison, a decision can be made that whether the hypothesis is to be rejected or not to be rejected.

4.4.1 Variables

The term *variable* represents a quantity that can take different values. For example, height, weight, age, attitude, and IQ are variables because there are different heights, weights, ages, races, attitudes, and IQs. A quantity that cannot change is referred to as a *constant*. Identifying relevant variables is an important step in a research programme. There are different categories of variables. A researcher should be aware of the difference between these categories of variables and he should be able to choose the right variables for his study.

- *Independent variables*: The independent variable is the factor that is manipulated or controlled by the researcher (Geoffrey Marczyk et al., 2005. p. 42). These are the factors that influence a research study. A researcher is mostly interested in examining the effect of the independent variable. For example, if we want to study the effect of a medicine, one group is subjected to the medicine and the other group is not. In this case, the independent variable is *medicine* and the researcher is interested in examining the effect of the medicine. Sometimes a research study may include more than one independent variable. For example, a researcher may want to study the effect of temperature and pressure on the rate of a chemical reaction. Here the temperature and pressure are independent variables.
- *Dependent variables*: Dependent variable is the phenomenon under study. It is influenced by the independent variable. In the above example, the dependent variable is the rate of the chemical reaction because it depends

on the temperature and pressure which are the independent variables. Thus in an experiment, a researcher studies the relationship between the independent variable and the dependent variable. Independent variables are the causes and the dependent variables are the effects. The selection of the independent and dependent variables based on the research problem and the hypotheses.

- *Categorical variables*: Variables that can take on specific values only within a defined range of values are called *categorical variables*. Variables such as gender, marital status, hair colour, etc., are categorical variables.
- *Continuous variables*: The quantitative phenomena that can take different values even in decimal points are called *continuous variables*. Weight, height, income, age, etc., are examples of continuous variables. When compared with categorical variables, continuous variables can be measured with a greater degree of precision. Also, the statistical techniques employed by the researcher are different for the categorical and continuous variables.
- *Quantitative variables*: A variable that varies in amount is called a quantitative variable. In a research programme, we heavily depend on quantitative variables.
- *Qualitative variables*: Variables that vary in kind are called qualitative variables. Ratings such as attractive, helpful, consistent are qualitative variables. They vary in kind, not in amount.
- *Extraneous variables*: In some cases, some independent variables that are not related to the research study but may affect the dependent variable are called extraneous variables. It distorts the association between the independent variable in the study and the problem under study, as it is related to both. Any effect of such variables is technically termed an *experimental error*. In the study of the effect of fertilizer on a crop, soil fertility is an extraneous variable. A study must be designed in such a manner that the effect upon the dependent variable is attributed entirely to the independent variables, not to an extraneous variable.
- *Confounding variables*: Extraneous variables that vary with the levels of the independent variable are the most dangerous type in terms of challenging the validity of experimental results. Such variables are called confounding variables. For example; an instructor may test the new strategy in the gifted classroom and test the control strategy in a regular class instead of randomly assigning students. Certainly, the *ability* would most likely vary with the levels of the independent variable. In this case, pre-knowledge would become a confounding extraneous variable.

It is to be noted that when dealing with a single variable, it may fall under more than one category. The variable *weight,* as we see, is a continuous variable at the same time it is a quantitative variable. Hair colour is a categorical as well as a qualitative variable.

4.4.2 Correlation and Causality

The main focus of the scientific study is to analyze the cause-effect relationship between variables. Cause and effect is a much discussed and complex topic in science. We say that two events C and E are connected as cause and effect if they are contiguous in space and time if C precedes E and if C is always followed by E. James Mark Baldwin (1901)gives a broad definition of cause and effect as: "cause and effect are correlative terms denoting any two distinguishable things, phases, or aspects of reality, which are so related to each other that whenever the first ceases to exist, the second comes into existence immediately after, and whenever the second comes into existence, the first has ceased to exist immediately before" [11]. In an experiment, we usually attempt to establish some cause-effect relationship between two or more variables.

Causality is a relationship between events processes or entities in a time series such that when one occurs the other follows invariably. The basic principle of causality is determining whether the results and trends seen in an experiment are actually caused by the variables under consideration or whether some other factor may underlie the process. Consider an experiment where a study of the effect of depression upon alcohol consumption is made and the researchers found that people who suffer from higher levels of depression drink more, and announce that this correlation shows that depression drives people to drink. But there is a problem with this conclusion. There is an equal possibility that the heavy consumption of alcohol makes people more depressed. This is a type of 'chicken and egg' puzzle and in such situations establishing causality is very difficult even though one could find a correlation. Thus the problem with causality is that a researcher cannot always guarantee that their particular manipulation of a variable was the sole reason for the perceived trends and correlation.

'One should not assume that there is a causal connection unless a well-founded theoretical explanation can be offered for the relationship' (Jeffrey A Lee, 2000, p. 32). In the physical sciences, such as physics and chemistry, it is fairly easy to establish causality, because a good experimental design can eliminate the effect of confounding variables. For example, suppose we study the variation of the terminal velocity of a ball with the density of the fluids. Since we know from the basic principles, that the fluid density determines the resistance (viscosity), we can confirm that the fluid density can cause a change in the terminal velocity. In non-physical sciences, it is impossible to establish a complete causality. We can correlate the academic performance of the students in a class with the teaching ability of a teacher, but it may not be the only reason. So we cannot guarantee a cause-effect relationship in this case.

It is evident that causality needs correlation; correlation does not need causality. In other words "*correlation does not imply causality*". The challenge to scientists is to observe many correlations and to infer the few primary causalities.

4.4.3 Experimental and Control Groups

The main task in experimental research is the isolation of the effect of the independent variable. For this purpose, a researcher usually considers two identical groups of subjects and he calls them the experimental group and the control group. These two are identical except for the independent variable. The independent variable will be missing from the control group and will be present in the experimental group. The meaning is that among the two identical groups a researcher applies the treatment to only one group and that is the experimental group. The group which is not subjected to any treatment is the control group. If any difference is noticed in the test group results and the control group results, then possibly the independent variable under consideration is involved.

This can be made clear through a simple example. To study the effect of fertilizer on the production of some vegetables like tomatoes, take two groups of them. The groups are identical in all extraneous factors like soil fertility, amount of water given, amount of sunlight received, etc. One group (experimental group) is then given the fertilizer and the other (control group) is not given any fertilizer. If the experimental group produced more tomatoes than the control group, then it can be concluded that the application of fertilizer caused increased production. On the other hand, if the control group was not selected one cannot conclude that the fertilizer improved the yield because there is no possibility for comparison. Without a control group, it is easy to draw a conclusion but it may go wrong. Therefore, 'it is essential to use the control group in an experimental research to make the research meaningful' (Wilson, 1952).

The effect of extraneous variables on both groups can be controlled through randomization (see section 4.4.6).

4.4.4 Sampling

Science is a broad area where we study a number of phenomena. These phenomena involve a number of variables; both independent and dependent. In many situations, we can notice that there is a variation in the variable under study. For example, in a study about humans; we notice that all humans are not identical. They vary in height, weight, attitude, health, and many other factors. As a result, a study made on an individual cannot be generalized to all humans. Anyway, it is not possible to make a study on all humans. Therefore as a compromise, the researcher uses the *sample*. In the present case, we select a group of individuals who are supposed to represent the whole group.

'Sampling may be defined as the selection of some part of an aggregate or totality on the basis of which a judgment or inference about the aggregate or totality is made' (Kothari, p. 152). In the sampling technique instead of every unit of the universe only a part of the universe is studied and the conclusions are drawn on that basis for the entire universe [12]. *The universe* is a term used to represent the total of all items or groups involved in the proposed research area. A universe can be finite or infinite depending upon the number of elements present in it. For example, the stars in the sky represent an infinite universe whereas the population in a nation is a finite universe.

Again the universe can be *existent* or *hypothetical*. An existent universe is one that comprises tangible items or units like the number of books in a library. A hypothetical universe is one that does not exist. It is an imaginary universe. An example of a hypothetical universe is the moves on a chess board because all possible and conceivable moves of chess can never be realized. They can only be imagined.

The elementary units or the group or cluster of such units may form the basis of the sampling process in which case they are called sampling units. A list containing all such sampling units is known as a *sampling frame* (Kothari, p.153). A sampling frame consists of a list of items from which the sample is to be drawn. The plan for obtaining a sample from a sample frame is called *sample design*.

Thus a *sample*; then is a representative of the universe. It exhibits the characteristics of the universe. To select an appropriate sample statistical procedures are employed. This ensures random selection and freedom from biases and leads to reliable conclusions. Certainly, sampling is not a perfect method; certain inaccuracies in the collected information will always be present. This inaccuracy is usually termed *sample error* or *error variance*. In order to select a good sample for a study, the researcher should have knowledge about the various sampling distributions like mean, proportion, chi-square distribution, etc. Also, he should have an understanding of sample theories and tests of significance like Student's t-test, z-test, and Sandler's A-test. A detailed discussion of these distributions and tests are available in many textbooks on statistics.

Kothari (2004) pointed out the following characteristics of a good sample design [13].

- Sample design must result in a truly representative sample.
- Sample design must be such which results in a small sampling error.
- Sample design must be viable in the context of funds available for the research study.
- Sample design must be such so that systematic bias can be controlled in a better way.

- Sample should be such that the results of the sample study can be applied, in general, for the universe with a reasonable level of confidence.

A brief discussion of various probability sample designs is given below [14].

- *Simple Random Sampling*: Simple random sampling is an "equal probability of selection method". It is a method of sampling in which every sample element is selected only on the basis of chance, through a random process. In simple random sampling, particular sampling units are selected using random numbers. In this method, each item in the population has the same probability of being selected as part of the sample as any other item. It is a lottery procedure and so it is very easy to understand. Simple random sampling is most useful when the population of interest is relatively homogeneous. Random sampling can be done with or without replacement. If it is done without replacement, an item is not returned to the population after it is selected and thus can only occur once in the sample. In replacement sampling, each element is returned to the sampling frame after it is selected so that it may be sampled again.
- *Systematic Sampling*: Systematic sampling is a variant of the simple sampling method. In this method, every n^{th} element from the list is selected as the sample, starting with a sample element (n) randomly selected from the first k elements. This method is useful for drawing a random sample when the population elements are arranged sequentially. The number of cases from one sampled case to another in a systematic random sample is called the sampling interval.
- *Stratified Sampling*: Stratified sampling is used when representatives from each subgroup within the population need to be represented in the sample. The first step in stratified sampling is to divide the population into subgroups (strata) based on mutually exclusive criteria. These subgroups or strata are supposed to be more homogeneous. Random or systematic samples are then taken from each subgroup. This method is more efficient than drawing a simple random sample because it ensures the appropriate representation of elements across strata.
- *Cluster Sampling*: This sampling method is also called block sampling. In cluster sampling, the population that is being sampled is divided into groups called clusters. Instead of these subgroups being homogeneous based on selected criteria as in stratified sampling, a cluster is as heterogeneous as possible to match the population. A random sample is then taken from within one or more selected clusters.
- *Adaptive Cluster Sampling*: In adaptive cluster sampling, n samples are taken using simple random sampling, and additional samples are taken at locations where measurements exceed some threshold value. Several additional rounds of sampling and analysis may be needed. Adaptive

cluster sampling tracks the selection probabilities for later phases of sampling so that an unbiased estimate of the population mean can be calculated despite oversampling of certain areas.

- *Composite Sampling*: In composite sampling volumes of material from several of the selected sampling units are physically combined and mixed in an effort to form a single homogeneous sample, which is then analyzed. Compositing can be very cost-effective because it reduces the number of analyses needed. It is most cost-effective when analysis costs are large relative to sampling costs.

The following are some non-probability sampling designs:

- *Judgmental Sampling*: In judgmental sampling, the selection of sampling units is based on knowledge of the feature or condition under investigation and on professional judgment. Judgmental sampling is distinguished from probability-based sampling in that inferences are based on professional judgment, not a statistical scientific theory. Therefore, conclusions about the target population are limited and depend entirely on the validity and accuracy of professional judgment; probabilistic statements about parameters are not possible.
- *Availability Sampling*: In availability sampling elements are selected on the basis of convenience. This sampling method is also known as a *haphazard, accidental*, or *convenience sample*. It just consists of whoever or whatever is available, without any concern for its similarity to the population of interest. The availability sampling is usually typically, quicker, and uses smaller sample sizes than other sampling techniques. The main disadvantage of haphazard sampling is that since it is not statistically based, generalizations about the total population should be made with extreme caution. What makes availability sampling *haphazard* is precise that a great many things other than chance can affect the selection of cases, ranging from the prejudices of the research staff to the work schedules of potential respondents.
- *Quota Sampling*: In this type of sampling elements are selected to ensure that the sample represents certain characteristics in proportion to their prevalence in the population. Instead of considering whatever is available, we can set a quota like unemployed, salaried, married, male and female, etc.
- *Purposive Sampling*: In purposive sampling, each sample element is selected for a purpose, usually because of the unique position of the sample elements. Purposive sampling may involve studying the entire population of some limited group or a subset of a population. Purposive sampling does not produce a sample that represents some larger population, but it can be exactly what is needed in a study.

4.4.5 Experimental Design

An experiment is observation under controlled conditions. Thomas B Braker (1994) defined an experiment as "*a structured set of coherent tests that are analyzed as a whole to gain an understanding of the process*". The validity of the hypothesis is tested through the data generated from an experiment conducted for that purpose. Therefore the experimentation became an inseparable part of science and designing an experiment is an integral component of every research programme. *Experimental design is the conceptual structure within which an experiment would be conducted.* In an experiment, the causal relation between an independent variable and a dependent variable is studied.

The three essential components for a successful, efficient experiment are as follows:

- Knowledge of the process.
- A response variable.
- Clear goals and objectives.

In performing such experiments care should be taken to reduce any type of biases. Therefore it is essential that an experiment should properly be designed and executed. 'Experience has shown that if the data collection is properly planned, organized, summarized, and interpreted using statistical principles, one will be able to draw valid and meaningful conclusions from the results. The design of the experiment was found to be an excellent tool for effecting engineering development, quality improvement, process optimization as well as cost reduction' (*Research Methodology: A Self Learning Manual*, p. 146). The basic techniques involved in designing an experiment are the *principle of replication, the principle of randomization, the principle of local control (blocking),* and *orthogonality.*

4.4.6 Principles of Experimental Design

Principle of replication: 'Replication is the repetition of treatments under investigation to different experimental units' [15]. This increases the statistical accuracy. It is different from repeating the measurements. It provides a valid estimate of the experimental error. For example, suppose we want to study the side effects of a particular medicine we can apply the medicine to a number of patients and a comparative study can be made. On the other hand, if the medicine is applied to a single patient at different occasions, it is just repetition. The results obtained in such situations cannot be generalized. So the replicability is essential for an experiment whose result is to be generalized.

Principle of randomization: The principle of randomization provides the researcher protection against the effect of extraneous factors, when he undertakes any experiment. It prevents systematic and personal biases from being introduced into the experiment. The origin of the principle of randomization can be attributed to Sir Ronald Fisher, through the book, *The Design of Experiments* in 1935. As Roger E Kirk pointed out 'randomization helps to distribute the idiosyncratic characteristics of participants over the treatment levels so that they do not selectively bias the outcome of the experiment. Also, random assignment permits the computation of an unbiased estimate of error effects (those effects not attributable to the manipulation of the independent variable) and it helps to ensure that the error effects are statistically independent.

The randomization ensures independence of observations which is necessary for drawing valid inferences using statistical techniques. In Fisher's words, 'apart from the avoidable error from the experimenter himself introducing with his test treatments, or subsequently, other differences in treatment, the effects of which the experiment is not intended to study, it may be said that the simple precaution of randomization will suffice to guarantee the validity of the test of significance by which the result of the experiment is to be judged' [16]. The randomization procedure can be applied when the experimental units are homogeneous and is limited in quantity and a sizable fraction of the units is likely to be destroyed [17].

Principle of local control: It is essential that the experimental conditions should be rather varied and should be representative of those to which the conclusions of the experiment are to be applied. However, if we increase the scope of the experiments, consequently there will be an increase in the variability of response. This problem can be dealt with by local control (blocking). To block an experiment is to divide the observations into groups called blocks in such a way that the observations in each block are collected under relatively similar experimental conditions. Here we first divide the field into several homogeneous parts, known as blocks, and then each such block is divided into parts equal to the number of treatments. Then the treatments are randomly assigned to these parts of a block. Here the comparison of two or more treatments can be made more precisely than similar comparisons from an unblocked design [18]. 'Under local control, the extraneous factor, the known source of variability, is made to vary deliberately over as wide a range as necessary and this needs to be done in such a way that the variability it causes can be measured and hence eliminated from the experimental error' (Kothari, 2004, p. 40).

In blocking we place a set of *homogeneous* experimental units into groups in order that the different groups may be subject to different treatments. In different blocks, the number of experimental groups may be different. Thus by grouping, blocking, and assigning different treatments we get a balanced configuration.

Orthogonality: Orthogonality in experiment results in the factor effects being uncorrelated and therefore more easily interpreted. The factors in an orthogonal experiment design are varied independently of each other. The main results of data collected using this design can often be summarized by taking differences of averages and can be shown graphically by using simple plots of suitably chosen sets of averages. In these days of powerful computers and software, orthogonality is no longer a necessity, but it is still a desirable property because of the ease of explaining results [19].

4.4.7 Characteristics of Good Experimental Designs

Experimentation is an essential component of scientific research. Therefore an experiment is to be designed properly so that it should be able to complete the mission for which it is designed. A properly designed experiment should have the following characteristics [20].

- *Replicability*: It should be possible for anybody to repeat the experiment, with the same result.
- *Accuracy*: A good experiment should yield accurate and clearly measurable results.
- *Precision*: The results should be well defined. An answer must be very precise. But a precise answer may not be accurate and vice versa. For example, one could calculate the value of acceleration due to gravity (g) to the 9th decimal place from an experiment, but it might be totally wrong due to some errors.
- *Generalizability*: It would be very useful to have our experimental results have relevance to not only the particular case but to a whole class of phenomena.
- *Simplicity*: Other things being the same, simpler experiments are preferable to more complicated ones.
- *Focus*: It is desirable that experiments have clearly defined goals and results.
- *Absence of systematic error*: Sometimes we may have certain errors built into the experimental procedure itself due to wrong assumptions or omissions. A good experiment should guard against this.
- *Consistency*: A good experimental design should be consistent with generally accepted procedures and principles.
- *Utility*: An experiment should yield some useful results or conclusions.

These are the commonly accepted characteristics of a good experimental design. However, there are some other factors like a simulation of the reality through the experiment, proper selection of samples, etc., that may affect an experimental design. 'There are both epistemological and axiological factors affecting the validity and value of the experiment', (RVG Menon, p. 108).

4.4.8 Types of Experimental Design

Experimental design is a basic requirement and the success of the research work greatly depends on the proper design of the experiment. An experimental design identifies the independent, dependent, and nuisance variables and indicates the way in which the randomization and statistical aspects of an experiment are to be carried out [21]. It aims to establish a causal connection between the independent and dependent variables. There are two broad classifications of the experimental designs:

- Formal experimental designs
- Informal experimental designs.

Examples of the formal experimental design are *completely randomized design, randomized block design, Latin square design,* and *factorial designs.* Informal experimental design includes; *before-and-without control design, after-only with control design,* and *before-and-after with control design.* A brief discussion of formal and informal experimental designs is given below [22].

1) *Completely randomized design*: A completely randomized design involves the principle of replication and the principle of randomization. In this design, the subjects are randomly assigned to the experimental treatments. The selections of the subjects are completely random. This design is used when the experimental areas (subjects) are homogeneous. For example, suppose we want to study the effect of two fertilizers (A & B) on ten plants such that five is treated with fertilizer 'A' and another five is treated with fertilizer 'B'. The complete randomization provides an opportunity for the selection of any five plants from the group of ten. 'Analysis procedure required to analyze such design called *one-way analysis of variance*. When all possible variation due to uncontrollable experimental factors is included under *chance variation*, the design of experiment is known as completely randomized design' [23]. *Research Methodology: Self Learning Manual* (p. 160) points out the following advantages of a completely randomized design:
   - Complete flexibility is possible.
   - The numbers of replications can be varied at will from treatment to treatment.
   - It is possible to utilize all the experimental data.
   - Statistical analysis is easy even if several replications are not the same for all treatments.
   - The analysis remains simple even when results from some units or treatments are rejected. The relative loss of information due to such rejection is the smallest compared with any other design.
2) *Randomized block design*: This is the most familiar research design. This was originated from agronomic research where several variables or

treatments are applied to different blocks to study the effect of replication. The first step is to divide the subjects into groups called *blocks*. The variable selected for grouping the subjects is one that is believed to be related to the measures to be obtained in respect of the dependent variable. For example, in a study, suppose the plants are grouped according to their size. Then each group of constant sized plants forms a block. The number of subjects (plants) in a given block would be equal to the number of treatments and one subject (plant) in each block would be randomly assigned to each treatment. If every treatment occurs in every block the design is said to be *complete*. Each treatment occurs the same number of times in every block; the design is also said to be *balanced*. Incomplete designs where all treatments do not occur in each block also exist, but it requires a large number of blocks; also the experiments must be executed in controlled environments. The randomized block design is analyzed by the two-way analysis of variance (two-way ANOVA) technique (for a detailed discussion of the analysis of variance see *Research Methodology* by Suresh Chandra and Mohit Kumar Sharma (2013), Narosa Publishing House Pvt. Ltd.).

However, in this example, there are two sources of heterogeneity in each block; the environment and the plant size. Within each block, these are kept constant and are said to be *confounded*. As P. Allen Hammer and Douglas A Hopper observed 'confounding of sources of heterogeneity in constructing a block is an acceptable experimental practice, but confounding of treatments with the sources of heterogeneity is not allowed'. Because the observed difference in the growth of plants may be due to the known heterogeneity and not because of the treatments being studied.

3) *Latin square design*: In an experiment, as we have seen, there are some confounding sources of inhomogeneity. For example in an experiment with plants or some crops, unwanted heterogeneities that cannot be confounded effectively in the block are seed difference and the fertility of the soil. If we want to study the effect of *five* fertilizers on the yield of a particular crop; the fertility of the soil in each block should also be taken into consideration. Latin square design is frequently used in agricultural experiments. In such a design the field is divided into blocks whose number is equal to the number of varieties of fertilizers being used (here five). Then each block is again divided into as many parts as there are varieties of fertilizers (here five) in such a way that each of the fertilizer varieties is used in each of the blocks, either column-wise or row-wise only once.

Fertilizer Level

		1	2	3	4	5
Seed differences	X_1	A	B	C	D	E
	X_2	B	C	D	E	A
	X_3	C	D	E	A	B
	X_4	D	E	A	B	C
	X_5	E	A	B	C	D

Fig. 4.1 Latin square design (Adapted from Kothari)

The *Latin square design* assumes that there is no interaction between treatments and blocking factors. The analysis of the Latin square design is very similar to the two-way ANOVA technique. The advantage is that the differences in soil fertility in the field can be eliminated in comparison to the effects of different fertilizers on the yield of the crop.

4) *Factorial designs*: This type of research design is usually applied in economics and sociology where the effects of varying more than one factor are to be determined. The economical and sociological problems are very often affected by a number of variables. Factorial designs can be of two types: (i) simple factorial designs, and (ii) complex factorial designs. In a simple factorial design, we consider the effect of variation of two factors on the dependent variable. In complex factorial designs, the effect of variation of a number of factors on the dependent variable is investigated.
5) *Before-and-after without control design*: In this design, a single test group or area is selected and the dependent variable is measured before applying the treatment. The treatment is then introduced and the dependent variable is measured again after the treatment has been completed. From a comparison of the results obtained in both cases, one can draw inferences about the effect of treatment on the test group or area. One difficulty associated with this design is that the results are subjected to extraneous variations with the passage of time.
6) *After-only with control design*: In this type of design two groups or areas are selected. One is called the test group and the other is called the control group. The treatment is introduced into the test group only. The dependent variable is then measured in both groups at the same time. The impact of the treatment on the dependent variable can be obtained through a comparison of the values of the dependent variable from the control group and from the test group after the treatment. It is to be remembered that in this design

we assume that both groups are identical with respect to their behaviour towards the phenomenon considered. If not so, extraneous variations will affect the result of the treatment. The advantage of this design over the before-and-after without control design is that the passage of time will not introduce difficulty in collecting the data.

7) *Before-and-after with control design*: Here again two groups are selected, the control and test groups. The dependent variable is measured in both groups for an identical time period before the treatment. Then the test group is subjected to the treatment. Then the dependent variable is measured in both for an identical time period after the introduction of the treatment. The treatment effect is determined by subtracting the change in the dependent variable in the control group from the change in the dependent variable in the test group. This design is considered superior to both *before-and-after without control design* and *after-only with control design* because it avoids extraneous variation resulting both from the passage of time and from non-comparability of the test and control groups.
8) *Paired design:* In many cases, random assignment alone will not help to control the experiment from problems arising out of differences between subjects. For example, let a researcher test the effectiveness of new medicine to treat a specific disease. After giving the medicine to the subjects consisting mainly of young people, if the researcher observed no difference between the treatment and control, he cannot conclude that whether the drug is ineffective or whether the ineffectiveness was because of the age of the subjects. In similar situations, when there is an extraneous variable that the researcher believes might become a confounding variable, a paired design can be used. This is also called *matched pairs design* as the pairs are formed based on some matching variable. Pairing or grouping should be based on a character that is likely to influence the actual character under the study. For example, in the case of human subjects, pairs can be matched based on gender, health conditions, race, religion, etc. The researcher after measuring the matching variable creates groups of subjects who have the same value or level of that matching variable. The limitation of the paired design includes the difficulties involved in measuring the matching variables, creating matched pairs, and assigning the selected subjects to the given experimental conditions.

4.4.9 Implementation of Experimental design

A research experiment should be properly planned and executed with utmost care. Jacqueline K. Telford (2007) describes the following steps to implement an experimental design [24].

- State the objective of the study and the hypotheses to be tested.
- Determine the response variable(s) of interest that can be measured.
- Determine the controllable factors of interest that might affect the response variables and the levels of each factor to be used in the experiment. It is better to include more factors in the design than to exclude factors, that is, prejudging them to be non-significant.
- Determine the uncontrollable variables that might affect the response variables, blocking the known nuisance variables and randomizing the runs to protect against unknown nuisance variables.
- Design the experiment, remembering to randomize the runs.
- Perform a pro forma analysis with response variables as random variables to check for estimability of the factor effects and precision of the experiment.
- Perform the experiment strictly according to the experimental design, including the initial setup for each run in a physical experiment. Do not swap the run order to make the job easier.
- Analyze the data from the experiment using the analysis of variance method developed by Fisher.
- Interpret the results and state the conclusions in terms of the subject matter.
- Consider performing a second, confirmatory experiment if the conclusions are very important or are likely to be controversial.
- Document and summarize the results and conclusions, in tabular and graphical form, for the report or presentation on the study.

However, even if the experiment is designed with utmost care one should always expect unexpected complications that may arise as the experiment progresses. Such complications arise when working with new techniques, new equipment, or when the phenomenon in the question is not understood clearly. This problem is to be faced not only by beginners but experienced scientists also. Nobody can anticipate all the problems that may arise during research.

What would a researcher do if a change is needed? If the change is minor that will not affect the result he can make it and is to be noted in his notebook. On the other hand, any significant change made by the researcher should be reported. As an extreme case; if the changes alter the nature of the experiment itself then the entire research project should be started over with a revised experimental design.

4.5 DATA COLLECTION AND ANALYSIS

After proper planning of the research design, the next step is the collection of required data and its subsequent analysis. The data can be classified as *primary* and *secondary* data. 'The primary data are those which are collected afresh and for the first time, and thus happen to be original in character. The secondary data, on the other hand, are those which have already been collected by someone else and which have already been passed through the statistical process' [25]. Data collection and analysis is a time-consuming process and it demands conscious effort on the part of the researcher. A researcher has to identify the appropriate method of data collection from a variety of possible methods and he should have knowledge about the methods of statistical analysis of those variables. Thus data collection and analysis constitutes a major part of any research programme and requires a detailed understanding. The various methods of data collection, presentation, analysis, etc., will be discussed in detail in the next chapter.

4.6 INTERPRETATION OF THE RESULT AND GENERALIZATION

After collecting data and performing the analysis, the next task for the researcher is the interpretation of the results. Through interpretation, a researcher can explain the various relations and processes in his findings. If a hypothesis is tested and proved several times a researcher can arrive at a generalization.

Interpretation refers to the task of drawing inferences from the collected facts after an analytical or experimental study. The interpretation has got two major aspects [26].

1) The effort to establish continuity in research through linking the results of a given study with those of another.
2) The establishment of some explanatory concepts.

According to William Emory 'interpretation is concerned with relationships within the collected data, partially overlapping analysis. Interpretation also extends beyond the data of the study to include the results of other research, theory, and hypotheses (cited in Kothari, p. 344). The importance of interpretation lies in the fact that the usefulness and utility of research findings lie in proper interpretation. Kothari (p. 345) pointed out the following reasons for considering the interpretation as a basic component of a research programme.

- It is through interpretation that the researcher can well understand the abstract principle that works beneath his findings. Through this, he can link up his findings with those of other studies, having the same abstract principle, and thereby can predict the concrete world of events. Fresh inquiries can test these predictions later on. This way the continuity in research can be maintained.
- Interpretation leads to the establishment of explanatory concepts that can serve as a guide for future research studies; it opens new avenues of intellectual adventure and stimulates the quest for more knowledge.

- A researcher can better appreciate only through interpretation why his findings are what they are and can make others understand the real significance of his research findings.
- The interpretation of the findings of exploratory research study often results in hypotheses for experimental research and as such interpretation is involved in the transition from exploratory to experimental research. Since an exploratory study does not have a hypothesis to start with, the findings of such a study have to be interpreted on a post-factum basis in which case the interpretation is technically described as a 'post factum' interpretation.

A researcher must keep in mind that a negative result that does not support the hypothesis was also important. If a researcher got a negative result then the hypothesis is to be revised and a new experiment is to be performed to support the hypothesis. When a hypothesis passes all possible tests it becomes a theory and a generalization of theory is a law.

A theory has often been described by logicians as a hypothesis that has been verified and found to be true. A theory is an explanation of a set of related observations or events based upon tested hypotheses and verified multiple times by different groups of researchers. According to Welton, 'when a hypothesis is established as true; it is called a theory'. 'While a hypothesis is simply an assumption, a theory is often made up of several assumptions related to each other. It is a coordinated system of hypotheses and their corollaries having an internal unity. It is a conceptual framework, thus built up, to provide a new way of looking at the subject under consideration' [27]. Kerlinger (1986) defines theory as "a set of interrelated constructs (concepts), definitions, and propositions that presents a systematic view of phenomena by specifying relationships among variables, with the purpose of explaining and predicting phenomena" [28].

The process of theory development involves a continual refinement of the interaction between constructs and empirical observations in a manner that makes possible deduction leading to explanation and prediction. Robert S Michael (2002) pointed out the following roles of theory in the development of science.

- Theories organize the existing knowledge in particular areas. A theory pulls together and enables deeper meaning by connecting the results of individual studies.
- Theories explain the results of research studies. However, explanation in most areas is never complete and is subject to revision as more empirical evidence becomes available.
- They predict future phenomena. Although the theory is not an absolute truth, some theories are sufficiently developed to allow prediction with a high level of accuracy.

- They enable control of occurrences of phenomena. Difficulty in manipulating variables limits inquirers; nevertheless, a goal of science is to enable control through theory.
- Theories suggest new avenues for inquiry. Rather than being finished, theory generates ideas for further investigation.

One can never argue that a theory is absolutely right. A theory is valid as long as there is no evidence to dispute it. Therefore, theories can be disproven.

A law generalizes a body of observations. Scientific laws explain things, but they do not describe them. Scientific laws are otherwise called *Laws of Nature*. Scientific laws must be simple, true, universal, and absolute. Scientific laws can sometimes be expressed in terms of a single mathematical equation. Some scientific laws, or laws of nature, include the law of gravity, Newton's laws of motion, the laws of thermodynamics, Boyle's law of gases, the law of conservation of mass and energy, and Hook's law of elasticity.

As Bill Robertson (2009) opinioned if a theory is any good, it explains a law. For example, the kinetic theory of gases explains Boyle's law, but the theory does not evolve into the law. 'The highest award for a theory is that it is a good theory, not that it becomes a law. Scientific laws simply explain how things behave. They do not result from theories but are rather descriptions of how the physical world does what it does' [29].

4.7 CLASSIFICATION OF EXPERIMENTS

The primary aim of an experiment is the testing of a hypothesis. The experiments used to test a hypothesis can be classified as:

- True experiments
- Quasi-experiments

True Experiments: The main features of true experiments are [30]: (i) Randomly formed groups (ii) Manipulation of treatments and (iii) Measurement of dependent variables.

The random selection of groups ensures that the subject groups are statistically equivalent. However, it is different from random sampling. That is random assignment is a method for assigning the subjects but not a method for selecting the subjects. The manipulation of treatment gives freedom to the researcher to decide the treatment that a particular subject has received. It is the decision of the researcher and the treatment is said to be manipulated. Finally, the independent variables are measured for different groups and are then compared.

In science, laboratory experiments and field experiments are examples of true experiments. There are many features of the laboratory experiments. They are carried out under controlled conditions, and the chances of error are very less. In a laboratory experiment, usually, one or more independent variables are varied and the dependent variables are measured according to the variation of the independent variables. In this way, it is possible to eliminate extraneous variables.

Field experiments are usually conducted in applied science. In such experiments, the independent variables are varied and measurements on the dependent variable are made on experimental units in their natural setting.

Quasi-experiments: The difference between a true experiment and a quasi-experiment is that in true experiments, the groups are randomly assigned, whereas in quasi-experiments such random assignment is not possible. The remaining part of the study is similar to that of a true experiment. For example, random assignment is not possible in many situations related to human subjects. If a study is to be conducted at schools, colleges, hospitals, etc., then the random assignment is almost impossible. In such cases, the possibility is a quasi-experiment. Quasi-experimental research is commonly used in health and social research. In such disciplines, it is not possible to create a truly controlled group because of ethical reasons and therefore randomization is not possible.

4.8 REQUIREMENTS FOR A GOOD EXPERIMENT

A good experiment to test a hypothesis must be properly designed for minimum errors, high precision. Moreover, it should be simple to conduct and must allow statistical analysis. D. R. Cox [31] pointed out the following requirements for a good experiment. They are the absence of systematic error, precision, a wide range of validity, simplicity, and calculation of uncertainty. A brief discussion is given below.

- *Absence of Systematic Error*: A good experiment should be free from any systematic error. This is possible if experimental units receiving one treatment do not differ in a systematic manner from other experimental units given another treatment. This is achieved by randomization.
- *Precision*: A good experiment should have maximum precision. The indicator of precision is the magnitude of standard error, which should be sufficiently small. If there are no systematic errors and the estimates of treatments differ from their true value only by random errors, the experiment has more precision. The precision of an experiment depends on the intrinsic variability of the experimental material, the accuracy of the experimental work, the number of experimental units, and the design of the experiment. Therefore, it is very important to take the necessary care to attain maximum precision.
- *Range of Validity*: To the extent possible, an experiment should be valid in a wide range of conditions so that the researcher can have a

high degree of confidence in its results. An experimental technique that gives encouraging results in a certain setting may not give favourable results when conditions are different. All necessary precautions should be taken to avoid or reduce threats to both internal validity and external validity.

- *Simplicity*: This is an obvious requirement. If an experiment is a complicated one, it will be difficult to carry it out properly without the help of trained and qualified personnel. It is, therefore, desirable for the experiment to have a simple design, particularly when it is to be performed by relatively unskilled people. Also, it is desirable to use relatively simple methods in the analysis of data. Further, only those methods and techniques that one can do with the available resources and facilities must be included in the experimental design.
- *The Calculation of Uncertainty*: A good experiment should enable the researcher to calculate the uncertainty in the estimates of treatment differences. This would mean that he should be able to ascertain the statistical significance of the differences between the treatments. To carry out statistical analysis, the researcher must have some basic knowledge of statistics.

REVIEW QUESTIONS

1. What is the scientific method?
2. What are the characteristics of the scientific method?
3. Explain the elements that comprise the scientific method.
4. What is the empirical approach?
5. What is a hypothesis?
6. What do you mean by replicability?
7. What are the methodological principles of the scientific method?
8. What are the principles that act as a guideline for scientific research?
9. Explain hypothesis.
10. What are the functions of a hypothesis?
11. Explain the main sources of the hypothesis.
12. What are the different types of hypotheses?
13. What is a null hypothesis? Explain.
14. What is an alternate hypothesis? Explain.
15. What are the rules for deciding the statements as the null hypothesis and alternate hypothesis?

16. Explain the characteristics of a good hypothesis.
17. What are the guidelines for the evaluation of the quality of a hypothesis?
18. Why testing a hypothesis is difficult?
19. What are the different steps involved in testing of a hypothesis?
20. What are variables?
21. What are independent and dependent variables? Give examples.
22. What are categorical and continuous variables? Give examples?
23. What are qualitative and quantitative variables? Give examples.
24. What are extraneous and confounding variables? Give examples?
25. What are variables? Discuss the various types of variables and their importance in research.
26. What is correlation? Explain.
27. What is causality? Explain.
28. "Correlation does not imply causality". Justify this statement.
29. Briefly explain the experimental and control groups and their importance.
30. Define sampling. What is its application in research?
31. What are the characteristics of a good sample design?
32. Explain simple random sampling.
33. What is systematic sampling? Explain.
34. Write a note on stratified sampling.
35. What are cluster and adaptive cluster sampling?
36. Explain composite sampling.
37. What is non-probability sampling? Give examples.
38. Explain the judgment sampling.
39. What is availability sampling?
40. Explain quota sampling and purposive sampling.
41. What is sampling? Discuss the various probability and non-probability sampling.
42. What are primary and secondary data?
43. What do you mean by an experimental design?
44. Explain the principles of experimental design.
45. What are the characteristics of a good experimental design?

46. What are formal and informal experimental designs?
47. Write a note on a completely randomized design.
48. Explain the randomized block design.
49. Explain the Latin square design.
50. What is factorial design?
51. Explain the before-and-after without control design.
52. Write a note on after-only with control design.
53. Explain the before-and-after with control design.
54. What is paired design? Explain.
55. What is experimental design and what are the principles of experimental design?
56. Discuss the various types of experimental designs.
57. What are the steps involved in the implementation of an experimental design?
58. Why interpretation is considered a basic component of research?
59. What are the major aspects of interpretation?
60. What are the roles of theory in the development of science?
61. What is a scientific law?
62. Explain the classification of experiments.
63. What are true experiments? Explain.
64. What are quasi-experiments?
65. Explain the five requirements of a good experiment.

REFERENCES

[1] Jose T Molina, 2013, "*Global Scientific Method*", www.molwick.com, p. 33.

[2] Peter Kosso, 2011, "*A Summary of Scientific Method*", Springer, p. 2.

[3] Robert Nola and Howard Sankey, 2007, "*Theories of Scientific Method*", Aucmen, UK, p. 57.

[4] Judith Haber, "*Research Questions, Hypotheses and Clinical Questions*", p. 28.

[5] Richard D Jarrard, 2001, "*Scientific Methods*", Online Book, p. 13.

[6] "*Research Methodology: A Self-Learning Manual*", pp. 92-93.

[7] Suresh Chandra and Mohit Kr Sharma, "*Research Methodology*", 2013, Narosa Publishing House Pvt. Ltd, New Delhi, pp. 74-75.

[8] Hulley S. B., Cummings S. R, Browner W. S, Grady D, Hearst N, Newman T. B., 2001, "*Getting ready to estimate sample size: Hypothesis and underlying principles In: Designing Clinical Research: An Epidemiologic Approach*", Philadelphia: Lippincott Williams and Wilkins; pp. 51–63.

[9] Richard D Jarrard, pp. 159-160.

[10] Suresh Chandra and Mohit Kr Sharma, pp. 80-81.

[11] James Mark Baldwin (ed.), 1901, "*Dictionary of Philosophy and Psychology*". New York-London vol. I, p. 164.

[12] "*Research Methodology: A Self-Learning Manual*", p. 248.

[13] Kothari, C R, 2004, "*Research Methodology-Methods and Techniques*", New Age International p. 58.

[14] "*Guidance on Choosing a Sampling Design for Environmental Data Collection*", United Stated Environmental Protection Agency, 2002, pp. 12-16.

[15] V K Gupta and Rajender Prasad, "*Fundamentals of Design of Experiments*".

[16] Ronald A Fisher, 1971, "*The Design of Experiments*", Hafner Publishing Company, p. 21.

[17] "*Research Methodology: A Self-Learning Manual*", p. 149.

[18] V K Gupta and Rajender Prasad, "*Fundamentals of Design of Experiments*".

[19] Jacqueline K. Telford, 2007, "*A Brief Introduction to Design of Experiments*", Johns Hopkins APL Technical Digest, Vol. 27, No. 3, p. 225.

[20] R V G Menon, 2010, "*An Introduction to the History and Philosophy of Science*", Dorling Kindersely (India) Pvt, Ltd., p. 108.

[21] Roger E Kirk, "*Experimental Design*", p. 24.

[22] Kothari, pp. 41-52.

[23] "*Research Methodology: A Self-Learning Manual*", p. 158.

[24] Jacqueline K. Telford, pp. 227-228.

[25] Kothari, p. 95

[26] Ibid p. 344

[27] J R Lakshmana Rao, 1998, "*Scientific 'Laws', 'Hypotheses' and 'Theories' How are They Related?*", Resonance, December, pp. 55-61.

[28] Kerlinger, F. N. "*Foundations of Behavioral Research*" (3rd ed.). New York: Holt, Rinehart, & Winston (1986).

[29] Bill Robertson, "How does a scientific theory become a scientific law?", Reading in Science Methods K-8, NSTA, January (2009) pp. 56-58.

[30] C George Thomas, *"Research Methodology and Scientific Writing"* 2021, Springer. p. 106.

[31] Cox, D.R. 1958. Planning of Experiments. John Wiley & Sons, New York, p. 308.

Suggested Readings

1. Biju Dharmapalan, 2013, "*Scientific Research Methodology*", Narosa Publishing House Pvt. Ltd.
2. Kumar Ranjit, 2005, "*Research Methodology: A Step-by-Step Guide for Beginners*", Singapore, Pearson Education.
3. Dawson, Catherine, 2002, "*Practical Research Methods*", UBS Publishers Distributors, New Delhi.

4. Fowler, W. S., 1962, "*The Development of Scientific Method*", Pergamon Press, Oxford.
5. Wilson, E. B., Jr., 1952, "*An Introduction to Scientific Research*", McGraw-Hill, New York.
6. Freedman, P., 1950, "*The Principles of Scientific Research*", Public Affairs Press, Washington, D.C.
7. Hans Radder (ed), 2003, "*The Philosophy of Scientific Experimentation*", University of Pittsburg Press.
8. Klaus Hinkelmann and Oscar Kempthorne, 2007, "*Design and Analysis of Experiments: Introduction to Experimental Design*", John Wiley & Sons, Inc.
9. Barker, Thomas B, 1994, "*Quality by Experimental Design*", Marcel Dekker Inc.

Chapter 5

Data Collection and Analysis

Men are probably nearer the central truth in their superstitions than in their science.

—**Henry David Thoreau**

I don't know anything, but I do know that everything is interesting if you go into it deeply enough.

—**Richard Feynman**

5.1 INTRODUCTION

The collection of data through an experiment and its subsequent analysis is an important stage of a research programme. *Data are any information or observations that are associated with a particular research project, including experimental specimens, technologies, and product related to the inquiry.* For example, data of students of a college include count of the number of students, and separate count of number of various types of students such as, male and female or undergraduates and postgraduates, etc. The identification and collection of appropriate data for solving the research problem using suitable methods is an area where a researcher's ability and understanding play an important role. Several methods of data collection are in practice and any type of data collection method can be used with any approach to research. However, some types of data collection are more commonly used with the experimental approaches. The data are usually classified into two:

- Qualitative data, and Quantitative data.

Data which are mainly in words, sounds or images that deal with meanings are called *qualitative data* and data which are in numerical form are called *quantitative data*. The main advantage of quantitative data is that a wide range of statistical tests is available for analyzing these data. The qualitative data are analyzed qualitatively based on the experience of the individual participants and the quantitative data are analyzed using some descriptive statistics to summarize the pattern of findings. These descriptive statistics include measures of central tendency within a sample like *mean* and measures of the spread of scores within a sample like *range*. Since there are large amounts of raw data to deal with, and the researcher needs specialized computer programs to analyze qualitative verbal information. Depending upon the sources the data are further classified as:

- Primary data, and;
- Secondary data.

The *primary data* are those which are collected afresh and for the first time, and thus happen to be original in character. These are recorded as observed or encountered. No modification is made and hence can be considered as raw material. There are advantages for the primary data; they include [1]:

- Primary data is more accurate and gives detailed information according to the requirement.
- The explanation of terms, definitions, and concepts are incorporated in the primary data.
- Methods of collection, its limitations and other allied aspects are highlighted.
- They are more reliable and less prone to errors.
- It often includes a copy of the schedule and description of the procedure used in selecting the sample and collecting the data.

In spite of these advantages primary data have some disadvantages like time consuming, expensive and the requirement of expertise, etc.

The *secondary data*, on the other hand, are those "which have already been collected by someone else and which have already been passed through the statistical process" [2]. It is also called published data. The advantages of secondary data include [3]:

- Less costly as data are already available.
- They are faster to collect and process as compared to primary data.
- It provides valuable insights and contextual familiarity with the subject matter.
- It provides a base on which further information can be collected to update these and finally can be used for the purpose of research.

However, the identification and accessing of appropriate data sources is time consuming. Not only that the available resources may be too vast. The use of secondary data may go erroneous if the researcher does not know the purpose for which the data were originally collected. Nevertheless, secondary data are very useful in research and should be collected from reliable sources.

Depending on the research problem a researcher can decide which type of data are required and accordingly he can follow appropriate method of data collection. Data in general possess some characteristics. Some of them are listed below [4]:

- Data are aggregate of facts; this means unconnected figures cannot be used.
- Data are affected to a large extent by multiplicity of factors.
- Data are estimated according to reasonable standard of accuracy.
- Data are collected in a systematic manner for a predetermined objective.
- The data must be related to one another.
- Data must be numerically expressed; qualitative data must be scaled to express in numeric terms.

5.2 NEED FOR DATA COLLECTION

Data collection is essential for the conduct of a research programme. According to Yogesh Kumar Singh (2006) data are needed to serve the following purposes in research [5]:

- Collection of data is very essential in any educational research to provide a solid foundation for it.
- It is something like the raw material that is used in the production of data. Quality of data determines the quality of research.
- It provides a definite direction and definite answer to a research inquiry. Data are very essential for a scientific research.
- The data are needed to substantiate the various arguments in research findings.
- The main purpose of data collection is to verify the hypotheses.
- Statistical data are used in two basic problems of any investigation:
- Estimation of population parameters, which help in drawing generalization.
- The hypotheses of any investigation are tested by data collection procedure.
- The qualitative data are used to find out the facts and quantitative data are employed to formulate new theory or principles.
- Data are also employed to ascertain the effectiveness of new devices for its practical utility.
- Data are necessary to provide solutions of the problems.

Realizing the importance of data collection one should take necessary precautions in collecting data. Any arbitrary data collected without proper care will not serve the purpose. The collected data must possess some qualities. The following points can be kept in mind when collecting data for a research work [6].

- The data must be relevant to the research problem.
- Standardized research tools must be used for data collection
- The data should be such as these can be subjected to statistical treatment easily.

- Minimum measurement error should be ensured in recording the data.
- It is essential that the data be tenable for the verification of the hypotheses.
- Completeness and comprehensive nature of the data should be ensured.
- The procedure for data collection should be objective.
- Data should be estimated for inferential purpose.
- The data should be accurate and precise.
- The data should be reliable and valid.
- The data should be such that these can be presented and interpreted easily.
- The scoring procedure of the research tool should be easy and objective.

5.3 METHODS OF DATA COLLECTION

Data collection techniques include direct observation, personal interview, forms, schedules or questionnaire method, documented sources of data, case study method, telephone interviews, mail questionnaires, etc. It is obvious that the data collection methods are different for primary and secondary data. Primary data are collected during the course of the experiment whereas the secondary data can be collected in advance. It should be noted that all methods of data collection cannot be applied in scientific research. Majority of the books on research methodology discusses data collection methods that are particularly used in survey and descriptive research. We can notice that majority of these methods have a subjective element in it and it is against the objective nature of scientific knowledge. Therefore, such methods are not used in scientific research. However, brief discussion of some important observation methods is presented below.

Observation method: This is the most commonly used method for data collection especially in behaviour sciences. Here the researcher observes directly what is happening to the *subject* and based on his observation he can elucidate a cause-effect relationship between the subject and an independent variable. "Observation becomes a scientific tool and the method of data collection for the researcher, when it serves a formulated research purpose, is systematically planned and recorded and is subjected to checks and controls on validity and reliability" (Kothari, 2004, p. 96). This method has the following advantages [7]:

- This method is very simple to use.
- In this method subjective bias is eliminated, if observation is done accurately.
- The information obtained under this method relates to what is currently happening; it is not complicated by either the past behaviour or future intentions or attitudes.

- This method is independent of respondents' willingness to respond and as such is relatively less demanding of active cooperation on the part of respondents as happens to be the case in the interview or the questionnaire method.
- The conclusions of observations are more reliable than non-observation conclusions, because they are based on first-hand perception by the eyes and can be verified by any one by visual perception.

However, there are some events like emotions, likes and dislikes that cannot be object of observation. In such cases we cannot employ this method for data collection. Further, in observation the atmosphere tends to become artificial and hence the individuals who are being observed become self-conscious. This will definitely affect the observation and data suffer some variations. In addition to these, as already mentioned, the final results of observation depend upon the interpretation and understanding of the observer. Thus, there is a possibility for subjectivity and bias to play in recording the observed data.

The observation method is further divided into *controlled* and *uncontrolled observations.* "If the observation takes place in the natural setting, it may be termed uncontrolled observation, but when observation takes place according to definite pre-arranged plans, involving experimental procedure, the same is then termed controlled observation' (Kothari, 2004, p. 97). The uncontrolled observations do not use precision instruments and are liable for subjective interpretation.

Interview Method: A thorough interview is a form of self-report that is a relatively simple approach to data collection. In interview method there is a face-to-face interaction between the researcher and the person from whom the information is to be obtained. Interview can be structured or unstructured. In structured interview predetermined questions are asked and standardized techniques are used for recording. Unstructured interview is characterized by flexibility in questioning. The researcher can ask supplementary questions as well as he can omit some questions if the situation demands so. An interview can cover any number of content areas and is a relatively inexpensive and efficient way to collect a wide variety of data that does not require formal testing [8].

The effectiveness of the interview depends on how it is structured. All the participants should be asked the same questions in the same order. Moreover the researchers who are conducting the interview should be trained properly to avoid any mistake in collecting the data. The interview method is further classified as:

- *Focused interview:* Here the attention is given to the experience of the respondent and its effects.
- *Clinical interview:* It is concerned with broad underlying feelings or motivations or with the course of individual's life experience.

- *Non-directive interview:* In this case the interviewer encourages the respondent to talk about the given topic with a bare minimum of direct questioning.

The characteristics of interview methods are:

- It is a close interaction including dialogue between two or more persons.
- There is a definite object of interview, such as knowing the ideas and views of others.
- There is face-to-face contact or primary relationship between the individuals.
- This is the most suitable method of data collection for business and economic problems.
- Interview method possesses a number of merits. They are [9]:
- This method offers greater flexibility.
- More information in greater depth can be obtained through an interview.
- Personal information about the respondent can as well be obtained easily under this method.
- Samples can be controlled more effectively as there arises no difficulty of the missing returns; non-response generally remains very low.
- Observation method can as well be applied to recording verbal answers to various questions.

In spite of these advantages, the interview method has some disadvantages also. It is quite natural that the respondent may not give adequate information in an open conversation. There may be several matters which may be written in private only. Moreover the response may be one sided. Sometimes the prejudices of the interviewer also affect the information collected. Further, "the data obtained may not be appropriate for extensive statistical analysis because they simply describe a construct rather than quantifying it" (Geoffrey Marczyk et al., 2005, p.117). A variation of personal interview is the *telephonic interview*; in which the respondent is contacted through a telephone. This method is used when a quick response is required and is more economical than personal interview.

Questionnaire Method: This is a very popular method of data collection. In this method researcher prepares a number of questions in a particular sequence and approached the respondent with these questions in printed form. The respondents are asked to read, understand and respond to the question. Usually, space will be provided against each question for marking the answer. The questionnaire can be of different types [10]:

- *Structured questionnaire:* Structured questionnaires are those in which a question is presented the respondents with fixed response categories.
- *Unstructured questionnaire:* Here every question is not necessarily presented to the respondent in the same wording and does not have fixed responses. Respondents are free to answer the question the way they like.

- *Mixed questionnaire:* This is a questionnaire which is neither completely structured nor unstructured. It consists of both the types of questions.
- *Disguised typed questionnaire:* Here the questions are direct and therefore respondents may give answer to certain sensitive questions whose accuracy may be questioned.

Kothari (2004, pp.100-101) identified the following advantages of the questionnaire method.

- There is low cost even when the universe is large and is widely spread geographically.
- It is free from the bias of the interviewer; answers are in respondents' own words.
- Respondents have adequate time to give well thought out answers.
- Respondents, who are not easily approachable, can also be reached conveniently.
- Large samples can be made use of and thus the results can be made more dependable and reliable.

However, the strength of this method depends mainly on the attitude of the respondent. Sometimes the respondents may not return the questionnaire duly filled. The subjectivity of the respondents also will affect the answer marked by them. This method can only be used among respondents who are educated and cooperative. Possibility for omitting some questions as well as providing ambiguous answer is another demerit of this method.

Schedule Method: "Schedule is the name usually applied to a pro forma containing a set of questions which are asked and filled by an interviewer in a face situation with a respondent" [11]. Through this method data can be collected in an objective manner. This method is almost similar to the questionnaire method, but the difference is that schedules are being filled in by the enumerators who are specially appointed for the purpose. The enumerator approaches the respondents with these questions and put to them the questions from the pro forma in the order the questions are listed and record the replies in the space provided for the same in the pro forma. Schedules are mainly used in business research. There are different types of schedules.

- *Observation schedule:* In observation schedules the observer records the activities and responses of the subject under specific conditions. The main purpose of the observation schedule is to verify information.
- *Rating schedule*: In rating schedule the value and the trend of qualities like attitudes, opinion, preferences, etc., are measured on a rating scale. This is mainly used in business and psychological research.
- *Document schedule*: The schedules of this type are used to obtain data regarding return evidence and case histories from autobiography, diary, case histories or final records of governments, etc.

The above methods are used in the collection of primary data. The methods for collecting secondary data are different from these methods. Secondary data are already available and hence the only task is to locate the correct source. Most of the secondary data are available in published form. The sources of secondary data include:

- Research journals
- Books, magazines and newspapers
- Proceedings of conferences
- Data collected by research agencies
- Public records and statistics, historical documents, and other sources of published information.
- Publications of the central and state governments, government of foreign countries or international bodies, etc.

5.4 PRINCIPLES FOR ACCESSING RESEARCH DATA

The research data are valuable and utilizing the data obtained by one research group by various other research groups is liable to certain policies and principles. Data access and sharing among researchers, research institutions, and national research agencies are promoted, while at the same time, the various national laws, research policies and organizational structures of member countries is to be taken into account. Access to such data is subject to a range of considerations. Moreover, in some instances, access to or use of data may be restricted to safeguard the privacy of individuals, protect confidentiality, proprietary rights or national security. The effective sharing of research data has the following implications [12]:

- Reinforces open scientific inquiry;
- Encourages diversity of analysis and opinion;
- Promotes new research;
- Makes possible the testing of new or alternative hypotheses and methods of analysis;
- Supports studies on data collection methods and measurement.
- Facilitates the education of new researchers.
- Enables the exploration of topics not envisioned by the initial investigators;
- Permits the creation of new data sets when data from multiple sources are combined.

The principles of sharing research data are meant for the practice of effective utilization of available research data from various sources. It is to improve the efficiency and effectiveness of the global science system. Organization for Economic Co-operation and Development (OECD), France; in their guidelines observed the following principles in sharing research data [13].

- *Openness:* Openness, in this context, means access on equal terms for the international research community at the lowest possible cost.
- *Flexibility:* Rapid and often unpredictable changes in information technologies, the characteristics of each research field and the diversity of research systems, legal systems and cultures of each country is to be taken into account.
- *Transparency:* Information on research data and data-producing organizations, documentation on the data and specifications of conditions attached to the use of these data should be internationally available in a transparent way, ideally through the Internet.
- *Legal conformity:* Data access arrangements should respect the legal rights and legitimate interests of all stakeholders in the public research enterprise. The accessibility of certain data is restricted because of national security, privacy and confidentiality, trade secrets and intellectual property rights, etc.
- *Protection of intellectual property:* Data access arrangements should consider the applicability of copyright or of other intellectual property laws that may be relevant.
- *Formal responsibility:* The arrangements for research data access should promote explicit, formal institutional practices, such as the development of rules and regulations, regarding the responsibilities of the various parties involved in data-related activities. These practices should pertain to authorship, producer credits, ownership, dissemination, usage restrictions, financial arrangements, ethical rules, licensing terms, liability, and sustainable archiving.
- *Professionalism:* Arrangements for the management of research data should be based on the relevant professional standards and values embodied in the codes of conduct of the scientific communities involved in the research activities.
- *Interoperability:* Technological and semantic interoperability is a key consideration in enabling and promoting international and interdisciplinary access to and use of research data. Access arrangements, should pay due attention to the relevant international data documentation standards. Member countries and research institutions should cooperate with international organizations that are charged with developing new standards.
- *Quality:* The quality of research depends mainly on the quality of data. Data managers, and data collection organizations, should pay particular attention to ensuring compliance with explicit quality standards.
- *Security:* The data, along with relevant meta-data and descriptions, should be protected against intentional or unintentional loss, destruction,

modification and unauthorized access in conformity with explicit security protocols.

- *Efficiency:* One of the main objectives of promoting data access and sharing is to improve the overall efficiency of scientific research to avoid the expensive and unnecessary duplication of data collection efforts.
- *Accountability:* The performance of data access arrangements should be subject to periodic evaluation by user groups, responsible institutions and research funding agencies. Such evaluations should help to increase the support for open access among the scientific community.
- *Sustainability:* The administrators should take necessary care to guarantee the permanent access of the data. For this long term retention of data is required. Research funding agencies and research institutions should consider the long-term preservation of data at the outset of each new project, and in particular, determine the most appropriate archival facilities for the data.

5.5 DATA PROCESSING

The procedure next to data collection is the processing, analysis and interpretation of the collected data. "Processing implies editing, coding, classification and tabulation of collected data so that they are amenable to analysis" (Kothari, 2004). The analysis for the interpretation of the data enables the researcher to decide whether the hypothesis is to be rejected or not. "Analysis of data includes comparison of the outcomes of the various treatments upon the several groups and the making of a decision as to the achievement of the goals of research. Data relevant to each hypothesis must be assembled in quantitative form and tested to determine whether or not there is a significant difference in the results obtained from the controlled groups" (Yogesh Kumar Singh, 2006). According to G.B. Giles "in the process of analysis, relationships or differences supporting or conflicting with original or new hypotheses should be subjected to statistical tests of significance to determine with what validity data can be said to indicate any conclusions" (cited in Kothari, p. 122). Data analysis utilizes many statistical techniques such as analysis of variance. It is not possible to have a data analysis without using statistical principles. Before going into the details of data analysis let us consider the various operations involved in the data processing.

Editing: The collected data is called the *raw* data. Editing involves examining the raw data for some errors or omissions and to make corrections when possible. "Editing is done to assure that the data are accurate, consistent with other facts gathered, uniformly entered, as completed as possible and have been well arranged to facilitate coding and tabulation" (Kothari, p.122).

Coding: "Coding refers to the process of assigning numerals or other symbols to answers so that responses can be put into a limited number of categories or classes" (Kothari, p. 123). Through coding several responses can be reduced to a small number of groups which contains critical information required for analysis. It is to be noted that each group should be defined in terms of only one concept.

Classification: The essential idea of classification is to form small groups with some meaningful relationships from a large volume of raw data. "Data having a common characteristic are placed in one class and in this way the entire data get divided into a number of groups or classes" (Kothari, pp.123-124). Depending upon the nature of the phenomena involved, the classification can be made either according to attributes (where the qualitative data are classified according to some *statistical attributes*) or according to class intervals (where the quantitative data called *statistical data* are classified on the basis of class intervals).

For determining the size of the class interval Sturges's formula can be used. According to this formula;

$$\text{Class interval; } i = \frac{R}{1+3.3\log N} \tag{5.1}$$

where, i is the class interval, R is the range and N is the number of items to be grouped. It is to be noted that the class interval is of equal magnitude and in practice we may have ten to fifteen classes.

Tabulation: Tabulation means the arrangement of the collected data according to some concise and logical order. Tabulation is the process of summarizing raw data and displaying the same in compact form for further analysis (Kothari, p.127). In tabulation the classified data are put in the form of tables. The advantages of tabulation can be observed through the following points [14].

- It conserves space and reduces explanatory and descriptive statement to a minimum.
- It facilitates the process of comparison.
- It facilitates the summation of items and the detection of errors and omissions.
- It provides a basis for various statistical computations.

Recently after the introduction of computer the handling of large quantity of data is not at all a problem. Classification and tabulation can be achieved within a fraction of a second with greater accuracy.

5.6 DATA ANALYSIS

If the collected data is to be translated into meaningful information, data must be managed and analyzed in an appropriate fashion. There is no single method for analyzing data. The form of analysis is in accordance with the particular project's functions and needs. Additional considerations in choosing a particular method for analysis might include the research setting or the type of research. Knowledge about data analysis can help a researcher interpret data for the purpose of providing meaningful insights about the problem being examined. The statistical procedures used for data analysis can be broadly classified into two: descriptive and inferential [15].

- *Descriptive statistics:* This allows the researcher to describe the data and examine relationships between variables.
- *Inferential statistics:* This allows the researcher to examine causal relationships.

5.6.1 Descriptive Statistics

Descriptive statistics is used to describe the data collected in research studies and to accurately characterize the variables under observation within a specific sample (Geoffrey Marczyk et al., p. 209). In data analysis, the descriptive statistics is used to describe the distributions of certain variables accurately. The advantage is that it provides necessary information for the other researchers who wish to replicate the study. The basic methods of descriptive statistics include:

- Frequency distribution
- Central tendency (mean, median and mode)
- Dispersion Measures of association

Frequency distribution: It is a complete list of all possible values of a variable along with the number of times that each value appears in the data set. The number of repetitions is the frequency of the value. This information can be represented in a table called the *frequency table.* An illustration of a frequency table is given below.

Table 5.1 Frequency distribution table

Value	Frequency	Cumulative frequency
22	2	2
24	3	5
25	3	8
29	4	12
30	6	18
35	2	20
39	2	22
42	1	23

This process could be more informative if we use grouped frequency distribution table. Here, the values are represented as a range as illustrated below.

Table 5.2 Grouped frequency distribution table

Value	Frequency	Cumulative frequency
21–25	8	8
26–30	10	18
31–35	2	20
36–40	2	22
41–45	1	23

The information contained in the above tables can be represented graphically and it is called a *histogram*. A histogram representing Table 5.2 is given below in Fig. 5.1.

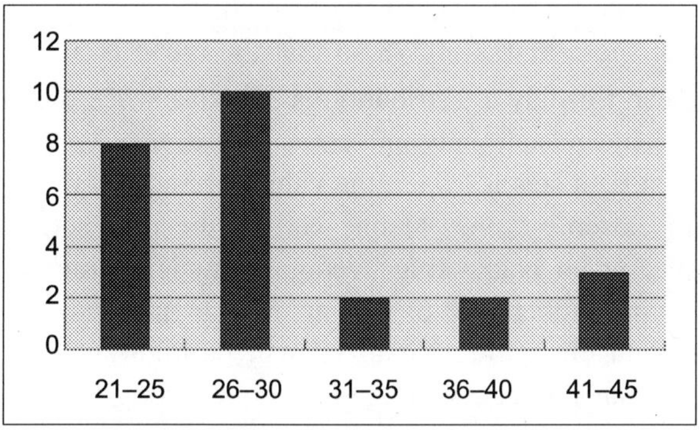

Fig. 5.1 Grouped frequency histogram

The major disadvantage of the frequency distribution table or a histogram is that it provides only an overview of the distribution. The other methods provide more precise way of describing the distribution of data.

Central tendency: For a given distribution we can define the central tendency *as the number that represents the typical or most representative value in the distribution*. In technical language the measures of central tendency tell us the point about which items have a tendency to cluster. This is also known as the statistical average. This allows the researcher to characterize the data set with a single value. The measures of central tendency include *mean, median* and *mode*.

Mean or *average:* It is the most widely used measure of central tendency. The calculation of mean of a distribution is very simple. Add all the numbers in the distribution set and the divide the sum by the total number of elements in the distribution.

Let there are n elements in a distribution; say: X_1, X_2, X_3, ..., X_n. Then the mean $\overline{X}$, of this distribution can be represented as:

$$\overline{X} = \frac{\sum_{i=1}^{n} X_i}{n} = \frac{X_1 + X_2 + X_3 + ... + X_n}{n} \tag{5.2}$$

In the case of frequency distribution the mean can be calculated as:

$$\overline{X} = \frac{\sum_{i=1}^{n} f_i X_i}{n} = \frac{f_1 X_1 + f_2 X_2 + f_3 X_3 + ... + f_n X_n}{n} \tag{5.3}$$

where $f_1, f_2, f_3, ...f_n$ are the frequencies of $X_1, X_2, X_3, ...X_n$, respectively. Also, $n = f_1 + f_2 + f_3 + ...f_n$.

As an example consider the marks obtained by ten students in a particular subject. The marks are:

46, 23, 42, 38, 35, 49, 40, 32, 39, 38

Then the sum of these marks is 378 so that the mean is 382/10 = 38.2.

The usage of mean to study a distribution is effective when the distribution is uniform. Usually, the mean is strongly influenced by extreme values. Therefore, it may be misleading in data sets in which the values are not normally distributed, or where there are extreme values at one end of the data set. This happens usually for example, when calculating average annual income of a group. If majority of the group members are of low annual income and; if a few report very high income, then the average obtained will be high and the conclusions made based on such an average will definitely go wrong. In such cases the usage of median or mode will be more meaningful.

Median (M): It is the value of the middle item of series when it is arranged in ascending or descending order of magnitude. Then the series is divided into two halves; in one half all items are less than the median, whereas in the other half all items have values higher than the median. Let us arrange the marks in the above example in the ascending order.

23, 32, 35, 38, 38, 39, 40, 42, 46, 49

In this particular example we have to mid-values: 38 and 39. Hence, the median is the average of these two, i.e., 38.5. "Median is not useful where items need to be assigned relative importance and weights. It is not frequently used in sampling statistics" (Kothari, p. 133). Since the median is a positional average the extreme values do not influence it.

Mode (Z): It is the value that occurs most frequently in a set of values. To find the mode, simply count the number of times (frequency) that each value appears in a data set. Like median, mode is a positional average and hence it is not affected by the extreme values. Once again listening to the above distribution,

23, 32, 35, 38, 38, 39, 40, 42, 46, 49

The value 38 occurs two times and hence the mode of this distribution is 38. The limitations of mode are:

- It is not amenable to algebraic treatment;
- It remains indeterminate when we have two or more model values in a series: andit is not suited when relative importance of items is to be considered.

Apart from mean, median and mode *geometrical mean* and *harmonic mean* are also used under certain conditions.

Geometrical mean is defined as the nth root of the product of the values of n times in a given series. This can be represented as:

$$GM = \sqrt[n]{\pi X_i} = \sqrt[n]{X_1.X_2.X_3.......X_n} \qquad (5.4)$$

Geometrical mean is usually used in the determination of average per cent of change.

Harmonic mean is defined as the reciprocal of the average of reciprocals of the values of items of a series. Mathematically this can be represented as:

$$HM = \frac{1}{\frac{1}{n}\left(\frac{1}{X_1}+\frac{1}{X_2}+\frac{1}{X_3}+.................+\frac{1}{X_n}\right)} \qquad (5.5)$$

The application of harmonic mean is limited, particularly, to cases where time and rate are involved. The disadvantage is that it gives largest weight to the smallest item and smallest weight to the largest item.

Dispersion: A measure of central tendency is a representative of a series. It is the most probable value. However, it cannot explain the variation of the values. Especially, it fails to give any idea about the scatter of the values of items of a variable in the series around the true value of average (Kothari, p.134). Measure of dispersion is used to calculate the measure of the spread of the values around the average value. Various measures of dispersion are *range*, *mean deviation*, *variance* and *standard deviation.*

Range: The *range* of a distribution tells us the smallest possible interval in which all the data in a certain sample will fall. In other words the range is the difference between the largest and smallest values of a distribution. Hence, the calculation of range is very easy; just take the difference of the largest and smallest value in the distribution. In the example presented above, the range can be obtained as:

$$49 - 23 = 26$$

Even though it gives an idea of the variability very quickly, one can see that the rage is not an effective representative of the dispersion because it depends only on two values; the largest and smallest. A more precise measure of dispersion, or spread around the mean of a distribution, is the *mean deviation.*

Mean deviation: Mean deviation is the average of *difference* of the values of items from some average of the series. Such difference is the deviation. In calculating mean deviation we ignore the minus sign of deviations while taking their total for obtaining the mean deviation. Mean deviation can be obtained from the arithmetic mean, median and mode. Thus the mean deviation can be represented as:

Mean deviation from mean,

$$\delta_X = \frac{\sum |X_i - \overline{X}|}{n} \tag{5.6}$$

Mean deviation from median,

$$\delta_M = \frac{\sum |X_i - M|}{n} \tag{5.7}$$

Mean deviation from mode,

$$\delta_Z = \frac{\sum |X_i - Z|}{n} \tag{5.8}$$

"When *mean deviation* is divided by the *average* (mean, median or mode) used in finding out the mean deviation itself, the resulting quantity is described as the *coefficient of mean deviation.* Coefficient of mean deviation is a relative measure of dispersion and is comparable to similar measure of other series" (Kothari, p. 135). It is a more effective representation of the measure of variability than range as it takes into consideration the values of all items of a series.

Variance: Variance is a precise measure of dispersion, or spread around the mean of a distribution. The variance can be calculated in the following steps. First subtract the mean of the distribution from each of the values. Then square each result and add all of the squared results. Finally, divide the result by the number of values minus 1. This can be expressed mathematically as:

$$\text{Variance} = \frac{\sum_{i=1}^{n} \left(X_i - \overline{X}\right)^2}{n-1} \tag{5.9}$$

The variance of a distribution gives us an average of how far, in squared units, the values in a distribution are from the mean, which allows us to see how closely concentrated the scores in a distribution are (Geoffrey Marczyk et al., 2005, p. 215).

Standard Deviation: Standard deviation is defined as the square-root of the average of squares of deviations, the deviations being obtained from the arithmetic mean. Thus,

$$\text{Standard deviation } \sigma = \sqrt{\frac{1}{n}\left\{\sum_{i=1}^{n}\left(X_i - \overline{X}\right)^2\right\}} \tag{5.10}$$

Standard deviation is the square root of variance. Technically the replacement of *n–1* with *n* will not produce any significant effect. However, keeping *n-1* in the expression provides a better representation of the error especially when *n* is small.

In the case of frequency distribution;

$$\text{Standard deviation } \sigma = \sqrt{\frac{1}{n}\left\{\sum_{i=1}^{n}\left(X_i - \overline{X}\right)^2\right\}} \tag{5.11}$$

The quantity obtained when the standard deviation is divided by the arithmetic mean of the distribution is called the *coefficient of standard deviation.* When this coefficient of standard deviation is multiplied by 100, the resulting figure is known as *coefficient of variation.* Standard deviation is widely used for data analysis in research studies because it provides a satisfactory measure of dispersion in a distribution. Also it will not be affected much by the fluctuations in sampling.

Measures of association: The main task of descriptive statistics is to describe the shape of variable distributions. However, it also aims to examine and describe the relationships or associations between variables. The most basic measure of association between two or more variable is the *correlation.* It is expressed in a single number called *correlation coefficient (r).* Correlation is expressed in a scale ranging from +1.0 to –1.0. It provides information about the direction of the relationship (either positive or negative) and the intensity of the relationship. Depending upon the type of variables there are a wide variety of correlations.

The direction of correlation tells us whether there is a *positive (direct)* or *negative (inverse)* relationship between the variables. A positive relationship implies that when there is an increase in the independent variable a corresponding increase is present in the dependent variable. For example, as the temperature increases the resistance of a metallic wire also increases. In a negative correlation, On the other hand, the situation is different. An increase in the independent variable causes a decrease in the dependent variable. Well known example is the Boyle's law, as the pressure on a gas increases its volume decreases. As already indicated the correlation coefficient ranges from +1.0 to –1.0. Therefore, a correlation coefficient +0.55 implies a positive correlation and –0.55 implies a negative correlation. Figure 5.2 given below shows the positive and negative correlation graphically.

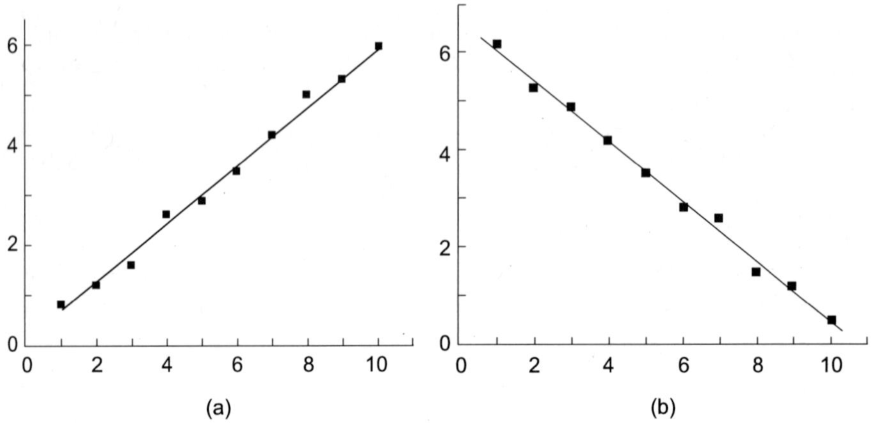

Fig. 5.2 (a) Positive and (b) Negative correlation directions

Correlation coefficient is an indication of the strength of the relation. If $r = \pm 1.0$, the relationship is very strong. In general a correlation coefficient 0.01 to 0.3 represents weak correlation, 0.3 to 0.7 are considered to be moderate, 0.7 to 0.9 are strong and 0.9 to 1.0 are very strong.

The square of the correlation coefficient is known as the *coefficient of determination* (r^2). It is used to determine the proportion of variance accounted for by the association. When the coefficient of determination is multiplied by 100 we get the *percentage of variance*. For example, if the correlation coefficient is 0.6, then the coefficient of determination is 0.36 and the percentage of variance is 36.

There are a number of correlations used in data analysis. The most widely used one is the *Pearson product-moment correlation* (*Pearson r*). It is used to examine associations between two variables that are measured on either ratio or interval scales. For example, the correlation between the days of exercise and kilogram loss of weight can be examined through *Pearson r*. Some other types of correlations are [16]:

- *Point-biserial* (r_{pbi}): The relationship between a variable measured on a naturally occurring *dichotomous* nominal scale and a variable measured on an interval (or ratio) scale is examined through this. For example, the correlation between the gender (dichotomous) and the score of a test (interval) can be measured through this.
- *Spearman rank-order* (r_s): The relationship between two variables measured on ordinal scales can be examined through this. For example, the correlation of class rank (ordinal) and socioeconomic status (ordinal) can be analyzed through this method.

- *Phi (Φ)*: This is used to examine the relationship between two variables that are naturally dichotomous. An example is the study of a correlation between gender (nominal) and marital status (nominal-dichotomous).
- *Gamma (γ)*: This is used to examine the relationship between one nominal variable and one variable measured on an ordinal scale. For example, a correlation of ethnicity (nominal) and socioeconomic status (ordinal) can be analyzed through this.

5.6.2 Inferential Statistics

When researchers conduct research on a large population, they select representative samples to draw inferences about the population. The analyses used to examine these inferences are referred to as *inferential statistics*. "Inferential statistics help us to draw conclusions beyond our immediate samples and data" (Geoffrey Marczyk et al., 2005, p. 219). Through a study among a small group of employees we can get information about the job satisfaction of the entire employees in the firm. From the conclusions drawn from the study of the sample a generalization regarding the population under study is possible. "Inferential statistics could be used to infer, from between-group differences in a particular study sample, how effective a new treatment or medication may be for a larger population" (Geoffrey Marczyk et al., 2005, p. 220).

The primary requirement of the inferential statistics is the random sampling. Inferences begin with the formulation of a hypothesis followed by tests to determine whether it is to be rejected or not to be rejected. For this researchers can choose appropriate inferential statistics from among a group of inferential statistics. This again based on the nature of research question and the types of variables involved. Widely used inferential statistics are *t-test, analysis of variance (ANOVA), chi-square*, and *regression*. A brief discussion of these is given below.

T-test: The mean differences between two groups are tested using *t*-test. It requires a single independent variable and a single continuous dependent variable to be present in both the experimental and control group. When a researcher wishes to compare the average (mean) performance between two groups on a continuous variable, he should consider the *t*-test.

Analysis of Variance (ANOVA): It is also a test of mean comparisons. It is used when a comparison of means across more than two groups is required. In *t*-test the comparison is between two groups. The ANOVA works by comparing the differences between group means rather than the differences between group variances. There are further modifications for ANOVA. A *one-way ANOVA* is used to compare the means of two or more levels of a single independent variable. When a study involves two or more independent variables *multifactor ANOVAs* can be used.

Chi-Square (χ^2) test: The *t*-test and ANOVA are appropriate only when the independent variables are continuous (interval or ratio). This need not be the situation always. Data can be nominal or ordinal. In such situations *chi-square statistics* can be used to test the hypothesis. "It does this by testing whether one set of proportions is higher or lower than one would expect by chance. Chi-square summarizes the discrepancy between observed and expected frequencies. The smaller the overall discrepancy is between the observed and expected scores, the smaller the value of the chi-square will be. Conversely, the larger the discrepancy is between the observed and expected scores, the larger the value of the chi-square will be" (Geoffrey Marczyk et al., 2005, p. 223). It can also be used to examine between-group differences on categorical variables, such as gender, employed, or grade level.

Regression: The method of predicting the value of some dependent variable from the given values of one or more independent variables is called *linear regression.* The main purpose of regression is prediction. There are two types of linear regressions: *simple regression* where the prediction of a single dependent variable is made with a single independent variable and *multiple regression* in which a researcher may use a number of independent variables to predict the dependent variable.

Another type of regression is the *logistic regression.* It has the ability to predict *dichotomous variables* (variables that are completely opposite and different from each other), such as the presence or absence of a specific outcome, based on a specific set of independent variables.

A detailed discussion of these techniques can be found in *Research Methodology*, by Suresh Chandra and Mohit Kumar Sharma (2013).]

5.6.3 Parametric and Non-parametric Tests

Statistical tests are mainly classified as parametric and non-parametric tests. The measured data are known as parametric data. The term 'parameter' refers to a function of the sample. Parametric statistics assumes that all the data are 'normally' distributed. For example, if we take the height of people, it could be seen that most people are of average height, with a few extremely tall or short cases; that is, the heights of people are normally distributed. Although individual values vary, the generality of this type of curve among populations is so strong that the parameter is taken as a constant. For applying parametric tests, the following three conditions must be satisfied.

- The observations must be independent.
- The samples must have equal or nearly equal variances.
- The variables involved are expressed in interval and ratio scales.

Parametric tests include the z-test, the t-test, and the analysis of variance (ANOVA). In certain cases, samples do not behave in the form of a normal bell curve. The data which not take the shape of a normal curve is called nonparametric data. Non-parametric data can be either counted or ranked. In such cases, we apply the chi-square test for statistical analysis.

A brief discussion of various statistical tests is given below.

T-test: The mean differences between two groups are tested using a t-test. It requires a single independent variable and a single continuous dependent variable to be present in both the experimental and control group. When a researcher wishes to compare the average (mean) performance between two groups on a continuous variable, he should consider the t-test.

Analysis of Variance (ANOVA): It is also a test of mean comparisons. It is used when a comparison of means across more than two groups is required. In the t-test the comparison is between two groups. The ANOVA works by comparing the differences between group means rather than the differences between group variances. There are further modifications for ANOVA. A *one-way ANOVA* is used to compare the means of two or more levels of a single independent variable. When a study involves two or more independent variables *multifactor ANOVAs* can be used.

Chi-Square (χ^2) test: The t-test and ANOVA are appropriate only when the independent variables are continuous (interval or ratio). This need not be the situation always. Data can be nominal or ordinal. In such situations, *chi-square statistics* can be used to test the hypothesis. 'It does this by testing whether one set of proportions is higher or lower than one would expect by chance. Chi-square summarizes the discrepancy between observed and expected frequencies. The smaller the overall discrepancy is between the observed and expected scores, the smaller the value of the chi-square will be. Conversely, the larger the discrepancy is between the observed and expected scores, the larger the value of the chi-square will be' (Geoffrey Marczyk et al., 2005, p. 223). It can also be used to examine between-group differences on categorical variables, such as gender, employment, or grade level.

Regression: The method of predicting the value of some dependent variable from the given values of one or more independent variables is called *linear regression*. The main purpose of regression is prediction. There are two types of linear regressions: *simple regression* where the prediction of a single dependent variable is made with a single independent variable and *multiple regression* in which a researcher may use a number of independent variables to predict the dependent variable.

Another type of regression is *logistic regression*. It has the ability to predict *dichotomous variables* (variables that are completely opposite and different from each other), such as the presence or absence of a specific outcome, based on a specific set of independent variables.

A detailed discussion of these techniques can be found in *Research Methodology*, by Suresh Chandra and Mohit Kumar Sharma (2013), Narosa Publishing House Pvt. Ltd., New Delhi.

5.7 NULL HYPOTHESIS IN STATISTICAL ANALYSIS

In statistical analysis, it is important to form a null hypothesis. A null hypothesis (H_0) states that there is no significant difference between the mean of a sample and the mean of a population from which the sample is selected. The null hypothesis is written as a negative statement. Then an alternate hypothesis or the research hypothesis (H_1) is also formed. Generally, the alternative hypothesis is a positive statement. Then, if the null hypothesis is rejected, then the alternate hypothesis is to be accepted and vice versa. In a null hypothesis even if a difference or an association is observed, it should be assumed that it is due to chance until it is confirmed by statistical analysis. However, if the statistical test rejects the null hypothesis, the alternate hypothesis is accepted by elimination. It is to be noted that the null hypothesis cannot be proved, because, anyone result that differs from the hypothetical result is sufficient to disprove it. The experimental data are analyzed statistically, and the probability of occurrence of the null hypothesis is calculated. If the probability is lower than the level of significance you have fixed, the null hypothesis is rejected.

5.7.1 Type I and Type II Errors

In a research study, usually, a sample is studied as the representative of a population. The study of the population as a whole is never possible. Due to this reason, there may be a possibility for the occurrence of errors, while rejecting or accepting the null hypothesis. After the data analysis, the researcher usually sets an arbitrary value for the rejection or acceptance of a null hypothesis based on some *level of significance*. The level of significance is usually a predetermined level of statistical significance and is expressed as a percentage. It represents the probability that a difference or an association as large as the one observed could have occurred by chance alone. Then, the null hypothesis is rejected if the value of the level of significance is less than the predetermined level of statistical significance. For example, we can set a level of 5% as the level of significance. This allows for a 5 per cent probability that the outcome could have occurred by chance. This indicates that a difference in means as large as that between the experimental and control groups would have occurred from sampling error in less than 5 out of 100 items of the experiment.

- The rejection of the null hypothesis while it is true is called a type I error. At the same time, failing to reject the null hypothesis while it is false is called a type II error.

"The maximum probability of rejecting the null hypothesis when it is actually true or the chance of committing type I error is known as the level of significance [17]. The two types of errors are unavoidable, and it is not possible to eliminate them simultaneously. When one type of error decreases, the other type increases, and therefore, the best option is to fix the probability of one of them and try to minimize the other. As a type I error is more serious, we usually fix the probability of type I error and try to minimize the probability of type II error.

5.7.2 Degree of Freedom

The concept of degree of freedom is very important while using statistical tools for data analysis. The concept of degree of freedom is basically discussed in classical mechanics while considering the motion of a particle. In a simple sense, the degree of freedom of a particle is the number of possible directions in which the particle can move. However, in statistical analysis, the degree of freedom has a different perspective. "The concept of degrees of freedom is based on the notion that if the total items in each set of measurements are fixed, then all the measurements minus one are free to have any value. In other words, the number of degrees of freedom in distribution is the number of items that are independent of each other, which cannot be assumed from each other" [17, p. 156]. This can be illustrated as follows.

Suppose we want to add two numbers so that their sum is 10. Then, if we fix one number as 6, then, the other number automatically becomes 4. Therefore, we have only one degree of freedom to select the number. Consider another example; that we need to add four numbers to get the sum equal to 50. Then, in this case, we have three degrees of freedom as we can select three numbers arbitrarily, and then the fourth number is fixed by the sum. That is, if we choose, 18, 15, and 8 are the first three numbers, then the fourth number can be 9 only since their sum must be equal to 50. In general, we can write, if the number of items is N and constraints k, then the degrees of freedom = N-k. The precision of experiments would be reduced if the number of degrees of freedom for error was less. Therefore, it is essential

5.8 PRESENTATION OF DATA

The communication of the analyzed data to the scientific community is achieved through the presentation of data. For easy understanding the analyzed data are presented in some visual form. A large amount of data presented in a textual form are difficult to understand. Therefore, a researcher uses non-textual methods for presenting his data in front of the scientific community as it readily attracts the attention of the readers. Most commonly used data presentation methods are: tables, graphs, histograms, frequency polygon, pictographs, pi-charts, flow charts, etc.

Tables: Tables are the most common method of presenting data. A large number of numerical data can be included in a table in a small place. It is more convenient than a lengthy explanation. It presents data in an understandable manner. In a table the exact texts are not repeated. Appropriate abbreviations can be used in a table and their meanings should clearly be defined in the caption, headnote or a footnote. "For clarity and understanding, a table may repeat information in more than one form, that is, in addition to the absolute values presented, it may give percentage, totals, means, averages, or ratios of their values" [18].

The major components of a standard table can be briefed as follows [19]:

- *Table number:* Tables are numbered in the order of appearance in the text using Arabic or Roman numbers.
- *Title:* The title of the table should be brief and should provide a clear idea of the contents of the table. Subheadings, if any, can be included on a new line inside parenthesis.
- *Column headings:* It indicates the nature of data in that column.
- *Horizontal rules:* Horizontal rules are drawn above and below the column headings.
- *Stub:* The leftmost column is called the stub and it describes the contents of each row.
- *Body of table:* Numerical data presented in the table should be decimal aligned. Numbers without decimal digits are aligned along the implied decimal point. Any missing data should be indicated by an em dash (—), two hyphens or three or more spaced dots (. . .).
- *Footnotes*: Footnotes appear at the bottom of a table. Each footnote begins on a new line. Instead of using letters as footnote symbol, typographical symbols like asterisk (*), number sign (#), section mark (§), dagger (†), and double dagger (‡) can be used.

Tables should be relatively simple so that it is easy to understand and the columns should have a clear relationship to each other. Column titles should be as brief as possible, consistent with clarity. Footnotes are needed in some cases to provide further explanation of the headings.

Graphs: There is no single statistical tool that is as powerful as a well-chosen graph. An enormous amount of quantitative information can be conveyed through graphs. It is one of the easiest ways of presenting the data visually. Most of the graphs falls under two categories: displaying either the data themselves or quantities derived from the data. Usually, the first type of display is used when we are exploring the data and are not fitting models, and the second is used to enhance numerical statistical analyses that are based on assumptions about relationships in the data. The graphical methods enable the researcher to explore data thoroughly, to look for patterns and relationships, to confirm or disprove the expected, and to discover new phenomena.

It represents a relationship between two variables. A graph consists of two mutually perpendicular axes; commonly designated as x-axis (horizontal axis) and y-axis (vertical axis). Usually, the independent variable is represented along the x-axis and the dependent variable along the y-axis. The point of intersection of these two axes is known as the origin of the graph. Each point on this graph can be characterized by two co-ordinates; x and y usually enclosed in parenthesis as (x, y). Then the co-ordinate of the origin is (0,0).

Points plotted on a graph may be interconnected or represented by a smooth curve. The points can be interconnected for nonfunctional relationships and for functional relationships the graph should be represented by a smooth curve.

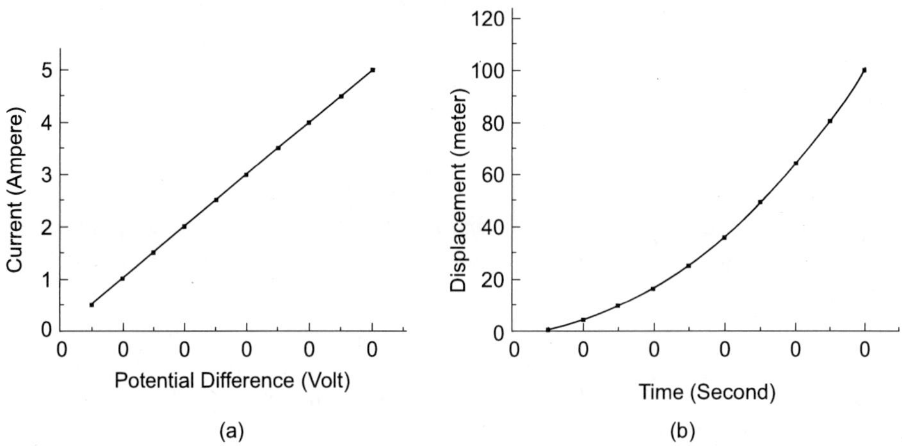

Fig. 5.3 (a) Linear dependence (b) Non-linear dependence

For example, consider the graphical representation of Ohm's law. In Ohm's law, as the potential difference (independent variable) across the ends of the conductor increases, the current (dependent variable) passing through it increase. This is a linear relationship and is represented by a straight line graph (Fig.5.3 a). The relation between the independent and dependent variable need not be linear always. In such cases the graph will deviate from straight line (Fig.5.3 b).

Another form of graph is the bar graph. A bar graph can be drawn either vertically or horizontally.

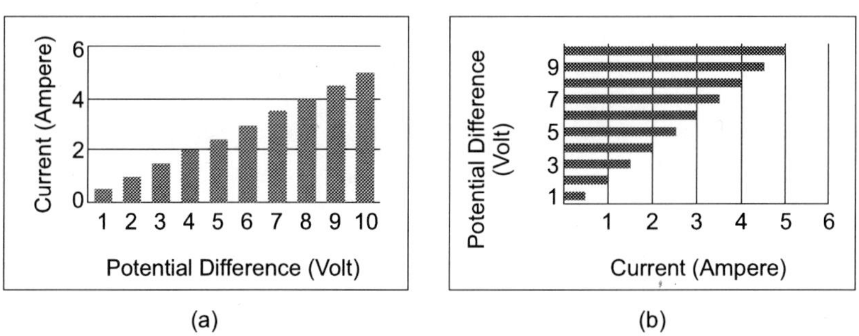

Fig. 5.4 Bar graphs (a) vertical and (b) horizontal

When using an orthogonal coordinate system the following points should keep in mind.

Correct labelling of the axes, including the quantity presented and its units.

- The text should remain legible, even if the editor reduces the size of the diagram.
- Use as much of the area available inside the diagram for presenting the data.
- Transform the variables and the corresponding axes to establish a simpler relation. For example, take logarithms to linearize an exponential.
- Transform one or both variables to show the difference between the data and the regression curve over the whole range. For example, logarithmic presentation of a large difference exists between the smallest and the largest data value, or zero point suppression that must be indicated.
- Self-explanatory captions must be given under the diagrams.

One major advantage of a graph is that it provides a means for interpolation of intermediate values, provided that the scale and rulings of the graph have sufficient resolution.

Histogram is another form of graphical representation which is used to represent discrete or continuous data that are measured on an interval scale.

[For details see section 5.6.1.]

Frequency polygon is obtained from a histogram by joining the midpoints on the top of the histogram bars. This is used when data from continuous variables are shown in a histogram.

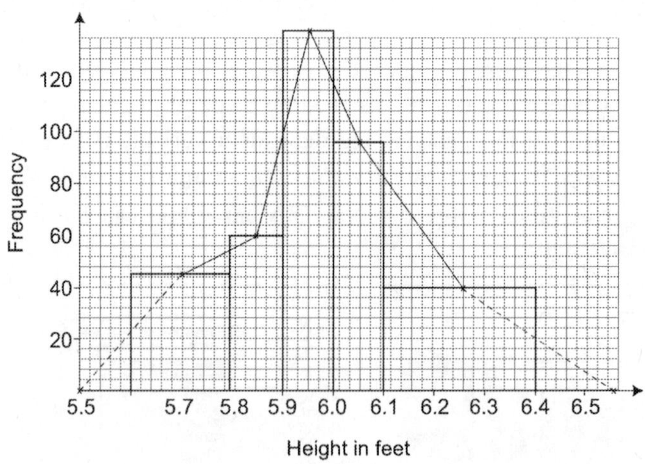

Fig. 5.5 Histogram with an overlaid frequency polygon

In a histogram there is an abrupt change in the values and this can be smoothened in a frequency polygon.

Pi-chart is a very popular way of representing a set of categorical data or displaying the different values of a given variable. In *pi*-charts a circle is divided into a number of segments; each representing a particular category. The area of the segment is the same proportion of a circle as the category is of the total data sheet. *Pi*-charts are quite popular as it is simple and easy to understand.

It provides a visualization of the whole.

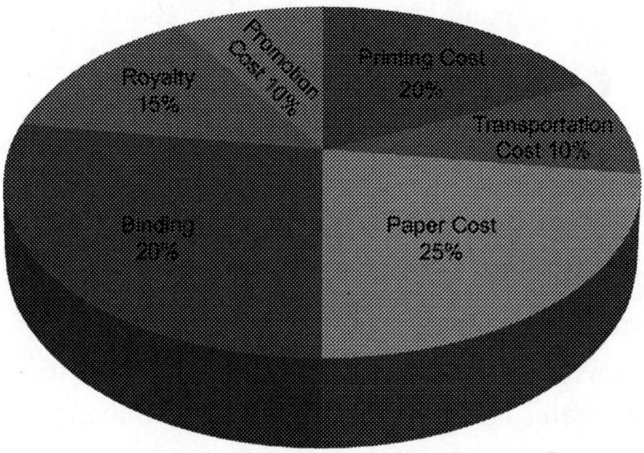

Fig. 5.6 *Pi*-chart depicting the expenditure for a publisher

Usually, the use of *pi*-charts is restricted when not more than six components are present. This is because a large number of components introduce much complexity in the display. It is not advised to use the *pi*-chart when the values of each component are similar; because it will be difficult to see the difference between the segments then.

5.9 ERROR ANALYSIS

"To err is human; to describe the error properly is sublime."

—Cliff Swartz, ***Physics Today*** 37 (1999), 388.

In science we try to explain reality using some theories. Since the reality is very complex any theory or model could only be a simplification of that part of reality. According to Popper even if we find that there is no difference between reality and the model, this statement itself is uncertain. Now listen to *Manfred Drosg's* words: who says that "due to the theory that underlies uncertainties an infinite number of data values would be necessary to determine the true value of any quantity. In reality the number of available data values will be relatively small and thus this requirement can never be fully met; all one can get is the best estimate of the true value" [20]. The essence of this statement can be summarizes in simple words as *a measurement can never be completely free from uncertainties*. This is because everything we define in terms of models not in reality. Thus, the uncertainty in measurements can be considered as a consequence of science.

The term *error* is used to represent the uncertainty in a measurement. In a measurement we assign a specific numerical value to a variable under investigation. One of the best ways to minimize the error in a measurement is to repeat the measurement a number of times and then take an average value. But this may not be possible in many situations. Therefore if a proper analysis is not made at various stages of the experiment the errors may accumulate, propagate and combine in composite functional relations.

5.9.1 Types of Errors

Several types of errors are possible in a measurement. A broad classification includes the following three:

1. Mistakes,
2. Systematic errors and
3. Random errors.

Mistakes: Mistakes can be accidental or intended. Accidental mistakes can be avoided by careful checking and rechecking. Intended mistakes such as selecting data that suit the purpose (or data manipulation) are to mislead the reader and this belongs to a scientific crime.

Systematic errors: This type of errors has a non-random character. A number of sources generate systematic error. They are [21]:

- Erroneous calibration or a lack of proper calibration of a measuring instrument;
- Careless measurements (uncorrected parallax, uncorrected zero-point deviations, time measurements uncorrected for reaction time, etc.);
- Tmpurities in materials;
- Bias of the experimenter. The experimenter might consistently read an instrument incorrectly, or might let knowledge of the expected value of a result influence the measurements.
- Many causes that the experimenter is not aware of.

The systematic error will be present in all measurements of a particular variable under given conditions. It is to be noted that the systematic error will not average to zero if we average many measurements. In other words we cannot estimate systematic error using statistical methods. This can be understood through a comparison of the results produced from different laboratories.

Random errors: Random errors are unpredictable (both the occurrence and the sources); the major sources being the limited precision of the instrumental reading and the environmental changes (physical noise). The presence of random errors can easily be identified as a spread of the observed values around an average value. Thus a possible way to minimize the random error is to make a large number of measurements and then take an average of these measurements.

A. F. Mills and B. H. Chang suggested that the following rules can be followed while analyzing the errors involved in a measurement [22]:

- As soon as an error from a particular source is seen to be significantly smaller than other errors present, it is given no further consideration.
- The major concern of most error analyses is the quantitative estimate of systematic errors and correction of data accordingly when possible.
- Whenever feasible, random errors should be estimated from repeated tests or from observed scatter in graphed results.
- In planning an experiment where it appears that significant systematic errors will be present, an *effort* should be made to ensure that random errors are much smaller.

Determining the source of uncertainty and the magnitude of this uncertainty is often difficult. Some errors or uncertainties are caused by natural fluctuations or irregularities. These effects cannot be eliminated. To estimate these uncertainties one frequently uses mathematical methods like averages and deviation. Error analysis mainly consists of two components:

- estimating the errors in all primary measurements, and
- estimating the propagation of the error at each step of the calculation.

The error in the final result is an indication of the precision of the measurement and, therefore an important part of the result. The error analysis also tells us which primary measurement is causing more error than others and thus indicates the direction for further improvement of the experiment.

5.9.2 Presentation of Errors

The possible errors in a measurement are to be expressed mathematically, so that further analysis and processing can be made. There are three different forms in which errors can be expressed: *absolute error, relative error* and *percentage error.*

Absolute error: Absolute error in a measurement is the difference between the measured value and the true value. Thus, for each measured value, A, there is an estimated error, ΔA. The complete result is given by $A \pm \Delta A$. This means that the "true value" probably lies between a maximum value of $A + \Delta A$ and a minimum value of $A - \Delta A$. Note that the absolute error has the same dimension as the measured quantity.

We usually speak about the minimum and maximum possible values of the absolute error, which is known as the *limiting error. When the measured value is expressed up to n^{th} place, the maximum possible absolute error is half a unit in the nth place.* This can be expressed as:

$$(\Delta A)_{\text{Max}} = \tfrac{1}{2}\,(10^{-n}) \tag{5.12}$$

For example, in a measured quantity $A = 2.58$; $n = 2$, then according to the above equation, the maximum possible error is $(\Delta A)_{\text{Max}} = 0.005$.

Relative error: Relative error in a measurement is defined as the ratio of absolute error to the measured value of the quantity. Thus, the relative error can be expressed as:

$$\text{Relative error} = E_r = \frac{\Delta A}{A} \tag{5.13}$$

In a measurement if the measured quantity is 10.25 and the absolute error is 0.3; then the relative error is $\frac{\Delta A}{A} = \frac{0.3}{10.25} = 0.02927$.

Percentage error: The relative error expressed in percentage is called the percentage error. This can be expressed as:

$$\text{Percentage error} = E_p = \frac{\Delta A}{A} \times 100\,\% \tag{5.14}$$

In the above example the percentage error is 2.927%.

The relation between the relative error E_r and absolute error E_a of a physical quantity A can be written as:

$$E_a = E_r \times N \tag{5.15}$$

5.9.3 Propagation of Errors

Let us calculate the error associated with a function *f*. We assume that *f* is a function of the measured quantities *a, b* and *c*.

$$f = f(a, b, c) \tag{5.16}$$

From differential calculus, using Taylor's series expansion we can write:

$$df = \frac{\partial f}{\partial a} da + \frac{\partial f}{\partial b} db + \frac{\partial f}{\partial c} dc \tag{5.17}$$

This represents the differential increments in *f* resulting from the differential increments in *a, b* and *c*. These increments are usually very small compared to the values of *a, b* and *c*. Let us denote these increments as δa, δb and δc. Thus equation (5.17) becomes;

$$\delta f = \left|\frac{\partial f}{\partial a}\right| \delta a + \left|\frac{\partial f}{\partial b}\right| \delta b + \left|\frac{\partial f}{\partial c}\right| \delta c \tag{5.18}$$

In this equation the modulus signs have been put because errors in *a, b,* and *c* are independent of each other and may be in the positive or negative direction. Therefore, the maximum possible error will be obtained only by adding absolute values of all the independent contributions. The propagation of errors under mathematical operations of addition, subtraction, multiplication, division and powers are illustrated below with examples.

1. *In addition or subtraction, the absolute errors are added.*

   Let $f = a + b - c$

   The total error in the quantity f can be written as:

   $$\delta f = \delta a + \delta b + \delta c \tag{5.19}$$

2. *In multiplication and division, the fractional (or percent) errors are added.*

   Let $f = \dfrac{ab}{c}$; then using equation (5.18) we can write;

   $$\left|\frac{1}{f}\right|\delta f = \left|\frac{1}{a}\right|\delta a + \left|\frac{1}{b}\right|\delta b + \left|\frac{1}{c}\right|\delta c \tag{5.20}$$

3. *When raised to constant powers, including fractional powers, the fractional error is multiplied by the power.*

   Let $f = a^{1.5}$, then;

   $$\left|\frac{1}{f}\right|\delta f = \left|1.5 \times \frac{1}{a}\right|\delta a \qquad ...(5.21)$$

4. *In mixed calculations, break up the calculation into simple parts.*

   Let $f = \dfrac{a}{b} - c^{4/3}$, then; $\delta f = \delta\left(\dfrac{a}{b}\right) + \delta(c^{4/3})$

Performing differentiation as in equation (5.18) we get;

$$\delta f = \left|\frac{1}{b}\right|\delta a + \left|\frac{a}{b^2}\right|\delta b + \left|\frac{4}{3}c^{1/3}\right|\delta c \tag{5.22}$$

5.9.4 Significant Figures and Rounding Off

The accuracy of a physical measurement is properly indicated by the number of figures used to express the numerical measure. Conventionally, only those figures that are reasonably trustworthy are retained. These are called significant figures. In determining the significant figures in a numerical measure, the following rules will be helpful.

1. Any nonzero digit is a significant digit. Examples: 5.86 has three significant digits and 523.56 has five significant digits.
2. Zeros in between nonzero values are significant digits. Example 80.2 has three significant digits.
3. Zeros to the left of the first nonzero digit are ***not*** significant. Example: 0.0062 has two significant digits.
4. For numbers with decimal points, zeros to the right of nonzero digits are significant. Example: Both 5.0 and 0.0050 have two significant digits.

5. For numbers without decimal points, zeros to the right of nonzero digits may or may not be significant. Example: 520 may have two or three significant digits. To avoid confusion, use a decimal point: **520.** has three significant digits, **520** has two.
6. The last significant digit in a number should have the same order of magnitude as the uncertainty. Examples: 4.4455±0.0001; 0.4±0.3
7. The uncertainty should be rounded to one or two significant digits.
8. When adding/subtracting, the answer should have the same number of decimal places as the limiting term. The limiting term is the number with the least decimal places. Example: 5.23 + 10.213 = 15.443; the answer should be rounded to 15.4 since the limiting term (5.23) has only one decimal place.
9. When multiplying/dividing, the answer should have the same number of significant figures as the limiting term. The limiting term is the number with the least number of significant figures. Examples: 1.23 × 2.224 = 2.73552, the answer should be rounded to 2.74 since the limiting term (1.23) has three significant figures.
10. When converting a number, the answer should have the same number of significant figures as the number started with. Example: $13.6\text{ in} \times \dfrac{1\text{ ft}}{12\text{ in}} = 1.133333$ should be rounded to 1.13, since the converting number (13.6) has three significant figures.
11. For exponents, the number of significant figures is the same as in the mantissa (the number to the right of the decimal point in the exponential). For example $10^{1.23} = 17$ or 1.7×10^1; this has 2 significant figures because there are 2 significant figures in the mantissa (.23).
12. For logarithms, retain in the mantissa (the number to the right of the decimal point in the logarithm) the same number of significant figures as there is in the number whose logarithm you are taking. For example: log (12.8) = 1.107. The mantissa is .107 and has 3 significant figures because 12.8 has 3 significant figures).

In reporting a numerical answer, one needs to know how to "*round off*" a number to include the correct number of significant figures. Even in a series of operations leading to the final answer, one must "round off" numbers. The widely accepted rules are given below:

1. If the figure to be dropped is less than 5, simply eliminate it. Example: 5.562 should be rounded to 5.56.
2. If the figure to be dropped is greater than 5, eliminate it and raise the preceding figure by 1. Example: 5.568 should be rounded to 5.57.
3. If the figure to be dropped is 5, followed by nonzero digits, raise the preceding figure by 1. Example: 5.5653 should be rounded to 5.57.

4. If the figure to be dropped is 5, not followed by nonzero digit(s), and preceded by an odd digit, raise the preceding digit by one. Example: 5.535 should be rounded to 5.54.
5. If the figure to be dropped is 5, not followed by nonzero digit(s), and the preceding significant digit is even, the preceding digit remains unchanged. Example: 5.525 should be rounded to 5.52.

5.10 PACKAGES FOR STATISTICAL ANALYSIS

Many statistical software packages are available for the statistical analysis of experimental data in research work. Many of them are free software and can be easily used by a researcher who has some basic knowledge of statistical principles. Some of the common statistical tools for analysis are given below.

Statistical Analysis System/Software (SAS): SAS is an integrated system of software for solving many tasks. SAS was developed by SAS Institute for data management, advanced analytics, multivariate analysis, business intelligence, criminal investigation, and predictive analytics. It helps the programmers to perform the following activities.

- Information retrieval and data management
- Report writing and graphics
- Statistical analysis, econometrics, and data mining
- Business planning, forecasting, and decision support
- Operations research and project management
- Quality improvement
- Applications development
- Data manipulation like extraction transformation, loading
- Platform independent and remote computing

SAS is a software suite that can mine, alter, manage and retrieve data from a variety of sources and perform statistical analysis on it. Its ability to help a business develop models and do predictive analytics has made SAS an irreplaceable business management tool. It performs popular statistical techniques such as Hypothesis Testing, Linear and Logistic Regression, Principal Component Analysis ANOVA, ANCOVA, T-Test, etc. SAS is not a free software; however, some features are available for free for students and researchers. For interested people, various training programmes are available on the SAS website. The main advantages of SAS are [23]:

- *Data Security:* Because of unparalleled/uncompromising data security provided by SAS software. Hence, SAS is leading the analytics software industry in BFSI sector.
- *Tech Customer Support:* SAS provides one of the best tech support personnel in the world. If you are stuck in anything related to SAS whether

it is an installation related issue or clarity in any of SAS functions and procedures, they have both online and offline communities to give support.

- *Detailed Documentation:* SAS documentation is very detailed as compared to open-source software like R and Python.
- *Memory Management:* One major advantage of SAS software is that it can store datasets on the hard drive and process bigger dataset than the size of your RAM.
- *Stable Software and Cost of Software Licence:* All the codes that contain functions and procedures of previous software versions are supported in new SAS versions. The cost of a software licence is very less to a bank or pharmaceutical company.
- *Legacy /Tradition System:* Many banks have been using SAS for the last 20–35 years and they have automated the whole process of analysis and have written millions of lines of working code. Making it one of the best reporting software.

MaxStat: MaxStat is a software package for statistical analysis specifically designed for students and researchers who do not have much knowledge of statistics. MaxStat is an easy-to-use and affordable statistics software. The Lite version is even free to use for non-commercial purposes. You can complete your statistical analysis in three easy steps within a single dialog box. MaxStat is perfect for students and young researchers because it guides users through each step of the analysis. The Lite version includes the features like descriptive statistics, hypothesis testing (t-tests, chi-square, 1-way ANOVA with post-hocs, non-parametric tests), distribution testing, linear regression, and correlation. The Pro version can handle advanced analyses, 2-way ANOVA, logistic and nonlinear regression, principal component analysis and multidimensional scaling, time series, power/sample size, factorial design, etc. The price for the Pro version is about 80 US dollars.

Stata: According to StataCorp, Stata is "a complete, integrated statistical software package that provides everything you need for data analysis, data management, and graphics". Basically, Stata is a software that allows you to store and manage data, undertake statistical analysis on your data, and create some really nice graphs. This was created in 1985. It helps professional researchers in the field of economics, sociology, education, epidemiology, institutional research, political science, and more explore, visualize, and manipulate data. Stata is also a paid software and the pricing is about 50 US dollars per year.

MSTAT-C: MSTAT or MSTAT-C is a computer-based statistical software package developed by the Crop and Soil Sciences Department of the Michigan State University of the United States. The software was written by Dr. Russel Freed, Professor, and Director of Crop and Soil Sciences Department of Michigan State University. This is widely used by researchers in the field of life sciences especially for the analysis of the results of scientific research. The main features of MSTAT-C include:

- It can analyze the field data from the single factor of multi factors experiment.
- It can calculate the mean, standard deviation range, and other dispersion value.
- Probability analysis can also be done by MSTAT software.
- It can calculate a complete ANOVA with probability as well as mean tales with their coefficient of variance.
- ANOVA can be checked before analysis.
- ANOVA of several parameters can be obtained at a time.
- Data obtained from any kind of design - CRD, RCBD, Latin Square of Factorial can easily be analyzed by MSTAT-C.
- Mean comparison of different factors can be performed.
- Correlation and regression analysis among the parameters can also be done.

MSTAT-C is free to download and use. It can be downloaded from the MSTAT-C website https://mstat.software.informer.com/

Microsoft Excel: Microsoft Excel is one of the most popular but very less used data analyzing application. It is commonly known as the spreadsheet. Since most people use Microsoft Office and Excel is included in the office the Excel spreadsheet is easily available. A large number of training programmes are available on the data analysis using Microsoft Excel.

5.11 SCIENTIFIC MODELS

Empiricists often draw a sharp distinction between scientific theories (representational devices) and natural world (which is being represented). 'Models and simulations are customarily placed among the representational tools, whereas experiments are considered parts of the natural or social world that have been carefully designed in order to answer some specific question' [24]. Models and simulations mediate between theory and reality. According to *semantic view* of theories (which is also called theories of scientific theories), the model is a fundamental component of theory. It is a structure that consists of a set of objects with properties and relations among them and/or their parts. They satisfy the linguistic component of the theory. 'The models must be put at work by means of a "theoretical hypothesis", stating that they stand in a certain relation (of similarity, isomorphism, analogy, etc., depending on which version of the semantic view one subscribes to) with real-world entities or systems' [25].

A model can be defined as a representation of the construction and working of some system of interest. It is a simplified version of the original. It enables the scientist to predict the effect of some changes made on the system. For a successful prediction the model must be close to the real system. Also it should not be so complex that it is impossible to understand and experiment with it. "A good model is a judicious trade-off between realism and simplicity" [26].

Scientific models can be classified as:

- static models and
- dynamic models

Stephan Hartmann (1996) distinguishes these two in the following way. 'A static model can only represent a system at rest. A dynamic model can also represent the time-evolution of the system' [27]. Therefore, a dynamic model can be in different states at different times, and usually each state will correspond to a specific combination of values taken by the variables in the model.

The various steps involved in modeling can be summarised as follows [28]:

- *Analyze the problem:* This is done to identify the problem precisely and understand its fundamental questions clearly. At this stage, we determine the problem's objective and decide on the problem's classification, such as deterministic or stochastic. Only with clear and precise problem identification can we translate the problem into mathematical symbols and develop and solve the model.
- *Formulate a model:* In this stage, we design the model, forming an abstraction of the system is modeling. Relevant data are collected to study about the system's behaviour. In formulating a model, owe should attempt to be as simple as reasonably possible.
- *Model abstraction:* Following problem formulation, the analyst abstracts relevant features of the system to be represented in the model. Depending on the goals of the simulation study, the analyst decides on the appropriate level of detail or granularity of the model. Then appropriate system boundary can be drawn for limiting the scope. Bounding the system is a very important analytical step because it defines internal and external factors as well as inputs and outputs. The process of bounding the system can help to reduce the level of complexity by reducing the size of the system in some cases.
- *Determine variables and units:* In this step we identify the independent and dependent variable and the system of units to be used. To simplify the model, we may decide to neglect some variables (such as air resistance), treat certain variables as constants. At this stage we establish a relation between the variables and determine the equations and functions.
- *Solve the model:* This stage implements the model. Before moving to this stage we should have a thorough understanding of the problem and of designing the model. Otherwise, we might waste much time, which can be most frustrating. We employ techniques like algebra, calculus, graphs and computer programs.

- *Model implementation:* The conceptual model generated during the earlier phase must be implemented. Recently, simulation techniques are widely used for the implementation of models. Various simulation languages such as SIMAN or GPSS, or standard programming languages such as C, C++, Java, or special simulators tailored for specific applications, etc., are used in simulations.
- *Verify and interpret models solution:* We should carefully examine the results to make sure that they make sense (verification) and that the solution solves the original problem (validation) and is usable. The process of verification determines if the solution works correctly, while the process of validation establishes if the system satisfies the problem's requirements. The solutions are to be tested to see that the predictions agree with the real data. We should analyze the model's solution to determine its implications. If the model solution shows weaknesses, we should return to initial steps to determine if it is feasible to refine the model.
- *Report on the model:* A proper reporting of the model is necessary to ensure its utility. A report on modeling involves the following components: analysis of the problem, model design, model solution, results and conclusions.

The modeling process is a creative, scientific endeavour. As such, a problem we are modeling usually does not have one correct answer. The problems may be complex, and many models provide good solutions. Thus, modeling is a challenging, open-ended, and exciting venture.

5.12 SIMULATIONS

In recent years simulations gained much attention in scientific research as routine tool for scientific inquiry. Computer simulations are used at an intermediate level between theory and experiments. Though simulation one can demonstrate how a model of a system works. In a simulation it is easy to reconfigure a system; usually it is not possible in models or otherwise too expensive or impractical. The properties concerning the behaviour of the actual system or its subsystem can be inferred through a simulation study. Simulation involves the development of descriptive computer models of a system and exercising those models to predict the operational performance of the underlying system being modeled.

Mario Bunge (1969) defined simulation as a relation between two entities, *x* and *y*, where *x* simulates *y* if (1) there exists a correspondence relation between the parts or the properties of *x* and *y*; and (2) the analogy is valuable to *x*, or to another entity (*z*) that controls *x* [29]. Francesco Guala (1998) strongly disagree with the first part of the definition; as it leads to include things that we would not intuitively call 'simulations' at all whereas the second part captures some important intuitions and connotations of the term simulation [30]. The simulation demands some dynamic aspects for the system under consideration. Guala made it clear through a simple example of drawing a map. If it has been drawn adequately, there exist some correspondence relation between its parts and parts of the territory it is aimed at representing. However, this is not a simulation. If the map is drawn together with a set of flags and miniaturized soldiers and tanks. If the toy-flags, mini-soldiers and mini-tanks are moved on the map according to the appropriate rules, we can say that a battle or a military campaign is being simulated. Thus only a dynamic system can simulate. In words of (Hartmann, 1996, p. 83), "a simulation imitates one process by another process".

According to Herbert Simon (1969), simulations rely on a process of abstraction from the fundamental principles governing the behaviour of the simulating and the target systems. If similar 'organizational properties' arise at a given non-fundamental level from different substrata, it is possible to abstract from the substrata and simulate the behaviour of a system *A* by observing the behaviour of another system *B* which happens to display those non-fundamental properties [31]. This enables us to create a demarcation between experiments and simulation. In experiments the correspondence holds at a '*deep*, and '*material*' level, whereas in the simulation the similarity is admittedly only '*abstract*' and '*formal*'. For example, we can simulate the propagation of light wave from its correspondence with the ripples on the surface of water. Here the similarity in only at an abstract level.

In a simulation experiment we perform a test or a series of tests in which meaningful changes are made to the input variables of a simulation model so that we may observe and identify the reasons for changes in the performance measures. As in the case of scientific models, simulation also involves a series of procedures to be executed with care. As Anu Maria (1997) explains, simulation involves the following processes [32]:

- Identification of the problem
- Formulation of the problem
- Formulation and development of a model
- Validation of the model
- Documentation of the model for future use
- Selection of appropriate experimental design
- Establishment of the experimental conditions for running the simulation
- Performance of simulation
- Interpretation and presentation of the result

One should keep in mind that all experimental situations do not demand the use of simulations. Simulations are usually used when we need to analyze the randomness in a system. Some situations that demand the use of simulation are given below (Anu Maria, 1997):

- When a process is impossible or extremely expensive to observe in the real world. For example, next year's cancer statistics, performance of the next space shuttle, and the effect of Internet advertising on a company's sales.
- Problems in which mathematical model can be formulated but analytic solutions are either impossible as in higher-order differential equations or too complicated systems like the stock market.
- When it is impossible or extremely expensive to validate the mathematical model describing the system. For example, due to insufficient data.

Simulation is a computer based program. A wide variety of simulation programs are available for the uses in various fields. Even though a simulation model can be built using general purpose programming languages which are familiar to the analyst (available over a wide variety of platforms, and less expensive) most simulation studies are implemented using a simulation package. Such packages have the following advantages:

- reduced programming requirements;
- natural framework for simulation modeling;
- conceptual guidance;
- automated gathering of statistics;
- graphic symbolism for communication;
- animation; and
- flexibility to change the model.

One difficulty that an analyst may face is the high cost of simulation programs. The average cost of a simulation program is as high as $15,000 or more.

Simulation packages are of two types: *simulation languages* and *application oriented simulators*. Even though simulation languages offer more flexibility it requires a programming expertise. Application oriented simulators, on the other hand are easier to learn and have modeling constructs closely related to the application. As far as scientific research is concerned we rely mainly on the second type of packages. The commonly used scientific and mathematical simulation programs include: Math lab, Mathematica, Schrodinger, Chemlab, Comsol Multiphysics and many others. For a physicist the most appropriate simulation software, as the author felt; are Mathematica and Comsol Multiphysics. A huge number of publications, books as well as journal articles are available on the application of these software in physics.

Simulation models are to be used judiciously in research. Anu Maria (1997) identifies the following advantages for simulation models:

- It is possible to obtain a better understanding of the system by developing a mathematical model of a system of interest, and observing the system's operation in detail over long periods of time.

- It is possible to test hypotheses about the system for feasibility.
- It helps to compress time to observe certain phenomena over long periods or expand time to observe a complex phenomenon in detail.
- It enables to study the effects of certain informational, organizational, and environmental and policy changes on the operation of a system by altering the system's model; this can be done without disrupting the real system and significantly reduces the risk of experimenting with the real system.
- It experiments with new or unknown situations about which only weak information is available.
- It identifies the *driving* variables—ones that performance measures are most sensitive to—and the inter-relationships among them.
- It employs a systems approach to problem solving.
- It is possible to develop well designed and robust systems and reduce system development time.

Even though the simulation has a number of advantages it has some difficulties or pitfalls also. While designing and performing a simulation experiment one should safeguard against (Anu Maria, 1997):

- Unclear objective.
- Using simulation when an analytic solution is appropriate.
- Invalid model.
- Simulation model too complex or too simple.
- Erroneous assumptions.
- Undocumented assumptions. This is extremely important and it is strongly suggested that assumptions made at each stage of the simulation modeling and analysis exercise be documented thoroughly.
- Using the wrong input probability distribution.
- Replacing a distribution (stochastic) by its mean (deterministic).
- Using the wrong performance measure.
- Bugs in the simulation program.
- Using standard statistical formulas that assume independence in simulation output analysis.
- Initial bias in output data.
- Making one simulation run for a configuration.
- Poor schedule and budget planning.
- Poor communication among the personnel involved in the simulation study.

We have seen that conceptualization of a problem and its computational solution is widely used in scientific research. It allows many complex systems to be analyzed in detail which is otherwise impossible. Also, 'it allows the understanding and visualization of cause-effect through simulations where initial and boundary conditions are prescribed specifically to gain insight' [33]. I would like to quote Henry A Ronald:

"It is a common error which young physicists are apt to fall into to obtain a law, a curve, or a mathematical expression for given experimental limits and then applies it to points outside those limits. This is sometimes called extrapolation. Such a process, unless carefully guarded, ceases to be a reasoning process and becomes one of pure imagination especially liable to error when the distance is too great.

In temperature our knowledge extends from near the absolute zero to that of the sun, but exact knowledge is far more limited. In pressures we go from the Crookes vacuum still containing myriads of flying atoms, to pressures limited by the strength of steel, but still very minute compared with the pressure at the center of the earth and the sun, where the hardest steel would flow like the most limpid water. In velocities we are limited to a few miles per second".

(APS News, January 2008, p. 8).

Our perception and knowledge is limited; we need simulation. Axelrod (1997) says that: "Simulation is a third research methodology. Like deduction, it starts with a set of explicit assumptions. But unlike deduction, it does not prove theorems. Instead, simulation generates data that can be analyzed inductively. Unlike typical induction, however, the simulated data come from a specified set of rules rather than direct measurement of the real world. While induction can be used to find patterns in data, and deduction can be used to find consequences of assumptions, simulation modeling can be used to aid intuition" [34]. This statement invited criticisms. However, simulation can be effectively used to support the deductive approach to research. 'It is not important if it does or does not prove theorems: it is only a virtual laboratory useful to verify hypothesis' [35].

REVIEW QUESTIONS

1. What are qualitative and quantitative data? Explain.
2. Explain the features of primary data.
3. What are secondary data? Explain.
4. What are the advantages of secondary data?
5. Mention some general characteristics of data.
6. Give a summary of the needs of data collection.
7. What are the important characteristics of data?

8. What care must be taken while collecting data?
9. Explain the observation method of data collection.
10. What are the advantages of the observation method?
11. Mention some drawbacks of the observation method.
12. What are controlled and uncontrolled observations?
13. Explain the features of the interview method.
14. Give the classification of the interview method.
15. What are the characteristics of the interview method?
16. Point out the merits and demerits of the interview method.
17. What is the questionnaire method? Explain.
18. What are the different types of questionnaires?
19. What are the advantages of the questionnaire method?
20. What is the schedule method of data collection?
21. Explain the different types of schedules.
22. What is data collection? Discuss the various types of data collection in detail.
23. What are the main sources of secondary data?
24. Brief the implications of the effective sharing of data.
25. Explain the various principles of data sharing.
26. Explain the various components involved in data processing.
27. What are the basic methods of descriptive statistics? Explain each.
28. What is the central tendency? Illustrate with an example.
29. What is mean? Illustrate with an example.
30. What is the median? Illustrate with an example.
31. What is the mode? Illustrate with an example.
32. Explain geometrical and harmonic mean.
33. Explain dispersion and range with suitable examples.
34. What is the mean deviation? Explain with an example.
35. What is a variance? Illustrate with an example.
36. Explain the standard deviation with a suitable example.
37. What are positive and negative correlations? Explain.
38. What are the different types of correlations?
39. What is inferential statistics?

40. What are parametric and non-parametric tests?
41. Explain various parametric tests in detail.
42. Explain the chi-square test.
43. What is regression?
44. Explain the importance of the null hypothesis in statistical analysis.
45. What are type I and type II errors? Explain.
46. What do you mean by the degrees of freedom in data analysis?
47. Discuss the various methods to present data.
48. What are the characteristics of a table?
49. Explain the use of graphs in presenting the data.
50. Explain histogram and frequency polygon. Illustrate.
51. What are the different types of errors in measurement? Discuss in detail.
52. What is a systematic error? What are the sources of systematic error?
53. What are the rules to be obeyed in error analysis?
54. Discuss the various methods for the presentation of error in detail.
55. What is absolute error? Explain with a suitable example.
56. Explain relative and percentage errors.
57. Explain the propagation of errors when there is more than one variable involved in measurement.
58. What are the rules for finding significant figures?
59. Explain the rules for the rounding off of a number.
60. Give an account of various software available for the data analysis.
61. What are the features of the Statistical Analysis System/Software?
62. What are the features of MSTAT-C?
63. What is a scientific model? Where the different types of scientific models?
64. Explain the various steps involved in modeling.
65. What is simulation? What are the various processes involved in a simulation?
66. Give some instances where the simulation is required.
67. What are the advantages of simulation packages?
68. What are the different types of simulation packages? Explain.
69. What are the main advantages of the simulation models?
70. What are the precautions that must be taken in simulating an experiment?
71. Explain the factors from which a simulation experiment must be protected.

REFERENCES

[1] *Research Methodology–Self Learning Manual*, p. 184.

[2] C R Kothari, "Research-Methodology: Methods and Techniques", 2004, New Age International, p. 95.

[3] *Research Methodology–Self Learning Manual*, p. 186.

[4] Ibid, pp. 181-182.

[5] Yogesh Kumar Singh, *Fundamentals of Research Methodology and Statistics*, 2006, New Age International, p. 212.

[6] Ibid, p. 220.

[7] Kothari, p. 96.

[8] Geoffrey R. Marczyk, David DeMatteo, David Festinger, *Essentials of Research Design and Methodology*, 2005, p. 118.

[9] Kothari, pp. 98-99.

[10] *Research Methodology–Self Learning Manual*, pp. 206-207.

[11] Ibid, p. 220.

[12] OECD Principles and Guidelines for Access to Research Data from Public Funding', 2007, OECD Publications, Paris, p. 10.

[13] Ibid, pp. 15-22.

[14] Kothari, p. 127.

[15] Geoffrey R et al., pp. 210-225.

[16] Ibid., pp. 219-220.

[17] C George Thomas, "*Research Methodology and Scientific Writing,*" 2021, Springer. p. 163.

[18] Biju Dharmapalan, *Scientific Research Methodology*, 2012, Narosa Publishing House Pvt. Ltd, New Delhi, p. 7.11.

[19] Ibid., pp. 7.11-7.12.

[20] Manfred Drosg, *Dealing with Uncertainties: A Guide to Error Analysis*, 2007, Springer, p. 14.

[21] Herman J C Berendsen, *A Student's Guide to Data and Error Analysis*, 2011, Cambridge University Press, p. 18.

[22] A. F. Mills and B. H. Chang, *Error Analysis of Experiments*, 2004, p. 6.

[23] https://medium.com/@reddipillink/intro-to-sas-statistical-analysis-system-software-ff8e232bad6c

[24] Francesco Guala, *Models, Simulations, and Experiments.*

[25] Ibid.

[26] Anu Maria, Introduction to Modelling and Simulation, 1997, Proceedings of the 1997 Winter Simulation Conference, ed. S. Andradóttir, K. J. Healy, D. H. Withers, and B. L. Nelson.

[27] Hartmann, S., 1996, The world as a process: simulations in the natural and social sciences, in Modelling and Simulation in the Social Sciences from the Philosophy of Science Point of View, p. 82.

[28] Jagat Kumar, *System Simulation and Modelling*, pp. 16-22.

[29] Bunge, M., 1969, Analogy, simulation, representation, Revue internationale de philosophie 87:16-33

[30] Francesco Guala, *Models, Simulations, and Experiments*.

[31] Simon, H.A., 1969, *The Sciences of the Artificial,* MIT Press, Boston. pp. 15-18.

[32] Anu Maria, Introduction to Modelling and *Simulation*.

[33] Sidney Yip and Tomás Diaz de la Rubia, (ed), *Scientific Modelling and Simulations*, 2009, Springer, p. 1.

[34] Axelrod R, Advancing in Art of Simulation in the Social Sciences, in Simulating Social Phenomena, edited by Rosaria Conte, Rainer Hegselmann and Pietro Terna, 1997, Springer, pp. 21-40.

[35] Michele Sonnessa, *Modelling and simulation of complex systems, PhD Thesis*, p. 21.

Suggested Readings

1. Roger Sapsford and Victor Jupp, *Data Collection and Analysis*, 2006, Sage Publications, London.
2. Madhu Viswanathan, *Measurement Error and Research Design*, 2005, Sage Publications, London.
3. Ajai S. Gaur and Sanjaya S. Gaur, *Statistical Methods for Practice and Research*, 2006, Sage Publications Inc.
4. John K. Taylor and Cheryl Cihon, *Statistical Techniques for Data Analysis*, 2006, Chapman and Hall/CRC.
5. John M. Chambers et.al. Graphical Methods for Data Analysis, 1983, Wadsworth & Brooks/Cole Publishing Company.
6. R. Lyman Ott and Michael Longnecker, *An Introduction to Statistical Methods and Data Analysis*, 2010, Brooks/Cole, Cengage Learning.
7. Ronald A Fisher, *Statistical Methods for Research Workers*, 1954, Hafner Publishing Company Inc. New York.
8. David Livingstone, *A Practical Guide to Scientific Data Analysis*, 2009, John Wiley and Sons.
9. Philip R Bevington and D Keith Robinson, *Data Reduction and Error Analysis for the Physical Sciences*, 2003, McGraw Hill.

Chapter 6

Research in Practice

One does what one is; one becomes what one does.

—**Robert von Musil, Kleine Prosa**

Put it before them briefly so they will read it, clearly so they will appreciate it, picturesquely so they will remember it and, above all, accurately so they will be guided by its light.

—**Joseph Pulitzer**

6.1 RESEARCH PROPOSAL

Scientific research is an expensive activity. So, it is very important to get sufficient financial support to carry out the research without any delay. There are many governmental and non-governmental funding agencies that provide financial assistance to conduct research in various fields of their interest. To get such grants, a researcher needs to submit a proper research proposal. The goal of a research proposal is twofold: to present and justify the need to study a research problem and to present the practical ways in which the proposed study should be conducted. The design elements and procedures for conducting research are governed by standards of the predominant discipline in which the problem resides, therefore, the guidelines for research proposals are more exacting and less formal than a general project proposal. Research proposals contain extensive literature reviews. They must provide persuasive evidence that a need exists for the proposed study. In addition to providing a rationale, a proposal describes a detailed methodology for conducting the research consistent with requirements of the professional or academic field and a statement on anticipated outcomes and/or benefits derived from the study's completion [1].

Therefore, while submitting a research proposal sufficient care must be taken while finding the research problem, planning the proposal, and preparing and submitting the proposal. The proposal must be submitted in the correct format. The researcher must be aware of the evaluation procedure of the research proposals. This will help him to prepare the proposal effectively and to get the approval of the funding agencies.

6.1.1 Finding a Research Problem

Finding a research problem, in general, is not that difficult. However, if the researcher is looking for financial support from a funding agency, then the research problem itself matters. The funding agencies must feel the relevance and importance of the research problem, its social impact, its outcome, and application, etc. That is, finding a research problem itself is a challenging problem. Therefore, ultimate care must be taken while finding a problem to write a research proposal. There are many sources of research ideas and C. George Thomas suggested the following sources of research ideas [2].

- Experience
- Literature search
- Internet search
- Daily observation
- Call for proposals
- Field problems brought by concerned people
- New developments and discoveries
- Conferences/seminars/lectures/discussions
- Colleagues and guides

A brief discussion of the above points is given below.

- *Experience:* In many cases, some research ideas evolve out of the experience of the researcher. As a professional or teacher, one may confront several questions or problems pertaining to your specialization. The background, culture, education, and practical experience of researchers influence the ideas and the way in which they think. Such ideas must be noted down immediately. Further, it is to be decided whether the idea is relevant in the current scenario and to be suitable for research activity. Then, based on this idea a research proposal may be developed.
- *Literature Search:* Scientific literature is one of the main sources of research ideas in a specific field. This has the advantage that the researcher can understand the most recent research developments and new problems encountered by the researchers in the world. Then, the researcher can work on such problems or even the same problems under varied parameters or at different socio-economic circumstances. Recent textbooks and reference books may also be of immense help. Reading more and more literature will help the researcher to develop his knowledge in his field and it becomes easy to find more relevant research problems.
- *Internet Search:* Researchers most frequently use the Internet to get information about the research topic. A general search on the Internet is helpful for getting new ideas or developing the topics further. There are many general search engines such as Google, Bing, Yahoo and this

will lead to a wide range of suitable web pages. Wikipedia, the world's largest online encyclopedia is helpful to have a broad understanding. Databases such as PubMed, PubAg, and CAB Direct are also helpful. These sites are rich in scholarly content and contain abstracts and links to a variety of sources.

- *Daily Observations:* Nature is a very good source of research ideas. One can learn many things from nature. In many research areas such as atmospheric science, agriculture, and farming, etc., the observation of nature will help to find research topics. The observation must be active that "observation involves detecting something and making a note of its significance by relation to something already known". Suppose you have just observed a malformation in a crop. In agricultural research, "field visits and mixing with farmers and extension workers will provide many ideas for research"
- *Call for Proposals:* Many research units of industries often invite research proposals on particular areas that help industrial development. In such notifications, it will be specified that the research needs to address specific problems, the contexts in which it operates, the approach they would like to investigate the problem, and the proposals must be based on those problems. They also specify the amount that they wish to spend on the research. Therefore, the researcher is free from the task of finding a research problem and he will get sufficient funds if the proposal is accepted.
- *Field Problems Brought by Concerned People:* Common people who directly deal with the fields like agriculture, animal husbandry, and medicine, may experience a large number of problems in their area. They may not be able to solve such problems and such problems need proper scientific investigation. If the problems are new and need investigation, the researcher can proceed with the idea of problem-solving. For example, medical professionals and researchers can undertake research programmes to address new diseases and epidemics.
- *New Developments:* Technological advancements and social developments are creating new problems and opportunities. For example, food security, genetically modified crops, pesticide residues, environmental pollution, water crisis, wetland conservation, increasing popularity of certain crops, global warming, climate change, energy crisis, need for alternate fuels such as biodiesel, and new diseases like Covid-19 are some new developments, which need immediate research intervention.
- *Conferences/Seminars/Lectures/Discussions:* Many of the research institutes, universities, and industries, often conduct various conferences, seminars, and workshops for researchers. These give the researcher an opportunity to mingle with scientists, hear their views, and exchange

new ideas. The interaction and the insights from such meetings provide an excellent opportunity to seek clarifications or further enlightenment on the ideas.

- *Colleagues and Guides:* For young scientists and research students, the research topics are often suggested by their research guides. In some cases, a colleague could also be able to suggest some research ideas.

6.1.2 Planning a Research Proposal

The research proposal must be properly planned before start writing the proposal. The researcher must be aware of the various purposes and components of the research proposal so that he can include all the necessary details in the proposal. The purpose of the research proposal can be the following:

- To propose a research project that will result in a significant contribution to knowledge.
- To formulate a detailed plan of the project including methodological approach and theoretical framework.
- To ensure that the proposed research is achievable within the required time and with the available resources.
- To demonstrate that you have adequate expertise and experience to undertake the project.

The proposal can be written down once a clear idea about the research problem, the various procedures involved, equipment to be used, etc., are obtained. The following points may be noted while writing a good research proposal.

Interests of the Funding Agency: The research can be completed successfully only if there is enough financial support from the funding agency. Therefore, it is essential to make the authority convinced about the research problem and its relevance. Also, they should be confident about the future application of the research outcome. Then only the agencies will sanction a research grant. Therefore, before writing a research proposal, the various motives and interests of the funding agencies and their fields of interest must be understood. For any clarification, one can directly contact any such agencies. Further, it should be ensured that the research idea must be in accordance with the programme of the funding agency.

Focus on Contemporary Issues: Any funding agency is interested in addressing contemporary issues in various fields of science and humanities. For example, the outbreak of the Covid-19 disease became a challenging issue for many researchers in the field of social science, life science, computer modeling and simulation, medical science, etc. In the last year, hundreds of research articles were published on this single issue. Therefore, if the proposal is able to address these aspects, the chances of getting grants are very high. There are a large number of current issues that may be interesting to many funding agencies. Some of them are climate change, food security, population explosion, solid waste crisis, pollution, energy crisis, lifestyle diseases, genetically modified organisms, and pesticide residues.

Clarity of the Proposal: A researcher who is submitting a research proposal must have good writing skills so that he can easily convey the various aspects of the research to the funding agencies quite easily. He must be confident about the proposal and the research outcome. The use of positive language will be helpful. The usage of words like may be, perhaps, probably, etc., must be avoided in any part of the research proposal. Good quality English with present and future tenses are to be used while writing the proposal. No need to be shy in seeking the help of an English teacher.

Proper Time Management: Most of the funding agencies want to complete a research programme that addresses an issue in a given time span. Otherwise, it will become insignificant and may lead to a loss of the investment made on the project. With proper planning of the research programme, it is possible to complete the research programme in the stipulated time. So, before submitting the proposal, the researcher must be confident about the time available to him. Most of the research grants are available for three years and some agencies may extend it for another two years if the research project has a long-term impact. Therefore, careful planning is required before submitting a research proposal.

While preparing a good research proposal, the following common errors should be avoided [3].

- *Failure to be concise:* A research proposal must be focused and not be "all over the map" or diverge into unrelated tangents without a clear sense of purpose.
- *Failure to cite landmark works in the literature review:* Proposals should be grounded in foundational research that lays a foundation for understanding the development and scope of the issue.
- *Failure to delimit the contextual boundaries of the research* [e.g., time, place, people, etc.]**:** As with any research paper, your proposed study must inform the reader how and in what ways the study will examine the problem.
- *Failure to develop a coherent and persuasive argument for the proposed research:* This is critical. In many workplace settings, the research proposal is intended to argue why a study should be funded.
- *Sloppy or imprecise writing, or poor grammar:* Although a research proposal does not represent a completed research study, there is still an expectation that it is well-written and follows the style and rules of good academic writing.

- *Too much detail on minor issues, but not enough detail on major issues:* The proposal should focus on only a few key research questions in order to support the argument that the research needs to be conducted. Minor issues, even if valid, can be mentioned but they should not dominate the overall narrative.

Writing a research proposal engages a number of skills. These skills can be grouped into three clusters [4]:

- *Subject knowledge and research skills:* The proposal gives you an opportunity to exhibit your mastery of subject knowledge and familiarity with current research trends.
- *Critical thinking skills:* A good research proposal displays evidence of advanced analysis, evaluation, and synthesis skills, as well as creativity and the ability to combine vertical and lateral thinking.
- *Communication skills:* The proposal displays your ability to express yourself in precise and concise language.

Therefore, it is necessary to keep these skills in mind during the work on the proposal since the funding agencies or authorities evaluate the progress continuously.

6.1.3 Format for Submitting Proposals

After completing the plan for a proposal, the next step is to submit the proposal. Almost all funding agencies prefer a particular format for submitting a research proposal that may be issued by the agency itself. According to Calnan [5], a successful proposal must answer the following questions:

- Why do I want to do this project?
- What do I want to do?
- How am I going to do the project?
- What do I need?

A research proposal must contain the following important sections [2, pp. 178-183].

- *Title Page:* The title of the proposal gives the first impression. In a common format, the title page includes the title, name, and address of the person who is proposing the research and the date of submission. The title should be clear and should represent the central concept of the proposal. Contradictory, ambiguous, or misleading titles must be avoided. Title should not be too long or too short. One can use 10 to 15 words in the title. No need to add familiar phrases in the title. Low impact expressions such as, "A Study of......", "Investigation on....", "Research on...." in the title. Such words cannot give any additional impression as the intention of the proposal itself is to conduct a research programme.

Usually, abbreviations are not used in the title and if the abbreviations are very common such as DNA, they can be used.

- *Information Regarding the Investigators:* The names of the principal investigator and co-investigators with full addresses and contact numbers for correspondence must be included in the proposal. Basic information such as the name, designation, department, name of the institute, address, date of birth, sex (M/F), and telephone/mobile/e-mail are to be included.
- *Abstract:* The proposal should contain an abstract of the proposed research work. The abstract of a proposal must include the justification for the study, the hypothesis or the research question, and the methods in a brief format. Usually, 300 to 400 words can be used for writing the abstract. If the funding agency suggests the word limit for the abstract, then one must follow it.
- *Introduction:* The introduction part must be written in such a way that the reviewers must be able to understand the interest and importance of the researcher in the problem. The introduction must analyze the problem with a short review of relevant literature. The hypothesis and the assumptions must be indicated with all the background logic. The rationale and justification for taking up the study must be given. The benefits and scientific utility of the proposed work must be indicated in the introduction.
- *Aims and Objectives:* Very often, many people consider aims and objectives as the same. But these are not the same. "Aim is a broad statement of desired outcomes or the general intention of the research project. Objectives are the specified tasks researchers should undertake in response to the research hypothesis or question to accomplish the overall aim of the project". For a research proposal, the objective is more important than the aim. The objective must impress the reviewers. It is important to include only those objectives that can be managed or handled by the researcher.
- *Methodology:* The methodology describes the methods that the researcher follows to conduct the research programme. A proper description of the methodology helps the reviewer to understand the strengths of the methodology. A specific description that how the research is going to be conducted, can be included in the methodology. The description of the methods must be brief and the evaluation plan of results, statistical tests, and a quantitative and qualitative measure of data can also be included.
- *Time Schedule of Activities:* The time schedule of the proposed research activities is to be included in the format demanded by certain agencies. Since the time schedule is to be followed strictly, the researcher should have a well-defined plan of execution of the proposed research project, and the time schedule must be indicated accordingly. The funding agencies monitor the progress of the research from time to time.

- *Ethical Considerations:* This is more important in the field of medical research. If the subject is a human being, then they must be informed clearly about the study that is going to be conducted on them. Similarly, if the experiments are done on animals, no animal should be harmed during the experiment. An approval from ethical committee may be required in such cases.
- *Available Facilities:* While submitting the proposal, the researcher needs to highlight the available facilities in the institution to conduct the research programme. This will help to convince the reviewers. If more than one researcher or institution is submitting the same proposal, then the preference will be given to the institution where more research facilities are available.
- *Budget:* "Budget is an important part of the project proposal, and it must be prepared carefully taking into consideration all the costs involved". While preparing the budget, the original cost and expected expenditure must be included. The reviewers and the funding agencies are highly knowledgeable on costs. "Include all categories of cost such as salaries/ wages for project associates and helpers, equipment, contingencies, travel, consultants, and supplies. Yearly expenses, rate of increase in salaries, costs for unforeseen expenditure, etc., must be shown".
- *References:* References are the list of articles and other sources which the researcher is used to formulate the research problem and the methodology of the research. A maximum of ten references that are very important can be included. The references must be relevant. Also, they must be recent publications. The references must be typed out correctly in the required format. This will help the reviewer to identify the references easily.
- *Curriculum Vitae of the Applicants:* This is also very important as it helps the reviewers to judge the capacity and capability of the researcher. The curriculum vitae must include the educational qualifications, professional experiences such as posts held in various institutions, awards, and achievements, and a list of published works. Noticeable works can be highlighted.
- Before submitting the proposal, it must be checked thoroughly for any mistakes, omissions, repetitions, etc. "Edit the material to shorten, to make it clear, punchy, and appealing before taking the final submission".

6.1.4 Evaluation Procedure of Proposals

After submitting the proposal and the expert committee will be evaluating the proposal. A well-prepared proposal is easier to evaluate.Note that there is no universal rule for evaluating a research proposal. However, the funding agencies usually set some criteria for the evaluation purpose. These criteria strongly depend on the interests of the funding agencies. However, the following criteria can be used by the committee to evaluate research proposals.

- *Goals and Significance:* To what extent is the proposed research expected to significantly advance the scientific or technological field? How likely is the proposed work to produce high-quality publications?
- *Content and Design:* Does the proposed methodology fit the research problem? The researchers in various science streams cannot use a questionnaire to collect data. Instead, he needs to setup an experiment and conduct the experiment to obtain data.
- *Population and Sampling Methodology:* Who is the target population? Is the sampling methodology appropriate to the research question(s)? Has the same student cohort and/or sample been used in another recent research? Could a smaller sample be drawn to achieve the same objective? How will the researcher(s) gain access to the proposed participants? In social and medical researches, sample selection is very important.
- *Timing:* Is the experimental plan achievable at the foundry within the proposed project term? The timely completion of the research project is very important. Otherwise, the expenditure will increase and there may be a possibility to lose the significance of the research work itself.
- *Resources:* What resources will be required to conduct this research (e.g., instrument design, Web application development, mail and/or e-mail services, data entry, and analysis)? Is the request for foundry resources well justified? To what extent will the proposed work take advantage of the unique capabilities or combinations of capabilities/expertise?
- *Overall Impact:* What will be the impact of the study? Are there any conceivable negative impacts on society? Will the study overburden respondents? Overall, do the expected benefits of the study appear to outweigh the costs?

In addition to these points, the competence and ability of the researcher will also be evaluated before sanctioning the fund. If the proposal does not meet these requirements, it will get rejected.

6.1.5 Sources of Financial Support

There are a large number of governmental and other industry agencies that give financial support to conduct researches in various fields. Such agencies usually invite the proposals from time to time once or twice a year. Industries may invite research proposals on specific problems that they want to solve. Most government agencies support basic research in various disciplines. In India, there are many such agencies and some of them are listed below [6].

- *University Grants Commission (UGC):* UGC is formed to promote teaching and research in emerging areas in Humanities, Social Sciences,

Languages, Literature, Pure Sciences, Engineering & Technology, Pharmacy, Medical, Agricultural Sciences, etc., through major research projects. University and college teachers can submit their research proposals and on the approval of the proposal, UGC will be sanctioning the research grant.

- *All India Council for Technical Education (AICTE):* Research Promotion Schemes (RPSs) of AICTE aim to create a research ambiance by promoting research in technical disciplines and innovations in established and emerging technologies; and to generate Masters and Doctoral degree candidates.
- *Industry-Institute Interaction Schemes (IIIS):* The main objective of IIIS is to plan, coordinate and execute integrated R&D programmes at the national level by a group of institutions. The technical/financial/administrative deliverables are to be spelled out clearly by the networking institutions with the lead institution being an IIT/IISc/IIM/NIT. Research supports are provided in Engineering and Technology, Architecture, Town Planning, Management, Pharmacy, Hotel Management, and Catering Technology, Applied Arts and Crafts, etc.
- *Council of Scientific and Industrial Research (CSIR):* The major functions of CSIR include promotion, guidance, and coordination of scientific and industrial research in India; establishment or development of and assistance to the existing special institutions or departments for the scientific study of problems affecting particular industries and trades; award of fellowship; utilization of Council's R&D results for industrial development; collection and dissemination of S&T information; and technology generation, absorption, and transfer. CSIR supports the research schemes in the field of S&T including agriculture, engineering, and medicine. Multidisciplinary projects which involve inter-organizational cooperation are also considered. Preference is given to schemes that have relevance to research programmes of CSIR laboratories.
- *Defense Research and Development Organization (DRDO):* DRDO is dedicatedly engaged in the formulation and execution of programmes of scientific research, design and development, testing and evaluation leading to induction of state-of-the-art weapons and equipment which would compete and compare favourably with its contemporary systems available elsewhere in the world. It consists of a chain of laboratories/establishments situated all over the country, pursuing assigned scientific goals with delegated powers under the policy direction provided by the headquarters in New Delhi. DRDO also supports a substantial amount of extramural research in academic institutions and other laboratories on defense related problems through various grants-in-aid schemes and other sponsored projects.

- *Department of Atomic Energy (DAE):* The Department of Atomic Energy supports research programmes in Nuclear Science and Technology through the Board of Research in Nuclear Sciences (BRNS). It extends the research supports through various schemes under R&D projects in universities and other research institutions. Also, various fellowships such as DAE Young Scientists Research Award, Dr. K.S. Krishnan Research Associateship, Raja Ramanna Fellowship, Homi Bhabha Chair Professorship, DAE Graduate Fellowships for Ph.D., etc., are offered to talented researchers.
- *Department of Biotechnology (DBT):* The setting up of a separate Department of Biotechnology (DBT), under the Ministry of Science and Technology in 1986 gave a new impetus to the development of the field of modern biology and biotechnology in India. It provides research supports in vast areas such as Animal Biotechnology, Aquaculture, and Marine Biotechnology, Basic Research in Biotechnology, Biofuels, Bioinformatics, Biological Control of Plants Pests, Diseases, and Weeds, Bioprospecting and Molecular Taxonomy, Biotech Process Engineering and Industrial Biotechnology, Biotechnology of Medicinal and Aromatics Plants, Biotechnology of Silkworms and host-plants, Crop Biotechnology, Environment & Conservation Biotechnology, Food Biotechnology, etc.
- *Department of Coal (DOC):* The Central Mine Planning and Design Institute Limited (CMPDI) is the nodal agency for coordinating R&D activities in coal and lignite sectors. The Ministry of Coal normally supports the project of shorter duration (2-4 years). The findings of which, if completed successfully, can be used directly for commercial exploitation and benefit to the industry. It supports research activities in various fields such as Production, Productivity & Safety, Coal Beneficiation & Coal Utilization, Environment & Ecology.
- *Department of Ocean Development (DOD):* The development in Ocean Science & Technology is linked with achievements in other scientific and technological areas. The research efforts should lead to fundamental understanding and ensure predictive capabilities. Research support is provided in areas such as newly emerging and front-line areas of Marine Geology and Geophysics, Marine Coastal Ecology, Marine Biology, Marine Microbiology, Coastal Marine Culture Systems, Marine Benthos, Beach Placers, Ocean Engineering, and Underwater Robotics.
- *Department of Science and Technology (DST):* The Department of Science & Technology plays a pivotal role in the promotion of Science & Technology in the country. Science & Technology Policy-2003 states that "Special emphasis will be placed on equity in development so that the benefits of technological growth reach the majority of the population, particularly the disadvantaged sections, leading to an improved quality

of life for every citizen of the country". The DST supports various research schemes such as Deep Continental Studies (DCS), Himalayan Glaciology (HG), Indian Climate Research Programme (ICRP), Instrument Development Programme (IDP), International S&T Cooperation (ISTC), Joint Technology Projects under STAC/IS-STAC, Monsoon and Tropical Climate (MONTCLIM) & Agrometeorology, Natural Resources Data Management System (NRDMS), Pharmaceuticals Research & Development Support Fund (PRDSF) Programme. There are many more programmes supported by DST.

- *Department of Scientific and Industrial Research (DSIR):* The Department of Scientific and Industrial Research (DSIR) is a part of the Ministry of Science and Technology, which was announced through a Presidential Notification, dated January 4, 1985. The Department of Scientific and Industrial Research (DSIR) has a mandate to carry out the activities relating to indigenous technology promotion, development, utilization, and transfer. Various schemes of DSIR include Industrial R&D Promotion Programme, Technology Development and Demonstration Programme, Technopreneur Promotion Programme, Technology Management Programme, International Technology Transfer Programme, Consultancy Promotion Programme, and Technology Information Facilitation Programme.
- *Indian Council of Medical Research (ICMR):* The primary aim of the ICMR is to promote research in the fields of medicine, public health, and allied areas. The Council promotes biomedical research in the country through intramural research through institutes totally funded by ICMR and extramural research through grants-in-aid given to projects in non-ICMR institutes. It supports the setting up of Centers for Advanced Research in different research areas around existing expertise and infrastructure in selected departments of medical colleges, universities, and other non-ICMR research institutes.
- *India Meteorological Department (IMD):* IMD was established in 1875. It is the National Meteorological Service of the country and the principal government agency in all matters relating to meteorology, seismology, and allied subjects. It provides research support in topics related to the advancement of knowledge in the fields of meteorology and atmospheric sciences, with particular reference to the Indian region.
- *Indian Space Research Organization (ISRO) Department of Space:* The Indian Space Research Organization (ISRO) was established in 1969. The Indian space programme has the primary objective of developing space technology and application programmes to meet the developmental needs of the country. The Indian space programme includes the development of operational systems in the areas of satellite based remote sensing,

telecommunications, broadcasting, meteorology, and the development of suitable launch vehicles for putting the satellite in various low earth orbits and geostationary orbits. ISRO supports the research activities in space science, application, technology, space communication, earth resources survey, meteorology, and satellite geodesy.

- *Ministry of Environment and Forests (MOEF):* The Ministry of Environment and Forests is the nodal agency in the administrative structure of the central government for planning, promotion, and coordination of environmental and forestry programmes. The main activities of the Ministry are conservation and survey of flora, fauna, forests, and wildlife, prevention and control of pollution, afforestation and regeneration of degraded areas and protection of the environment. It supports research schemes such as Environmental Research Programme (ERP), Ecosystems Research Scheme (ERS), Eastern and the Western Ghats Research Programme (E&WRP), etc.
- *Ministry of Non-Conventional Energy Sources (MNES):* The Ministry promotes renewable energy technologies and creates an environment conducive for their commercialization through innovative policy initiatives and strategies. The range of its activities covers renewable energy resource assessment, research and development, demonstration, extension, and production in the areas of biomass energy, solar thermal and solar photovoltaics, wind energy, and small hydropower. It also promotes and supports studies and research in new technology areas such as tidal energy, geothermal energy, alternate fuels for transportation, hydrogen energy, and fuel cells.
- *Ministry of Water Resources (MOWR):* Ministry of Water Resources (MOWR) provides financial assistance to promote research work in the field of Water Resources Engineering. The assistance is provided by way of grants to academicians/experts in the universities, IITs, recognized R&D laboratories, Water Resources/ Irrigation departments of the Central and State Governments, and NGOs. Research proposals of applied nature, as well as basic research, are considered for MOWR support.

There are many other research institutes, both governmental and private, that provide research supports to conduct value research programmes in various disciplines. One can find many international agencies also supporting research activities in various fields.

6.2 LITERATURE REVIEW

The term literature refers to the information relevant to the topic of study of a researcher. Such information may be obtained from a variety of sources such as books, journal articles, reports, proceedings of conferences, etc. However, the focus is on scholarly published materials. The source material can be categorized as [7]:

- *Primary source:* Original research from journals, articles or conferences, original materials such as historical documents, or creative works such as art or literature.
- *Secondary source:* Evaluation, reviews or syntheses of original work
- *Tertiary source:* Broadly scoped material put together usually from secondary sources to provide an overview, e.g., a textbook.

In selecting the source material the following points should be kept in mind [8].

- *Relevance:* It should contribute to the development of the selected topic, clarify your position, provides an alternative point of view you wish to argue against or provide useful primary source material.
- *Authority:* It should be published in a reputed journal, has been critically evaluated, been used extensively as a source material, and been peer reviewed or be a recognized authority in the area.
- *Currency:* It should be recent research or still influential in the area.

As mentioned above a number of sources are available for collecting appropriate literature. Different sources vary in degree of usefulness. A brief discussion of various sources of literature is given below.

- *Journal articles:* These are good especially for up-to-date information. Keep in mind, though, that it can take up to two years to publish articles. They are frequently used in literature reviews because they offer a relatively concise, up-to-date format for research, and because all reputable journals are refereed (i.e. editors publish only the most relevant and reliable research).
- *Books:* Books are less up-to-date as it takes longer for a book to be published than for a journal article. Textbooks are unlikely to be useful for including in your literature review as they are intended for teaching, not for research, but they do offer a good starting point from which to find more detailed sources.
- *Conference proceedings:* These can be useful in providing the latest research, or research that has not been published. They are also helpful in providing information on which people are currently involved in which research areas, and so can be helpful in tracking down other work by the same researchers.
- *Government/corporate reports:* Many government departments and corporations or commissions carry out research. Their published findings can provide a useful source of information, depending on your field of study.
- *Theses and dissertations:* These can be useful sources of information. However there are disadvantages: 1) they can be difficult to obtain since

they are not published, but are generally available from the university library; 2) the student who carried out the research may not be an experienced researcher and therefore you might have to treat their findings with more caution than published research.

- *Newspapers/Magazines:* Often newspapers are more helpful as providers of information about recent trends, discoveries or changes. Specialized magazines may be more useful but usually magazines are not useful for your research except as a starting point by providing news or general information about new discoveries, policies, etc., that you can further research in more specialized sources.
- *Internet:* The fastest-growing source of information is the Internet. However, it should be noted that anyone can post information on the Internet so the quality may not be reliable. Information we find on the Internet may be intended for a general audience and so not be suitable for inclusion in the literature review.The refereed electronic journals (e-journals) that appear on the Internet are more reliable.

A literature review can be contained as an integral part of an article, research proposal, research report or dissertation. It describes, compares, contrasts and evaluates the major theories, arguments, themes, methodologies, approaches and controversies in the scholarly literature on a subject. It also compares and contrasts these arguments, themes and methodologies, etc., with the concerns of a proposed topic of research. A literature review presents an examination and evaluation of the previous achievements on the topic selected for study. It is not a summary of the historical background of the topic under study.

Literature review in a research has various purposes. A good literature review shows:

- that the researcher is aware of what is going on in the field;
- that there is a theory base for the work that he proposing to do;
- how the work fits in with what has already been done;
- that the research work has significance; and
- that the research work will lead to new knowledge.

6.2.1 Need for Literature Review

There are many reasons why a researcher should write a literature review. 'For professional researchers, literature reviews are useful reports that keep them up-to-date with what is current in the field. For scholars, the depth of the literature review emphasizes the credibility of the writer in his or her field. Literature reviews also provide a solid background for a research paper's investigation' [9]. Comprehensive knowledge of the literature of the concerned field is essential to most research papers. Roberts and Taylor (2002) pointed out the following reasons that make a literature review an important component of a research work [10]:

- to avoid making the same mistakes as others have done;
- to carry on from where others have reached;
- to increase your breadth of knowledge in your subject area;
- to identify key works, information and needs in your area;
- to identify and learn terminology;
- to position your own work in context;
- to identify opposing views;
- to demonstrate that you can access research in the field;
- to identify methods relevant to your project;
- to identify studies that are worth replicating or improving; and
- to find experts in your field whom you could contact.

Through a literature review a researcher usually cites another author. Since a literature review is information dense it is crucial that the work is intelligently structured to enable a reader to grasp the key arguments with ease. *Edith Cowan University* of Australia (Academic Tip Sheet) suggests the observation of the following **5C**'s in preparing a literature review.

1. Cite: keep the primary focus on the literature.
2. Compare various arguments, theories, methodologies, approaches and findings expressed in the literature: what do the authors agree on? Who employs similar approaches?
3. Contrast the various arguments, themes, methodologies, approaches and controversies expressed in the literature: what are the major areas of disagreement, controversy, debate?
4. Critique the literature: which arguments are more persuasive, and why? Which approaches, findings, methodologies seem most reliable, valid, or appropriate, and why? Pay attention to the verbs you use to describe what it is an author says/ does: e.g. asserts, demonstrates, etc.
5. Connect the literature to your own area of research and investigation: how does your own work draw on, depart from or synthesize what has been said in the literature?

A good literature review will analyze the research that has been done and synthesize or collect those elements that are similar or most pertinent to the topic that one has selected for his study. It provides an overview on the research topic.

6.2.2 Planning and Locating Relevant Literature

All research begins with a literature review which enables the researcher to understand the current progress in his field of research. This is essential since the new knowledge depends strongly on the existing knowledge. A thorough review of the literature will help the researcher to:

- understand their topic;
- develop and focus a topic;
- provide a clear rationale for, or better situate, their topic;
- fine-tune their research questions;
- identify well-vetted data collection and analysis methods on their topic;
- determine whether to replicate a previous study or develop a completely new study; and
- add rigour and validity to the research by validating the topic, methods, and significance.

The researcher must have a clear understanding of the scope of the literature review and he must be conscious about the time that must be dedicated to the literature survey. This is due to the fact that in many cases, the research is a time-bound process. Therefore, the researcher will get only a short time for literature review. Another important factor is the identification of the most appropriate keywords for searching the literature. Keywords with a broad meaning gives lots of search results and to find the most relevant article from the list is again a task. Choosing the correct keyword will help to refine the search results. The researcher must have knowledge about the various sources like a library, online database, etc., from where research articles can be obtained.

After completing a literature review, the researcher must be able to answer the following questions.

- What do you know about your topic?
- What do you not know about your topic?
- How does your research address the gap between what you know and what you don't about your topic?

After reviewing the literature, if the researcher is able to answer those three questions, he would have a very clear and well-rationalized justification for the inquiry. If he cannot answer these questions, then he should continue the literature review of related topics.

The first step in literature review is the identification of relevant literature. There are many sources and it is important to identify the significance of various sources for the study. The following steps may be followed for finding the relevant literature [2, pp. 267-268].

Selection of Appropriate Tools: A large number of tools and methods are available for literature search. In recent years, the most common search tool is the Internet. Huge online databases are available for research. In earlier years, the most commonly used sources were libraries. In good libraries, one can find a good collection of literature either in the form of printed journals or soft copies. Now, most of the libraries are digitized and the search for resources is easy. Libraries in universities and colleges, provide facilities like INFLIB NET for access to paid journals.

Trace Relevant Publications: The selection of the most relevant research articles related to the topic is very important. To get such articles while searching over the Internet, it is essential to identify the most appropriate keywords. One can use synonyms and alternative keywords also. For common topics, usually one can find a large number of articles. In such cases, the most appropriate keywords must be used. One can try by changing the search engine also. Google Scholar is the most widely used search engine to locate scholarly articles over the Internet.

Note Making: To keep the information obtained from the literature in an organized manner, note-making is compulsory. Note down any relevant point related to the topic and write in the notebook with the date and reference. This will be helpful while preparing the final article. The summary or comments of the relevant point must be written in your own words.

Keeping Bibliographic Record: During the literature review, a researcher refers to a large number of sources. From each source, he may get some useful information related to the research topic. While you write notes, it is important to ensure that bibliographic details such as the name of author/s, the title of publication, name of publisher, and page numbers are accuratelywritten for later use [11]. The following bibliographic details are to be included [2, p. 268].

- Name of author/s, editor/s, or organization
- Title of article, book, or chapter (if it is an edited book)
- Volume and issue numbers (for journals)
- Edition and reprint (for books)
- Name of the publisher (for books)
- Place of publication (for books)
- Year of publication
- Inclusive page numbers (for articles and chapter articles)
- Total pages (for books)

In the case of research articles, the following details are to be recorded.

- Name of author/s, editor/s, or organization
- Title of the article
- Title of the site (i.e., homepage)
- Date of creation of the site
- Date, last modified, or updated
- Date of access
- Digital Object Identifier (DOI) or full address of URL.

6.2.3 How to Manage Information

A meaningful literature review gives information related to the research topic. The details collected during the literature review are to be managed wisely and judiciously. The proper bibliographic note is to be maintained for future reference purposes. There are different methods like index cards, software for reference management, and electronic card files for managing the collected information.

Index Cards: This is an old system of managing references and information. This is particularly useful and easy to manage when the number of references is less. The index card contains the bibliographic details of the collected information. With the development of computer applications and software, this became less frequently used by researchers.

Software for Reference Management: With the development of computer applications, many software packages were developed for the management of references. Researchers find these software packages more efficient, convenient, and easy to use. There are free reference management software packages like EndNote, Mendeley, CiteULike, Refbase, and Zotero. Such packages can be installed on a computer and one can find an add-on in the Microsoft Word document. We can open these and add as many references as needed. These can be used while the manuscript is typing in Microsoft Word and the reference can be inserted at the point where it is required from the already added list. The advantage of such software is that the software itself provides the referencing format for various international journals and one can select the desired format based on the journal to which he wishes to submit his article.

6.2.4 Academic and General Search Engines and Online Searching Methods

With the development of information technology, a number of search engines are available for the online search of information. The most commonly used search engine is Google. However, there are many other search engines. Some of them are discussed below.

Google (www.google.com): Google is the most popular search engine. Google began in January 1996 as a research project by Larry Page and Sergey Brin when they were both Ph.D. students at Stanford University in California. Google indexes billions of web pages to allow users to search for the information they desire through the use of keywords and operators. Google has a unique database called Google Scholar (https://scholar.google.co.in/) for scholarly articles. Google Scholar provides information about journal articles, conference papers, abstracts, theses, books, and other such academic publications in various disciplines and sources.

Microsoft Academic (https://academic.microsoft.com/): Microsoft Academic is a free academic search engine from the Microsoft group. In addition to the search results and access to sources, it also gives citation information such as a number of citations, g-index, and h-index. Microsoft Academic has been posited as a competitor to Google Scholar, Web of Science, and Scopus for academic research purposes as well as citation analysis.

WorldWideScience.org (https://worldwidescience.org/): This provides, various information in various fields of science. WorldWideScience.org is a global science gateway—accelerating scientific discovery and progress through a multilateral partnership to enable federated searching of national and international scientific databases and portals. It allows a single query search in the national database of many countries. "WorldWideScience.org uses federated searching to offer its coverage of global science and research results. It provides access to 'deep web' scientific databases, which are typically not searchable by commercial search engines".

SciDev.Net (https://www.scidev.net/global/): SciDev.Net is the world's leading source of reliable and authoritative news, views, and analysis about science and technology for global development. SciDev.Net started in 2001 with an aim of bridging the sizeable gap in scientific knowledge between rich and poor countries. It engages with development professionals, policy-makers, researchers, the media, and the public to use science and technology for global development.

Directory of Open Access Journals (https://doaj.org/): The articles published in the open access journals are free to access by any person. Many of the open access journals charge an amount as open access fee from the contributors of the article and not from the person who is accessing the paper. DOAJ is an online directory that indexes and provides access to high-quality, open-access, peer-reviewed journals.

Bing (https://www.bing.com/): Microsoft Bing is a web search engine owned and operated by Microsoft. Microsoft has already introduced search engines like MSN Search, Windows Live Search and later Live Search; later these were discontinued and started the Bing search engine in 2009. Bing provides a variety of search services, including web, video, image, and map search products.

Yahoo (https://www.yahoo.com/): Yahoo is a web service provider based in America. It provides a web portal, search engine Yahoo Search, and related services. Yahoo! provides a lot of information including the latest news and entertainment programmers and helps user's instant access to other Yahoo! services such as e-mail, maps, finance, and group messaging facilities.

Wiki.com (www.wiki.com): A wiki is a server program that allows users to collaborate in forming the content of a Website. A wiki is essentially a database for creating, browsing, and searching through information.

SlideShare (https://www.slideshare.net/): SlideShare is a specialized search engine, which allows to search for slideshow presentations uploaded in the database. One can also search for e-books and documents. SlideShare also allows to save slides and the slideshow can be downloaded and saved on a storage device.

6.2.5 Writing a Literature Review

A literature review shows that how much work has already been done in the selected research area. It helps to highlight and provide a justification of the research proposal. In fact a literature survey helps the researcher to identify a thrust area in which he can continue his research work. Literature review is different from a research paper. While the main focus of an academic research paper is to support your own argument, the focus of a literature review is to summarize and synthesize the arguments and ideas of others. However, they contain some of the same elements.

From the point of view of a researcher the literature review:

- should discuss the theories and concepts that underlie the research;
- is a discussion of the area of research, the main issues relating to it;
- should also explain any terminology that the researcher uses in his research area and provide definitions of any areas that may not be obvious; and
- need to critically analyze the material that he is looking at.

After identifying credible, up-to-date and accurate information sources, the researcher need to start building up an outline of the subject area and the research that has been done on it. One has to search a range of resources to get a full picture. This should be followed by grouping of information according to relevance, making notes, identifying themes and performing a critical analysis. We can quote other authors, but it should be properly referenced in order to avoid plagiarism (Chapter 8 Section 8.4.1).

Similar to a research paper a literature review should contain three parts [12, 13]:

1. *The introduction:* In the introduction the researcher defines or identifies the general topic, issue, or area of concern. It gives the reader a quick idea about the research topic and the central theme or organizational pattern that the researcher will follow. Through the introduction the researcher points out overall trends in what has been published about the topic; or conflicts in theory, methodology, evidence, and conclusions; or gaps in research; or a specific problem or new perspective of immediate interest. It also explains the criteria used in analyzing and comparing the literature and the organization of the review.
2. *The body:* This part contains the discussion of various sources. It includes group research studies and other types of literature (reviews, theoretical

articles, case studies, etc.) analyzed over various common platforms such as qualitative versus quantitative approaches, conclusions of authors, specific purpose or objective, chronology, etc. It summarizes individual studies or articles according to its comparative importance in the literature.

3. *The conclusion and recommendations:* In this section the researcher posits what he has learned from the literature review, what questions it raised, how he plans to proceed, or what seems to be missing from the research. It summarizes major contributions of significant studies and articles to the body of knowledge under review, maintaining the focus established in the introduction.

Among these three, the major component is the body of the literature review. A researcher usually spends more time and effort in preparing the body. Several frameworks are followed in organizing the body of the literature review. Some of them are:

- *Chronological:* This could be a logical framework for organizing your literature review if the research in the sources appears to follow a chronological progression.
- *Trends:* Sometimes it is possible to identify some trends within the research. This may be similarly chronological, but certain trends may encompass greater historical eras and be geographical, cultural, etc.
- *Thematic:* Thematic reviews focus on central issues or topics. When writing the literature review, the researcher will use the sources to reinforce the theme that he has chosen as central to his paper.
- *Methodological:* This framework is organized around the methodologies used in the research. The researcher may present an analysis of research techniques according to certain criteria. This approach to literature review is often used by researchers who wish to justify their choice of one methodology over others.
- Current state: In this framework the researcher's intention is to present only the present status of the topic.

6.3 JOURNALS

Journals are publications that intended to publish scholarly and scientific articles, which are being selected through peer review. It is different from a magazine that publishes journalistic articles. Journals consist of various elements, and the main element of a journal is articles. In the case of academic journals, articles are contributed by academics, authors or researchers on a specific topic relevant to the subject of the journal. They are often the result of years of serious research work. Other elements include editorials; where the editor of the journal presents their own opinion or highlights relevant topics within the journal and reviews of books and other research work. Journals are published continuously in regular issues ranging from weekly to bi-annually. Journals after a certain period of time, usually one year, individual journal issues are collated together into a volume.

In the editorial of Journal of Biological Rhythms (Vol. 15 No. 6, December 2000) Martin Zatz wrote: "I believe the dignity of the Journal, insofar as that phrase has any meaning, derives from the quality of the science it presents. Yes, I want articles in the Journal to be cited 20 years from now, but I want, even more, that it be read today (and cited next year). I believe that the Journal should serve as archives.... and a forum for exchange of information and views and for the expression of our fields and our subscribers' vitality."

Journals are rich source of information. It is important to maintain the archival nature of the journal. There are several types of journals published with different purposes. The style and content of each journal will be different. The three main categories of journals are:

- *Academic Journals:* These journals are published for a special group, learned society or profession. They focus on research topics and feature peer-reviewed articles (see section 7.6.3. for details about Peer Review). Examples are *Nature*, *Journal of Pure and Applied Physics*, *Thin Solid Films*, *Journal of Raman Spectroscopy*, and *Journal of Solid State Chemistry*, etc.
- *Popular Journals:* These are often thought of as magazines. These types of journals include glossy, short articles written for general readers, alongside advertisements of general interest. Examples are *Resonance*, *National Geographic*, *PC Magazine*, *BBC History Magazine*, and *Electronics for You*, etc.
- *Trade and Professional:* These types of journals are service or industry-based. They include feature news items and job advertisements as well as advertisements for products relevant to that journal's field. Examples are *Investors Chronicle*, *Nursing Standard*, and *The Instrumentalist*, etc.

Journals are available both in printed and electronic forms. Publishing the research findings in high quality peer reviewed journals is considered to be a contribution to the knowledge field. The research findings are valuable to the community if they are published in journals and other scholars get an opportunity to access it. In recent years the availability of electronic journals over the Internet made the access of those articles much easier all over the world. With the advancement of technology the findings can be published in journals within a short period.

The major functions of a journal include [14]:

- *Registration:* establishing the author's precedence and ownership of an idea.

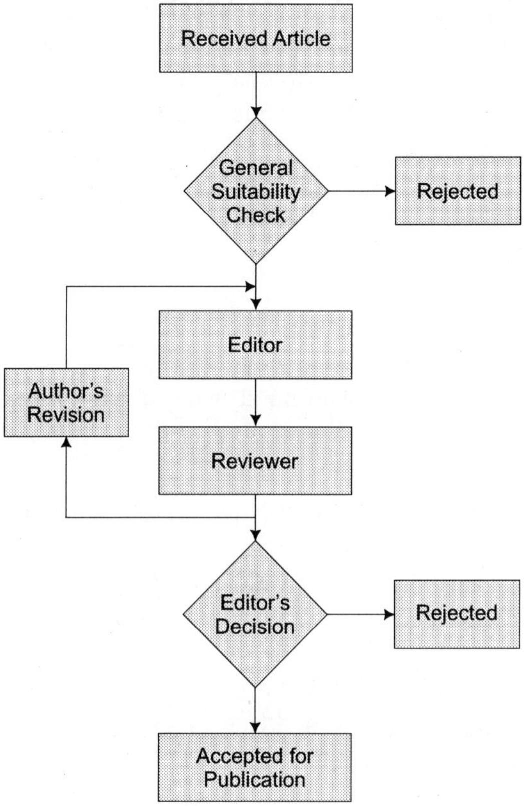

Fig. 6.1 Flowchart depicting the processes involved in the publication of an article. (Adapted from the release of the Japan Society of Applied Physics, May 2013)

- *Dissemination:* communicating the findings to its intended audience.
- *Certification:* ensuring quality control through peer review and rewarding authors.
- *Archival record:* preserving a fixed version of the paper for future reference and citation.

A journal will not publish all the research papers that they receive. Before accepting for publication the article will be subjected to a peer review by a group of experts in the concerned field. The reviewers then suggest some corrections or modifications. After making those corrections or modifications the author has to resubmit the article. After the acceptance the article will be published in the journal. The flow chart given above shows the various processes involved in the publication of a research article in a journal.

Publishing an article in an established journal promotes the acceptance and discussion of the findings by the scientific community. Such articles will be picked up by the major abstracting and citation services than are newer journals or journals of generally low repute. Abstraction and citation of articles published in a new journal is not so common. It usually takes some years before a new journal to reach the major stream. A journal which is not indexed and cited will be available in hard copy only, and the researchers using the Internet or electronic databases will ever encounter and cite it. However, newer journals may take a shorter time to review, accept, and publish the research work. A new journal can survive only if it presents valuable findings continuously before the scientific community.

Therefore, the selection of a journal for publishing the research findings is to be done carefully. As a first step one has to shortlist the potential journals and from the shortlisted journals an appropriate selection is to be made. The following questions should be kept in mind while making a selection process [15].

1. Is the journal peer reviewed?
2. Does the journal currently publish papers on subjects related to the research topic?
3. Which journals have the best reputation in the field of the selected topic? This information can be obtained from colleagues and read recent articles and judge their importance.
4. What is the journal's impact factor?
5. Which journals are most likely to be cited by others in the selected research field?
6. Is the journal published by a society? Society journals are usually the most prestigious and have the largest circulation.
7. Is the journal indexed in the major electronic databases such as Medline, Biological Abstracts, Chemical Abstracts, Current Contents, etc.?
8. Which journals have the kind of expertise that would ensure your paper is given a "fair hearing"?
9. Are there journals whose readership you need/want to influence?
10. How often is the journal published? What is the usual time lag between receiving and publishing papers? Using the "date submitted", the "date accepted", and the date of the issue of published articles one can estimate the length of the review process as well as the time from acceptance to publication in print.
11. Is the journal published in English? This is also important because English has become the language for international scientific communication.
12. What is the focus of the journal; is it broad or narrow? Which disciplines are represented? What is the journal's research orientation; is it basic, theoretical, or applied?

13. Do you like the appearance of published articles – the format, typeface, and style used in citing references? If relevant, does the journal publish short and/or rapid communications?
14. Do the figures published in the journal have the resolution that you need?
15. How frequently the journal is published? Monthly journals have a shorter lag time than quarterly journals.

'The audience for a journal is largely determined by the scope and aims of the journal, the journal's reputation and history of publishing in the field, and the accessibility of the journal to researchers' [16]. If the journal is expensive and if it does not provide any open access to the authors, then definitely the number of scholars using that particular journal will be limited. However, Internet access to journal titles, abstracts, and homepages has allowed by many journals. For research scholars in the universities, the article from various journals may be available at free of cost or for a nominal amount. The university libraries usually subscribe major research journal, in various disciplines.

6.3.1 Short Communication

Short communications are concise articles that contain limited, (approximately 2500), words. They aim to report new ideas, recent advances in various areas of knowledge. Short communications do not cover in detail background information about the problems treated or the applications, rather they provide key pointers to the reader. The work reported needs to be technically sound, innovative and significantly unique. Authors must include a statement about the word count in their submission. Since the new findings are to be reported at the earliest, the review time of the articles must be as short as possible. This is ensured in short communication, because the article is short and the reviewers need to spend only a small time in reviewing it.

A short communication may be used when something is discovered which is never explored in any literature. That is, one can use the rapid communication when there is a novelty in the discovery. The reviewers of short communications expect high quality, originality and proposal of a new theory or concepts based on the existing knowledge. A short communication also has a different structure than a regular article; usually the findings are expressed without separation between the sections. While performing research sometimes we may come across important information that dilutes the main information related to the research principal objectives. Therefore, we usually omit it from the published paper. If one thinks that the omitted information is important but cannot serve as the content of a full paper then it is worth to include that information in a short communication note.

6.3.2 Rapid Communication

The Rapid Communications section in a research journal is devoted to the publication of short papers presenting highly original and significant research findings. It has focused on the speedy dissemination of important findings in various research areas. A Rapid Communication should present the work that is notably important, interesting, or timely to those in a particular subfield. Authors submitting a rapid communication article should provide reasons why the work is urgent and requires rapid publication.

Most of the journals offer open access to the article published under the section Rapid Communication. When considering the article length an article under rapid communication can occupy four or five journal pages.

Many journals offer the following general criteria for submitting an article under the heading rapid communication.

- Articles must be distinctive, timely, and of special interest to the readership.
- Articles must describe innovative or pioneering research, or present unique and new concepts in the subject matter area.
- Articles must be enhanced by rapid publication in that the article has a high profile and is need-to-know information by the readership.
- Articles must be pre-reviewed by at least 3 experts in the subject matter area if possible.

The rapid communication section aims at encouraging the submission of preliminary, yet exciting findings, whenever appropriate. Reviews of these papers will focus not only on quality but also on whether early publication is justified. An important advantage of publication of short articles is to allow a larger number of papers per issue. Also submission and publication of short articles has the potential of making the journal more reader-friendly as well as reducing the burden on reviewers and editors.

6.3.3 Research Paper

Research papers or research articles are fully documented reports of original research and are always peer reviewed. Research articles are the backbone of any journal and 80% or more articles in a particular issue will be of this category. A research article addresses a clearly stated research question or hypothesis, which is tested empirically through the application of a solid research protocol. In a research paper a researcher presents his findings using technical language. A research paper should be presented in a prescribed format. The format may be different in different disciplines. However, the experimental research, especially scientific research follows almost a similar structure for writing a research paper. The most generally used format of a research paper includes:

- Abstract
- Introduction
- Materials and methods
- Results and Discussion
- Conclusion

A detailed discussion of each of these components is given in chapter 7. In addition to these; acknowledgements and a list of references are included at the end of the paper.

Writing a scientific paper is not an easy task. Majority of the journal demands the use of a technical language in preparing an article. The use of flowery language is discouraged in writing a scientific paper. The preparation of an article includes the creation of a first draft, revision and editing of the first draft, preparation of the second draft, correction of the second draft and finally the submission for publication. The particular format and style adopted for writing a research paper depend upon both the nature of the report and the journal or other publication in which the paper is to be published. Most of the journals issue guidelines for the authors annually in one of the issues. The same may also be available in its website throughout the year. A researcher becomes familiar with details of organization, section headings, methods of data presentation, and ways of citing and listing references by examining recent papers in any well-established scientific journal through his experience.

6.3.4 Review Paper

A review paper is basically different from a research paper; both in purpose and format. They are also published in peer reviewed journals. Review papers usually synthesize and summarize the work of a number of scientists in a particular sub-field. They do not report new results. Review articles often does not have the component *"materials and methods"*. The review paper will have an *"abstract"*, and it may have a section called *"introduction"*. The rest of the paper will be an overview of the topic, organized either historically or by different treatments of the problems connected with the topic. Review articles are very useful when a researcher wants to know what development occurred in a particular area in which he is interested.

The numbers of published research papers are increasing fast in recent years. A research scholar cannot collect and read all these papers. This increases the importance of review articles which summarise the publications made by a number of researchers. 'The main purpose of a review article is to examine the current state of the relevant publications on a given topic and to initiate a discussion about the research methodologies and the findings related to the said topic'[17]. A review article should contain a comprehensive list of supporting references cited in the text. There is no strict limitation concerning the review topics. Review articles can be classified as non-systematic and systematic [18].

- *Non-systematic reviews:* These types of reviews include a selection of a wide range of technical results related to a particular topic which reflects the professional experiences and interests of the authors. The quality of such papers depends on the competence and the sincere efforts of the authors to provide a good review of the subject matter.
- *Systematic reviews:* Systematic reviews require an advanced preparation and rely on a specific methodology being chosen in advance, where selected foundation studies are being analyzed in connection to a long-standing problem statement.

In a review paper the author transforms a collection of technical information presented by various researchers into a coherent script that presents, most often, an alternative point of view on the previously unknown or not well understood relations among different studies.

Sometimes a review of a single paper also appears in journals; such a review is known as a critical review. In a critical review the reviewer studies the problem stated in the article and its scientific content. Then he makes an analysis on the strength and weakness of the paper and finally discusses the research findings. Based on a critical evaluation an interpretation on the research findings is to be made. The critical reviews must be based on verifiable scientific facts, logic and reasoning.

The purposes of a review paper can be summarised as follows [19]:

- Give enough background so that those learning about the field can get started.
- Contain frequent citations of other work throughout the paper, not just in the Introduction and it will be very useful to the reader.
- Summarise knowledge and trends in the field.
- Criticise the results obtained by other scholars and scientists.
- Suggest avenues for future research: problems to solve, gaps in the knowledge to fill, ways to extend results to other materials or applications.
- Review paper provides a foundation for further research and experimentation.

Good uses of review papers help the researcher to create a strong foundation in his area of interest.

6.3.5 Interactive Journal

The numbers of scientific journals nowadays are very large. Thousands of articles are published every year through these journals. It is a common observation that a large proportion of scientific publications are careless, useless or false, and do not contribute to scholarly communication and scientific progress. As *Ulrich Pöschl* (2004), observed 'Many papers reflect a mentality of publishing just as much and as fast as possible, rather than participation in vital scientific exchange and discussion. The inflationary increase of scientific publications is fuelled by the habit of evaluating scientific productivity by the number of papers. In many research areas scientists have to spend an excessive amount of time to maintain an overview of the information dispersed and diluted in the increasing flood of publications' [20]. The reason for such a situation is the limitation of present process of peer review. The so called closed peer review is hampered by limited competence and the sometimes conflicting interests of editors and referees. This is not surprising because the research field need not be a narrow one; it includes enormous variety of research activities, institutions and interdisciplinary aspects. Even scientists within the same discipline may not be aware of all the details relevant for a critical and thorough review.

Also the best qualified referees for a particular manuscript may not be known to the editor and even if they are known sometimes those referees need not have enough time and motivation for a thorough evaluation of the manuscript. Ulrich Pöschlcontinues: 'incomplete, unbalanced or plainly unsustainable presentations are rarely complemented or corrected by public commentaries'. This is because the publication of a commentary in a journal requires almost the same effort as in the case of a regular paper.

The quality of the articles published in mainstream journals can be improved through adopting the concept of interactive journal. It is a two-stage publication process consisting interactive peer review and public discussion.

The first stage can be summarised as follows [21]:

- Manuscripts that pass a peer review or pre-selection by the editor are immediately published in a scientific discussion forum on the internet.
- They are then made available for full interactive peer review and public discussion, during which the comments of referees (anonymous or attributed), additional short comments by other scientists and scholars and the authors' replies are published alongside the discussion paper.
- The interactive comments are published without peer review and revision, but can be censored in case of abusive commenting (personal offence, etc.).
- To ensure publication precedence for authors and to provide a lasting record of scientific discussion, the discussion papers and interactive comments are permanently archived and fully citable.

The second stage can be summarised as follows:

- The completion of the peer review process.
- Publication of the final revised papers in the scientific journal.

Compared to the traditional journals the interactive journals have got many advantages. Some of them are [22]:

- The discussion paper offers *'free speech'* and rapid dissemination of novel results (authors' and readers' advantage).

- The interactive peer review and public discussion offer direct feedback and public recognition for high-quality papers (author's advantage).
- They prevent or minimise the opportunity for hidden obstruction and plagiarism (authors' advantage);
- They provide complete and citable documentation of critical comments, controversial arguments, scientific flaws and complementary information (referee's and reader's advantage).
- They prevent submissions of careless, useless and false manuscripts (referee's and reader's advantage).
- The final revised paper offers a maximum of scientific information density and quality assurance through full peer review, public discussion and final revision based on the referee's comments and additional comments from other interested scientists (reader's advantage).

A researcher or a scientist usually publishes as many papers as possible through journals with closed peer review and does not get direct public feedback and recognition for his work. The two-stage publication process involved in the interactive journals encourages scientists to prove their competence through producing high-quality results and papers and their discussion.

6.4 CONFERENCE PROCEEDINGS

Apart from journal articles, conferences are the second major form of formal communication among scientists. Conferences are frequently organized by various research organizations as well as universities and colleges. In a conference a group of scientists gather to present findings about their research. Some conferences have very broad topics (like annual meeting on the advancements in medical biotechnology) and some can be very specific like the (conference on Optics and Photonics for defence applications). In these conferences scientists address the group or audience and share their ideas or findings on a particular topic. Sometimes, the researchers write up a formal paper explaining the same thing they did at the conference in more detail. These papers can be published in book form referred to as the "Proceedings of Conference X". A conference proceedings includes abstracts also. The authors are usually requested to submit their write-ups well in advance for a review by an expert committee. Based on the suggestion of the expert committee the article may be included in the proceedings of the conference.

Usually, the proceedings are issued to the participants at the beginning of the conference itself. This creates an opportunity to go through the articles in advance and any doubt that may arise can be made clear when the author presents it in the conference. However, the number of articles included in the proceedings will be much larger than the actual number of presentations in the conference. This is because the organizers usually invite articles from the authors in the relevant fields and the number of articles received will be large in number. Then after the review by the expert committee most significant articles will be allowed in the conference, some others will be selected for poster presentation.

6.5 JOURNAL IMPACT FACTOR (JIF)

Journal impact factor, which provides a means to assess the quality of a journal, was introduced in 1963 by Garfield and Sher. It is a measure of importance or influence based on the number of citations during a given period of time. 'Journal impact factor' provides a systematic, objective means to critically evaluate the world's leading national and international journals, with quantifiable, statistical information based on citation data available in online and print mode' [23]. It is descriptive statistics computed from the data gathered by Thomson Scientific (earlier the Institute for Scientific Information; ISI) of Web of Knowledge data base. It is also known as *Garfield impact factor or journal citation rate.* It provides a quantitative tool for ranking, evaluating, categorizing and comparing journals.'It refers to the extent to which an article appearing in a given journal is likely to be cited anywhere in the journal literature within two years of the original article's appearance in print' [24]. The citation includes that appears in any journal in the next two years. This index provides a measure of the average recent use of articles in a given journal.

Journal impact factor can be calculated using the formula:

$$\text{Journal Impact Factor }(\text{Year}_x) = \frac{\text{Cites to recent articles }(\text{Year}_{x-1} + \text{Year}_{x-2})}{\text{Number of recent articles }(\text{Year}_{x-1} + \text{Year}_{x-2})}$$

In general, the higher a journal's impact factor, more scholars are utilizing that journal. Therefore, a work published in that journal will gain attention of a large group of scholars. Therefore, it helps the authors to decide where to submit their articles. As a general rule, the journals with high impact factors include the most prestigious.

Journal impact factor is useful to librarians, editors, publishers and researchers. It is basically used in the process of academic evaluation. Masood Fooladi et.al., pointed out the following advantages of journal impact factor [25].

- The primary advantage of impact factor is that it is very easy to measure.
- Another advantage of journal impact factor (IF) is to mitigate the absolute citation frequencies. Since large journals provide a considerable body of citable literature which cites to the small journals, IF discounts the benefits of large journals in favour of small journals.
- Journal impact factor is a useful tool to compare different research groups and journals. It is commonly applied to administer scientific library

collections. Impact factor is a measure to choose most frequently cited journals (Khan & Hegde, 2009, cited in Masood Fooladi et al.).

- Journal impact factor for indexed journals greatly available and easy to understand and use. Impact factor is more acceptable and popular than other alternative measures (Khan & Hegde, 2009, cited in Masood Fooladi et al.).

However, journal impact factor has some limitations also. While assessing the quality of a journal through the impact factor one should keep in mind the following points [26]:

- The journal impact factor is a retrospective metric. For example, the 2013 Impact Factor looks at citations to papers from 2011 and 2012, but is not available until 2014.
- Not all journals have impact factors as calculated by Journal Citation Report. For example, Journal Citation Report indices over 8000 journals while SCOPUS indices over 15,000.
- New journals do not have journal impact factors. A minimum of three years' data are required for calculating the impact factor.
- Usually non-English language journals are only included if they include at least a table of contents in English
- Self-citation can distort low journal impact. It is a common way to manipulate and increase the impact factor of journals.
- Journals publishing a large proportion of non-citable items such as editorials can have lower journal impact factors, regardless of the quality of the material, if these items are considered "articles" by Journal Citation Report.
- Review articles are often highly cited, so journals can raise their journal impact factor by publishing more of these.
- Citations to a paper may not reflect its quality or significance.
- The impact factor of a journal does not necessarily reflect the quality of papers published by a particular author in that journal.
- Journal impact factors vary between and even within academic disciplines, depending on how popular a particular topic is, or how narrowly a journal is focused.
- Journal impact factor carries little meaning without comparison to impact factors of other journals on the same topic.

While making decisions based on journal impact factor the above limitations are to be considered; otherwise it will be potentially misleading. Despite these limitations; the scientific community is widely using the impact factor as a tool to assess the journal. This situation was well explained by Hoeffel (1998) through the following words [27]:

"Impact Factor is not a perfect tool to measure the quality of articles but there is nothing better and it has the advantage of already being in existence and is, therefore, a good technique for scientific evaluation. Experience has shown that in each specialty the best journals are those in which it is most difficult to have an article accepted, and these are the journals that have a high impact factor. Most of these journals existed long before the impact factor was devised. The use of impact factor as a measure of quality is widespread because it fits well with the opinion we have in each field of the best journals in our specialty".

As mentioned earlier impact factors for science and social science titles are published as part of the ISI Web of Knowledge database, under the title Journal Citation Reports. This can be accessed via http://www.isiwebofknowledge.com/. It is possible to search for individual titles, and for the impact factors for groups of journals in a subject area.

There are other measurements to assess the influence of a journal on its field of research. They are [28]:

- *Journal Immediacy Index:* It is calculated as the number of citations to articles in the year with respect to the number of articles published in that year, giving a measure of how rapidly the average article in a given journal is used.
- *Journal Cited Half-Life:* It is calculated as the number of publication years from the current year that account for 50% of citations received by the journal, giving a measure of the longevity of use of the average article in a given journal.
- *Aggregate impact factor for a subject category:* It is calculated taking into account the number of citations to all journals in the subject category and the number of articles from all the journals in the category.
- *Median impact factor:* It is the median value of all journal impact factors in the subject category.
- *Year Impact Factor:* It is similar to the journal impact factor; but, instead of one year consideration the citations are counted to the previous five years and again divided by the source items published in the previous five years. This approach may be more appropriate for journals in certain fields because the body of citations in two years may not be large enough to make reasonable comparisons, as the publication schedules may be consistently late, or it may take more than two years to disseminate and respond to published works.

6.6 CITATION INDEX

"Citations are an acknowledgement of intellectual debt."

—Henry Small

Citation is a simple concept to understand. Citations are used in articles to support, to provide a precedent for, to illustrate or to elaborate what an author has to say [29]. Thus, citations form an explicit linkage between the current research and prior works stored in the archive of scientific literature. Citations imply that an author has found a particular theory, method or published data are useful in some way in his research work. The citation work (reference) can be a paper or a book. Usually, citations are mentioned in the bibliography or as footnotes of the citing work (source). One major advantage is that the citation helps the scholars to locate the related papers.

The concept of citation index is designed to provide information retrieval and dissemination using the source-reference connections [30]. A citation index is a reference tool that presents bibliographic data on published journal articles. Science citation index was first published in 1964 by Institute for Scientific Information (ISI) and this was followed by other citation indices. Science citation index covers major journals in basic and applied sciences (medicine, engineering, mathematics, computer sciences, etc.).

According to Eugene Garfield (1963) : *"A citation index is a directory of cited references where each reference is accompanied by a list of source documents which cite it. The most characteristic feature of the citation index is that the user begins a search with a specific known paper (target reference). From this starting point one is brought forward in time to subsequent papers related to the earlier paper"* [31].

Citation analysis, involves counting how many times a paper or researcher is cited. It assumes that influential scientists and important works are cited more often than others. 'It alphabetically lists the names of all the first authors whose works were cited during the period covered, and tells when and where those works were originally published. Citation index basically consists of authored cited items, anonymous cited items and patents cited' [32].

Swales and Feak (1994) pointed out the following importance of citation [33]:

- Citations recognise and acknowledge the intellectual property rights of authors. They are a matter of ethics and a defence against plagiarism.
- Citations are used to show respect to previous scholars.
- Citations operate a kind of mutual reward system. Writers 'pay' other authors in citations. (*Ravetz,* 1971)
- Citations are tools of persuasion; writers use citations to give their statements greater authority. (*Gilbert,* 1977)
- Citations are used to demonstrate familiarity with the field. (Bavelas, 1978)
- Citations are used to create a research space. (*Swales*, 1990)

The citation index has got a number of applications. Some of them are:

- It helps in discovering who is referencing your research and how it is being used to support current research.

- It helps in following the history or methodology of an idea from its first communication into the present day.
- It uncovers the early unpublished stages of research activity.
- It helps in locating current research based on earlier patents, reports, published works or findings.
- It creates a list of experts within a field of research.
- It helps in analyzing the impact of published research.

The citation index explicitly helps to find out how many times a paper has been cited in different articles. Also the citation index serves as an indicator in evaluating an author based on publications. In addition to these 'citation analysis helps to know the Journal Impact Factor (JIF), Author Self- Citation (ASC) i.e. number of times an author cites herself / himself, and Journal Self-Citation (JSC) i.e. number of times the journal cites itself, which forms a portion of total citations' [34]. Citation index has various components such as author, anonymous and patent cited index, source index, permuterm subject index, corporate index like geographic and organization index. A detailed discussion of these components with examples can be seen in reference 27.

A list of Web based citation analysis tools adapted from www.physicsweb. orgis given below: More than 100 new databases or tools that allow citation searching are available on web. These fall into three categories.

The first allows the user to search in the full-text field to determine whether certain items, authors or journals have been cited in a document. Examples of these databases or tools include the following.

- arXiv e-print server (arXiv.org)
- CiteSeer (citeseer.ist.psu.edu)
- Google Book Search (books.google.com)
- Google Scholar (scholar.google.com)
- The Institute of Physics Journal Archive (journals.iop.org/archive)
- Physical Review Online Archive (prola.aps.org)
- Elsevier's Scirus (scirus.com)

The second category of databases or tools allows the user to search in the cited-references field to identify relevant citations. These tools first became available in the late 1990s when subject-specific databases began adding cited-references information to their own records. Examples include the following.

- ASA's Astrophysics Data System Abstract Service (adsdoc.harvard.edu)
- MathSciNet from the American Mathematical Society (ams.org/mathsci. net)
- Elsevier's ScienceDirect (sciencedirect.com)
- SciFinder Scholar from the American Chemical Society (cas.org)

- Scitation/SPIN from the American Institute of Physics (scitation.aip.org)
- SPIRES-HEP at Stanford Linear Accelerator Center (slac.stanford.edu/spires)

The third category is databases that work exactly like Web of Science. The best example of this category is Scopus (scopus.com), which was launched in 2004 by Elsevier. Although it covers more refereed journals and conference proceedings than Web of Science, Scopus provides citation searching only from 1996 onwards, whereas Web of Science goes as far back as 1900.

6.7 *h*-INDEX

The *h*-index was originally introduced by US physicist Jorge E. Hirsch in 2005. He presented the definitionof *h*-index as:

"A scientist has index ***h*** if ***h*** of his/her ***N*** papers have at least ***h*** citations each, and the other (***N-h***) papers have no more than *h* citations each" [35].

EXAMPLE 1

Suppose an author has published 12 papers. Each paper got citations as indicated in the table below (after arranging in the decreasing order of citations).

Paper	1	2	3	4	5	6	7	8	9	10	11	12
Citations	52	18	12	9	8	5	4	2	2	0	0	0

In this example $N = 10$; now assume $h = 6$. Then as per the definition the first 6 papers should have at least 6 citations. But it is not so. The 6th paper has got only 5 citations. So the *h*-index cannot be 6. If *h* is assumed to be 5, then each of the first five papers have more than 5 citations. Consequently the *h*-index of the author is 5.

H-index gained the attention of the scientific community as it is simple to compute and it takes into accounts both the quantity and impact of the publications. In traditional methods these are measured using different indicators. *H*-index can be calculated from the citation data available through the Web of Science of the ISI Web of Knowledge. The advantages of *h*-index include that [36, 37]:

It is a better index than total number of publications or total number of citations alone.

- The *h*-index has been proven to be robust in the sense that it is insensitive to a set of lowly cited papers (Vanclay 2007).
- The difficulty of increasing the *h*-index grows exponentially as all the most cited papers of the researcher have to receive new cites to obtain a higher index.

- The *h*-index is insensitive to one or several outstandingly highly cited papers.
- Increasing the number of publications alone does not have an immediate effect on this index.
- In principle, any document type can be included.
- Publications that are hardly ever cited do not influence the *h*-index.

However, there are some factors that complicate the calculation of *h*-index. An author's *h*-index can vary with the number of his/her papers included in the data base used for calculating the *h*-index. Also it depends on the number of citations of those papers included in the same data base. The University of Auckland Library issued the following guidelines to keep in mind while considering *h*-index [38].

- Researchers in specialist areas of research may have lower *h*-indices than those in more popular or currently-topical areas.
- Researchers who produce a small number of influential papers with high citation rates will have low *h*-indices if citations to their other papers are low, even though their contribution to a discipline may be high.
- Researchers with short publishing careers might appear disadvantaged by low *h*-indices, even though their impact in a discipline may be high.
- Researchers may boost their *h*-indices by manually noting citations to their work that have not been recorded by Web of Science or Scopus. Average *h*-indices may vary significantly between disciplines.
- Self-citations may increase *h*-indices.
- Researcher metrics describe the number of citations to your publications (or those of your department or institution), rather than describing the journals in which they are published.
- Authors may have high *h*-indices for items published in low impact journals, and vice versa.
- Neither the *h*-index nor *g*-index decreases with time, so they may appear to undervalue the work of younger authors, but in context, a *g*-index is a measure of an author's contribution to the research environment.
- It is unfair to compare *g* or *h*-indices of two researchers working in different disciplines, e.g., a discipline that uses more references per paper will be expected to have higher indices.

As Hirsch suggested; the *h*-index for a scientist can be calculated for definite period (say for one year) whereas journal's *h*-index can be computed for a *life time contribution*. *H*-index is usually used to supplement the impact factor. It is not a substitute for other indicators. In spite of its wide acceptance *h*-index has got significant limitations. Hirsch himself pointed out some of them like the inability of *h*-index to compare scientists from various disciplines and the dependency of *h*-index on the researcher's age, etc. Egghe (2006) pointed out the major limitation of *h*-index and all later indices that depend on it as its focus on "*h-core*" papers [39]. (*H*-core papers are the papers of a particular author that are getting the highest number of citations relative to the rest of the author publications). Some other disadvantages that are specifically related to the *h*-index are [40]:

- The *h*-index, in its original setting, puts newcomers at a disadvantage since both publication output and observed citation rates will be relatively low. In other words, it is based on long-term observations.
- The index allows scientists to rest on their laurels since the number of citations received may increase even if no new papers are published.
- The *h*-index is only useful for comparing the better scientists in a field. It does not discriminate among average scientists.
- The *h*-index lacks sensitivity to performance changes: it can never decrease and is only weakly sensitive to the number of citations received.

As Martin (1996) opinioned since the *h*-index is easy to obtain, we run the risk of indiscriminate use, such as relying only on it for the assessment of scientists. Research performance is a complex multifaceted endeavour that cannot be assessed adequately by means of a single indicator [41]. According to Van Raan (2006), the use of the h-index could provoke changes in the publishing behaviour of scientists, such an artificial increase in the number of self-citations distributed among the documents on the edge of the *h*-index [42]. Therefore, ,*h*-index as a single indicator is inadequate for assessment of the performance of a scientist. It is to be complemented with other indicators for making an effective assessment.

6.8 *g*-INDEX

As an improvement over *h*-indexL. Egghe (2006) proposed the *G*-index for publications assessment. According to Egghe, "If a set of articles ranked in decreasing order of the number of citations that they received, the *g*-index is the unique largest number such that the top *g*-articles received together at least g^2 citations" [43]. This also means that the top $(g + 1)$ paper have less than $(g + 1)^2$ cites.

EXAMPLE 2

Suppose an author has published 12 papers. Each paper got citations as indicated in the table below (after arranging in the decreasing order of citations).

Paper	1	2	3	4	5	6	7	8	9	10	11	12
Citations *C*	52	18	12	9	8	5	4	2	2	1	0	0

Sum of C	52	70	82	91	99	104	108	110	112	113	113	113
SQR	1	4	9	16	25	36	49	64	81	100	121	144

In this example the total citations for the author is 113, which is greater than 100 and less than 121. Therefore, the g-index of the author is 10. Evidently $(g + 1) = 11$ papers have less than $(g + 1)^2 = 11^2 = 121$ citations.

'It was designed to provide more importance to the most cited papers of the author, as in the case of the h-index, it does not matter if a paper has more than h cites when computing the measure' [44]. It can be easily proved that $g > h$. G-index is a successful indicator in evaluating the production of a researcher incorporating the actual citations of his papers. But 'the g-index may be greatly influenced by a very successful paper' [45]. This can be made clear through the calculation of g-indices of two different researchers.

Let the first researcher has published 25 papers and only one of them got 300 citations. The rest of them did not get any citations. Evidently his g-index is 17, (172 = 289 < 300, the cites of the best 17 papers). The g-index cannot be 18 because 182 = 324 which is greater than 300; (the total citations of best 18 papers). Let the second researcher has published 30 papers and all of them have received 10 cites. In this case the g-index is 10, (102 = 100, the cites of best 10 papers). We cannot take the g-index as 11 because 112 =121 which is greater than the total citations of best 11 papers ($11 \times 10 = 110$).

In this example both authors got the same total number of citations. But the g-index of the first author is much larger than that of the second author. However, the publications by the second researcher has got more visibility and produced more interest among the scientific community.

Rodrigo Costas and Maria Bordons (2008) performed a comparative study of h-index and g-index and observed the following improvements for g-index as compared to the h-index [46]:

- The weighting of the citations received by the documents is considered in the g-index calculation.
- The g-index for a given scientist is not limited by his/her total number of publications.

These features make g-index more suitable for assessing selective scientists (scientists with homogeneous citation pattern) who are less likely to obtain high values of h-index.

6.9 *hg*-INDEX

In 2010 S. Alonso *et al.* presented a new index called hg-index to characterize the scientific output of the researchers. Both h-index and g-index measures different aspects of publications by an author. However, both are incomplete and together present a concise picture of a scientist's achievements in terms of publications and citations. The *hg*-index is proposed to utilize the benefits of both *h*- and *g*-indices and to minimize the drawbacks inherent with them.

According to S. Alonso *et al.* [47], the *hg*-index of a researcher is computed as the geometric mean of his *h*- and *g*-indices, that is:

$$hg = \sqrt{h \times g}$$

It can be demonstrated that $h \leq hg \leq g$ and that $(hg–h) \leq (g–hg)$, that is, the *hg*-index corresponds to a value nearer to *h* than to *g*. This property helps to avoid the problem of the big influence that a very successful paper can introduce in the *g*-index. In other words *hg*-index softens the influence of a high *g*-index when the *h*-index is low.

The *hg*-index can be interpreted geometrically as the area of a rectangle with sides *h* and *g* units. It has got the following benefits [48]:

- It is very simple to compute once the *h*- and *g*-indices have been obtained.
- It provides more granularity than the *h*- and *g*-indices. This is specially interesting when compared with the *h*-index. To increase the *h*-index is difficult (more when the *h*-index is high) and it is usual to find that many different researchers have the same *h*-index with a very different number of publications and cites. The *hg*-index provides a more fine-grained way to compare scientists.
- The *hg*-index is valued in the same scale as both *h*- and *g*-indices. Thus, the *hg*-index it is easy to understand and to compare with those existing indices.
- It takes into account the citations of the highly cited papers (the *h*-index is insensitive to highly cited papers) but it significantly reduces the impact of single very high cited papers (a drawback of the *g*-index), thus achieving a better balance between the impact of the majority of the best papers of the author and very highly cited ones.

The *hg*-index provides a more balanced view of the scientific productions than other indicators like *h*- and *g*-indices.

In addition to the indicators discussed here there are many other indicators frequently used by the scientific community to assess a paper, a journal or an author. They are generally called *h*-type indices and are modifications of *h*-index to minimise one or the other limitation of the *h*-index. Major examples are *H*(2) index, *W*-index, *A*-index, *R*-index, *E*-index, *M*-index, *AR*-index, *M*-quotient, Q^2-index, *K*-index, *HW*-index and *J*-index. A detailed discussion of all these indicators are available in the article "*A* Survey of Impact and Citation Indices: Limitations and Issues", by Majdi Maabreh and Izzat *M.* Alsmadi, in the *International Journal of Advanced Science and Technology* Vol. 40, March, 2012.

6.10 READING A SCIENTIFIC PAPER

Scientific research papers are the main source of information for a researcher. Research papers are available in different journals related to the particular subject area. A research paper may be relevant to a researcher in many ways; it can provide knowledge about the basic theories involved in his research area, it can supplement his knowledge in a particular area or it can provide new ideas for conducting research. Reading a research paper is not just like reading a textbook. Through the reading of a research paper one should try to answer the following questions.

- What problem(s) are they solving?
- Why are these problems important?
- What did they really do?
- What is the contribution of the work or what is interesting or new to the field?
- What methods are they using?
- Would you have solved the problem differently?
- Do all the pieces of their work fit together logically?
- What were the results?
- Did they do what they set out to do?

These questions are critical and answer to them is essential for understanding the paper. Inferences can be made based on the reader's experience and knowledge. Also noting down the points in a notebook is required for a thorough understanding. Reading research papers is partly a matter of experience and skill. The following steps can be followed while reading a research paper.

Skimming the article: As a first step skim the article by noting the headings, figures and structure of article. Majority of articles follow the basic structure that includes an abstract followed by an introduction, methods, results and discussion. Skimming the article usually takes a few minutes to provide an overview of the article.

Understanding the vocabulary: The second step is to read the paper line by line, highlighting every word and phrase that the reader does not understand. The meaning and definition of scientific vocabulary can be understood through textbooks and scientific dictionaries. An understanding of the meaning of words and phrases that are used to describe an experiment can be derived from the context in which they are used.

Comprehension of different sections: After understanding the meaning of scientific vocabulary and phrases, the paper is to be read once again. This time it is for a section-by-section comprehension of the whole paper.

- In the introduction section note the author's summary and comments on previous research, also the hypothesis of the paper and how it will be tested.

- In methods try to get a clear picture of what was done in each step. It is advised to make an outline of the procedures and instruments used.
- In results try to understand the figures and tables. It is a good practice to try an explanation for the figures by the reader.
- The discussion contains the conclusions drawn by the author from the data. Try to understand the interpretations given by the author. The author usually relates his findings with the other findings in the related field.

From each section note the questions that may arise while reading the paper.

Reflection and criticism: Now the reader can summarise the article and can answer some questions that were noted during the paper reading. Summarising the paper is one way to try to determine the scientific contribution of a paper. To assess the scientific merit, the reader must compare the paper to other works in the area. Are the ideas really novel, or have they appeared before? It should be noted that some papers offer new ideas; others implement ideas, and show how they work; others bring previous ideas together and unite them under a novel framework. Knowing other work in the area can help the reader to determine which sort of contribution a paper is actually making.

After reading and understanding a research paper one should write reading report. It has two essential parts:

1. A concise summary of the paper, providing an overview of what the authors actually did what methods they used, and what their results were.
2. A brief critique of the paper, giving a technical evaluation of the work, what things were unclear or not addressed, and the merits of the work.

The following guidelines can be followed while writing a reading report:

- In order to write a good reading report, the paper must be read critically.
- Write the report on a computer so that it can be edited easily and also for legibility.
- Use your understanding of the paper to write a cohesive summary rather than a *play-by-play* account of the paper.
- Be concise, but include some technical detail.
- Avoid unwanted phrases like *I would have liked to see . . . and I thought* . . .that are not really relevant in the report.
- The report should reveal that the reader understands of the key points of the paper.
- Do not simply copy choice phrases from the paper; this does not demonstrate that you have understood the paper, and it is not good writing.
- Write the report in such a way that it should be understood by others in the same field.

6.11 SEMINAR, CONFERENCE AND WORKSHOPS

Seminars, conferences and workshops provide a platform for the exchange of knowledge. Recently, such programmes are conducted very often at various levels like institutional, national and international in all disciplines of knowledge.

Seminars are educational events that feature one or more subject matter experts delivering information primarily via lecture and discussion. According to Jaques (1991) seminars are small group teaching and learning arrangements that use group interaction as a means of engaging participants. Although seminars usually begin with a presentation or mini-lecture to provide the basis for discussion, the word "seminar" also includes rather formal group discussions led by the teacher and focused on the content rather than on issues arising from participants. Usually, topics are selected for a seminar based on the relevance and interest of the participating community. The organising committee prepares the budget and identifies and ensures participation of eminent persons from the research fields related to seminar topic. The success of a seminar mainly depends on the quality of the resource persons. Therefore, the speaker should be someone at the top of their field or someone who possesses an in-depth knowledge of the particular area either academically or professionally. The following points should be considered while selecting the speakers.

- Expertise or strong research background
- Ability to convey knowledge to a large audience
- Unbiased and non-partisan. Speakers that are aligned with a particular group should not be invited.
- It is always advised to identify alternate speakers as a precaution.
- After the presentation by each resource person, time for an open discussion will be provided in a seminar. During this time the participants can interact with the speakers and share their idea or observations.

Workshops are teaching and learning arrangements, usually in small groups, that are structured to produce active participation in learning. Traditionally, workshops provide participants with the opportunity to practise skills and receive feedback. The need to conduct a workshop depends on the following factors [49].

Assessment of participating institutions or organizations: This is concerned with the needs of the institution or organization. Workshops are conducted in those areas in which the institution needs a support. The institution looks for gain from the workshop to support its functioning.

Assessment of workshop participants: This is concerned with the benefits of the participant of the workshop. It mainly depends upon the experience and expertise of the participants. The workshop must help to improve the abilities of the participants.

Resource persons: This consideration assesses the ability of the resource persons to lead a team. The resource person should be familiar with the content area of the workshop, preferably through both professional experience and educational background. He should have some prior training and facilitation experience.

Funding: This is an important factor. Main question to address is that what will be the approximate cost of the workshop and how and from where the fund can be obtained. The organization committee can find sponsors for the workshop who will provide necessary funding and sometimes resource persons also. Various government agencies also encourage the conduct of seminars and workshops in various disciplines by providing financial support.

Conference is an event which attracts the attention of a large number of scholars and scientists from various institutions and organizations. Conferences are usually conducted at national or international level. In a conference the organizing committee issues a "Call for Papers" and presenters are chosen from these abstract submissions. Conferences are platforms for presenting and exchanging innovative ideas and findings in the relevant field. Conferences often features keynote presentations delivered to all attendees, as well as multiple break-out sessions. Conference attendees have the opportunity to connect with internationally recognized scientists and other resource persons.

Presenting a paper or giving a talk in a seminar or a conference requires a serious preparation. The presenter should have a clear understanding about the content and area of the seminar or conference. The following points will help the presenter to plan and present a talk effectively.

- Have an understanding about the focus of the seminar/conference.
- Get an idea about the academic or professional level of participants.
- Have a glance on the topics being presented by other resource persons.
- Make sure that your topic is related to the contentment area covered in the seminar/conference.
- Try to connect your topics with other talks being present.
- Arrange the content of the talk orderly so that the talk can be developed in a sequence.
- Prepare a good power point presentation.
- Speak clearly and slowly so that all can understand you, especially if the language of the workshop is not the first language of the participants.
- If your presentation is being translated, spend some time with the interpreter prior to your presentation and give him/her your notes at least the day before, and speak even more slowly and clearly to facilitate more accurate translations.
- Speak loudly enough for everyone to hear.

- Participants are more comfortable and the presentations are more effective if the resource person is active, instead of sitting or standing still the entire time.
- Try to maintain your plan and timing as much as possible so as not to impose a time penalty on future presentations.
- Ask questions frequently including breaking every few minutes to ask participants if they have any questions to ask.
- Summarise the content of the talk at the end.

6.12 FEATURES OF A SCIENTIFIC CONFERENCE

A scientific conference is a gathering of scientists and other people interested in scientific activities. Scientists from various parts of the country and from other countries exchange their knowledge and their recent works to the delegates participating in the conference. The conference, as the name suggests is a formally structured meeting of several delegates at an arranged place and scheduled date, which is meant to 'confer' something about a subject, usually of common concern to all the persons present or attending it. The purpose of organizing a seminar is to impart knowledge about a specific aspect to the interested or concerned group of people. A conference is more about sharing opinions and thoughts on a topic of common interest. A conference brings those people together who have some shared interest in the topic which is going to be discussed. It is a form of participatory meeting, organized to discuss an issue, find out facts or solve a problem at hand. It usually continues for a few days, where deliberation or exchange of information takes place, on a specific subject or removal of differences takes place.

The following differences between the seminar and conference can be noted.

- Conference refers to a type of get-together conducted formally to convey certain information on a particular topic by way of presentations, speech, and deliberation. On the other hand, the seminar can be described as a long hour educational lecture provided to a particular group of people by one or more experts, for conveying information relating to a subject, or a message.
- Seminars are basically interactive sessions, wherein Socratic dialogue takes place between the expert and the attendees. Conversely, conferences are consultative in nature, arranged by an organization or committee, aiming to talk about or exchange their views on a specific matter.
- Seminars are held to inform, educate, counsel, guide, or train the participants, on a specific subject. As against, conferences are organized at the organizational, national or international level, which objects to present and discuss a certain matter with the attendees and get their opinion and views on that matter.

- In conferences, the participation of the attendees is somewhat limited. On the contrary, in seminars, the people attending it, take an active part in the discussion, by questioning the expert on their doubts, and the expert, in turn, provides a suitable answer to those questions to clear their doubts.
- Conferences are organized at the conference room of the organization, convention centre, or any other place such as hotel meeting space. In contrast, seminars are held at the institution itself, i.e., at the seminar hall if it is conducted on the institute level, however, if it is organized at a high level, a large space is required for sitting and so they are conducted in an auditorium or any other hotel space.
- When it comes to duration seminars last for a few hours only, whereas conferences continue for one or more days, wherein there are several breakout sessions.
- In conferences, the topic is presented by the speaker, who gives his views to resolve an issue or decide something. On the flip side, in seminars, a lecture is given by the guest speaker, who is an expert in the field to introduce the topic to the participants, after which further discussion takes place between the expert and the participants.

6.13 PRESENTING A RESEARCH WORK AT A CONFERENCE

Researchers usually consider a presentation at a conference as the easiest way to report their research findings. Both oral and poster presentations can be used in such conferences to report research findings. The advantage is that the researcher will get immediate feedback from the experts present in the conference and they can have an evaluation of their presentation as well as findings.

Usually, the organizers of the conference invite scientific papers or abstracts for possible presentations at the conference. Abstracts are a brief description of the research work in less than ten sentences. The screening committee of subject experts will verify the content of the submitted paper and abstracts and select the qualified papers for the presentation at the conference. If the number of participants is large, then the organizers may arrange a poster presentation also. Then, some very important submissions will be allowed to present orally at the conference and the remaining will be presented as posters. A separate time schedule is allowed for a poster presentation. The success of the oral and poster presentation depends on the confidence level and understanding of the presenter. The following points may be noted.

- Try to present the results in a concise manner at the conference.
- Do not attempt to present too much data and information in a short presentation.

- Convey only the main points and stimulate the interest of the audience so that they can seek out additional information from the speaker in the course of subsequent discussion.

The presentation should be such that it gains the attendance of the audience. Proper practice to deliver the talk is necessary for a beginner to present the paper without and any mistakes and within the allotted time. For oral presentation, one can use the PowerPoint presentation. The PowerPoint must be prepared by including all relevant points without explanation. The explanation of various points can be given at the time of presentation. Figures, charts, graphs, etc., can also be included in the PowerPoint presentation. The following points should be kept in mind while preparing for a presentation [2, p. 505-506].

- Before presenting the paper at a conference, it should be ensured that the findings are organized systematically and in an easy understanding way. The PowerPoint presentation is to be verified several times for any possible mistakes.
- After proper planning, a rehearsal can be done either in front of the research guide, research students, teachers, etc. Proper rehearsal exudes confidence, and during such rehearsals, time is checked, and incomprehensible slides are eliminated.
- Further, the presenter must know the audience before him. Considering the audience and their expectations will help you to choose the most suitable approach, the structure, the content, and the details of the presentation. If the audience have sufficient knowledge about the basic principles of your research work, then you can skip those slides . Otherwise, it will be boring for the audience.
- The presentation must be brief. It is not suited to include too many details in the presentation. Only the very important part of the findings and results can be included. The experimental details such as the working of certain instruments must be avoided. The sense of humour of the presenter will certainly help the audience to relax. Also, always try to deliver the speech fluently. One should not read the details by just looking at the presentation slides.
- Finally, the presenter must be conscious of the time allotted to him. The presentation must be finished within the allotted time. The long speech will create restlessness among the audience. This can be accomplished through proper planning and rehearsal.

6.14 VIDEOCONFERENCING PLATFORMS

Presently, most of the organizers prefer videoconferences. Online conferences and seminars have become very common. The outbreak of the pandemic situation of Covid-19 outspread made the organizers to conduct the seminars and workshops through online mode. There are many free and paid videoconferencing platforms that provide various features for easy presentation of the topic. Some of the most commonly used videoconferencing platforms are discussed below.

- Zoom
- Google Meet
- Microsoft Teams
- Dialpad Meetings
- TrueConf Online
- Skype
- FreeConference
- Facebook Live
- YouTube Live

Zoom: Zoom is a simple video conferencing application on Windows. Zoom allows for simple, easy setup meetings or regular meetings in your personal meeting room. Quickly create and share a private meeting link with an individual or team and connect remote team members through video. Zoom has the following characteristics such as free host up to 100 participants, 1-1 and group meetings, unlimited meetings, HD video and audio, screen sharing, scheduled meetings, private and group chat, host controls. The paid version permits the participation of more than a hundred participants.

Google Meet: Google Meet (formerly known as Hangouts Meet) is a video-communication service developed by Google. It is one of the most commonly used videoconferencing platforms. It is one of two apps that constitute the replacement for Google Hangouts, the other being Google Chat. Google meet is a simple way to connect with Google contactsand can be used for personal or business reasons. Use this tool for individual conversations, team meetings, and recorded demos. Google Meet is available for free and paid versions. The paid version allows the admission of more than a hundred participants.

Microsoft Teams: Microsoft Teams is a proprietary business communication platform developed by Microsoft. The main features of the Microsoft Teams include a chat function, online video and screen sharing, online meetings, and audio conferencing. Another important feature is document storage in SharePoint. Every team that uses Microsoft Teams will have a site in SharePoint Online, which will contain a default document library folder. All files shared across all conversations will automatically save to this folder. Permissions and security options can also be customized for sensitive information.

Dialpad Meetings: The free version of Dialpad Meetings is ideal for teams of 10 or less. The free version limits your video call duration to 45 minutes, so it's ideal for shorter meetings. Since it allows unlimited videoconferences, you can use this videoconferencing tool for recurring meetings, like 1-1s or monthly team debriefs.

TrueConf Online: TrueConf offers a free plan with up to three participants in video calls. This solution is great for small teams or individuals who need a simple tool to host video calls or collaborate with small groups. The collaboration features like remote desktop control are helpful for those in support who may need to help a client solve a problem.

Skype: The free version of Skype works well for small teams with less than 10 total members. It is a good tool if you need a simple way to video chat from your computer, phone, or tablet and tends to have smaller group meetings or 1-1 conversations. Skype also offers a tool called Skype in the Classroom specifically designed for online education.

FreeConference: FreeConference is ideal for smaller teams or meetings with five or fewer participants. One major advantage of this videoconferencing tool is its ease of use. No downloads are required to start using it and moderator controls allow the individual who's running the videoconference to keep the meeting on track.

Facebook Live: Facebook Live is a great fit for businesses, influencers, or individuals who are looking to broadcast demos, videos, or showcase their company's culture while streaming live. Your followers on Facebook can comment and chat live, one can select a custom audience for the video, and you can schedule videos ahead of time to gain excitement.

YouTube Live: YouTube Liveis a platform for demonstrating a product with live questions and action or interactions between the audience and the resource persons. Seminars and conferences can be live-streamed through YouTube live. Since there is no restriction to the number of participants, any person interested can attend seminars and conferences.

REVIEW QUESTIONS

1. What is a research proposal?
2. Discuss the various sources of research ideas.
3. What are the purposes of a research proposal?
4. Explain the various considerations in detail while preparing a research proposal.
5. Mention the errors that must be avoided while preparing a research proposal.
6. What is the classifications of skills involved in the research proposal?
7. Give a detailed account of the format for the submission of a research proposal.
8. Explain the various criteria based on which a research proposal is evaluated.

9. Give an account of some funding agencies in India that provides research support.
10. Explain the purpose of the literature review.
11. Explain the primary, secondary, and ternary sources of literature.
12. Give a detailed account of various sources used in the literature survey.
13. Why the literature review is considered an important component of research activity?
14. What are the 5C's in preparing a literature review?
15. Explain the importance of planning in the literature review.
16. What are the steps to be followed in locating the relevant literature?
17. What are the details that must be included in a bibliographic record of literature?
18. What is the importance of note-making?
19. Explain the proper management of information collected in the literature review.
20. Give the details of some academic search engines.
21. What are the components and purpose of a literature review from the point of view of a researcher?
22. Explain the three main components of a literature review.
23. Briefly explain some of the frameworks for preparing the literature review.
24. What is a journal? Explain the three main categories of journals.
25. What are the major functions of a journal?
26. Draw the flowchart showing the various processes involved in a journal's publication.
27. What are the points that should be kept in mind before selecting a journal for the possible publication of an article?
28. What are the features of short communication?
29. What is rapid communication? What are the criteria for publishing an article under rapid communication?
30. What is the structure of a research paper?
31. What is a review paper and what are its features?
32. What are systematic and non-systematic reviews?
33. What are the purposes of a review paper?
34. What is an interactive journal?

35. What are the stages involved in the publication of an article in an interactive journal?
36. What are the advantages of an interactive journal?
37. Write a note on conference proceedings.
38. Define and illustrate the journal impact factor.
39. What are the advantages of the journal impact factor?
40. How can we assess the quality of a journal through the journal impact factor?
41. Why scientific community uses the journal impact factor to assess the quality of a journal?
42. What is a citation index? How it is calculated?
43. What is the importance of the citation index?
44. Give examples for some tools and database for citation index.
45. What is H-index and how it is calculated?
46. What are the advantages of H-index?
47. Mention some disadvantages of H-index.
48. What is G-index and how it is calculated?
49. What is H-G index and how it is calculated?
50. Give some benefits of H-G index.
51. What is the importance of reading a scientific paper?
52. Explain the various steps of reading a scientific paper.
53. What are the essential components of a reading report?
54. What are the guidelines that must be followed while writing/reading a report?
55. What is a seminar and what are its advantages?
56. What is a workshop? How it is beneficial for researchers?
57. What are the differences between a seminar and a workshop?
58. What is a conference? Why it is important?
59. On what factors does the need to conduct a workshop depends?
60. What are the considerations that must be followed by a presenter while presenting a paper in a seminar?
61. Explain the features of some videoconferencing platforms.
62. What are the characteristics of a good presentation?

REFERENCES

[1] Krathwohl, David R. *How to Prepare a Dissertation Proposal: Suggestions for Students in Education and the Social and Behavioral Sciences*". Syracuse, NY: Syracuse University Press, 2005

[2] C. George Thomas, *Research Methodology and Scientific Writing*, 2021, Springer.

[3] https://libguides.usc.edu/writingguide/researchproposal

[4] https://www.monash.edu/rlo/graduate-research-writing/write-the-thesis/writing-a-research-proposal

[5] Calnan, J. 1976. *One Way to do Research: The A-Z for Those Who Mus*t. Heinemann Medical Books, London, p. 250

[6] https://www.nitt.edu/home/icsr/funding_agencies.pdf

[7] *"Literature Review-Academic Tip Sheet"*, Edith Cowan University, Australia.

[8] Ibid.

[9] *"How to Write a Literature Review"*, www.unc.edu/depts/wcweb/

[10] *"Writing a Literature Review"*, The Flinders University of South Australia, http://www.flinders.edu.au/SLC

[11] Barrass, R. 2002. "*Scientists Must Write: A Guide to Better Writing for Scientists, Engineers and Students"* (2nd Ed.). Routledge, London, p. 204.

[12] *"Literature Review"*, St. Kate's Libraries Guides, http://library.stkate.edu

[13] *"Review of literature"*, 2006, UW-Madison Writing Centre, http://www.wisc.edu/writing/Handbook/ReviewofLiterature.html

[14] *"An overview of scientific and scholarly journal publishing"*, The STM Report, 2009, International Association of Scientific, Technical and Medical Publishers, Oxford, p. 12.

[15] *"Selecting a Journal"*, Scientific, Medical and General Proofreading and Editing, www.sfedit.net

[16] Margaret Cargill and Patrick O'Connor, 2009, *"Writing Scientific Research Articles"*, Wiley Blackwell, pp. 69-70.

[17] Dobri Atanassov Batovski, *"How to Write a Review Article"*.

[18] Ibid.

[19] Mary Westervelt, *"Review Papers: Organization and Content"*, School of Engineering, University of Pennsylvania.

[20] Ulrich Pöschl, 2004, *"Interactive journal concept for improved scientific publishing and quality assurance"*, Learned Publishing (2004) 17, 105–113.

[21] Ibid.

[22] Ibid.

[23] *"Journal Inpact Factor 2012"*, Global Institute of Scientific Information, www.jifactor.com

[24] Bruse A Thyer, *"Preparing Research Articles"*, 2008, Oxford University Press, p. 23

[25] Masood Fooladi et al., 2013, *"Does Criticisms Overcome the Praises of Journal Impact Factor?"*, Asian Social Science; Vol. 9, No. 5.

[26] *"The H-Index and Journal Impact Factors"*, The University of Auckland Library, http://www.library.auckland.ac.nz/

[27] Hoeffel C, 1998, *"Journal impact factors"*, Allergy, 53:1225.

[28] Margaret Cargill and Patrick O'Connor, p.70.

[29] B M Gupta, A K Jha and P K Mishra, 2004, *"Citation Indices and Important Products of ISI"*, Analysis of Library and Information Studies, 51,1: 1-10

[30] Ibid.

[31] Garfield A, 1963, *"Science Citation Index"*, Science Citation Index 1961, 1, p. v-xvi.

[32] *"Using Citation Indices"*, University of Texas Libraries, http://www.lib.utexas.edu/chem/info/citations.html

[33] Swales, JM & Feak, 1994, *"Academic writing for graduate students"*, University of Michigan Press, Ann Arbor.

[34] Shailendra Kumar and Manisha Singh, 2011, "Structure and Functions of Citation Index", 8th International CALIBER-2011, Goa University, Goa.

[35] Hirsch, J. E, 2000, *" An index to quantify an individual's scientific research output"*, Proceedings of the National Academy of Sciences, 102, 16569–16572.

[36] S. Alonso, F. J. Cabrerizo, 2010, E. Herrera-Viedma and F. Herrera, *"hg-index: a new index to characterize the scientific output of researchers based on the h- and g-indices"*, Scientometrics, 82:391–400

[37] Ronald Rousseau, *"Reflections on recent developments of the h-index and h-type indices"*, Collnet journal of scientometrics and information management 2(1) June 2008.

[38] "The *H*-Index and Journal Impact Factors", The University of Auckland Library, http://www.library.auckland.ac.nz/

[39] Egghe, L., *"Theory and practice of the G-index"*. Scientometrics, Vol. 69, No. 1, (2006), pp. 131–152.

[40] Ronald Rousseau.

[41] Martin, B. R, 1996, "The use of multiple indicators in the assessment of basic research", Scientometrics, 36(3), 343–362.

[42] Van Raan, A. F. J, 2006, *"Comparisons of the Hirsch-index with standard bibliometric indicators and with peer judgment for 147 chemistry research groups"*, Scientometrics, 67(3), 491–502.

[43] Egghe, L.

[44] Alonso. S et al.

[45] Ibid.

[46] Rodrigo costas, María bordons, *"Is g-index better than h-index?: An exploratory study at the individual level"*, Scientometrics, Vol. 77, No. 2 (2008) 267–288.

[47] Alonso. S. et al.

[48] Ibid.

[49] *"How to Organize and Run Learning Workshops"*, UNICEF and United Nations Staff College.

Suggested Readings

1. Bruse A Thyer, "Preparing Research Articles", 2008, Oxford University Press.
2. Margaret Cargill and Patrick O'Connor, 2009, "Writing Scientific Research Articles", Wiley Blackwell.
3. Robert Barrass, 2002,"Scientists must Write", Routledge.
4. Harvey Gould, Jan Tobochnik, "An Introduction to Computer Simulation Methods Applications to Physical Systems", 1995.

Chapter 7

Scientific Writing

The opposite of a correct statement is a false statement. But the opposite of a profound truth may well be another profound truth.

— **Niels Bohr**

It is most important to have a beautiful theory. And if the observations don't support it, don't be too distressed, but wait a bit and see if some error in the observations doesn't show up.

— **Paul M Dirac**

7.1 SCIENTIFIC PAPER

The aim of scientific research is generation of knowledge. A research work that took many years may go waste unless its result is not communicated to the scientific community. Hence, writing and publishing scientific papers is considered an integral part of researcher's professional life. Scientists should be exposed to the new discoveries so that they get novel ideas and can work over it. Moreover by writing a research paper, a researcher will develop an in-depth knowledge in a particular area. Realizing the importance of writing and publishing the scientific discoveries, in the preface of the book "*How to Write and Publish a Scientific Paper*" Robert Day (1983) wrote that "Good Scientific writing is not a matter of life and death; it is much more serious than that".

Robert A Day defined a scientific paper as a written and published report representing original research report [1]. According to O' Conner and Woodford, (1976) "a scientific or research article or paper is a technical document that describes a significant experimental, theoretical or observational extension of current knowledge, or advances in the practical application of known principles" [2].

A scientific paper is the most common way of communicating the results of a scientific research to the scientific community. Writing a scientific paper requires an expertise as well as experience because for effective communication of the knowledge the report must be written in a technical language and should communicate what it is intended to communicate. Moreover it should be published in a peer reviewed journal in the respective field so that it gains the attention of the scientific community easily. The Council of Biology Editors

(CBE), suggested that "an acceptable primary scientific publication must be the first disclosure containing sufficient information to enable peers to assess observations, to repeat experiments, and to evaluate intellectual processes; moreover, it must be susceptible to sensory perception, essentially permanent, available to the scientific community without restriction, and available for recognized secondary services….." (Cited in Robert A Day p. 9).

The result of a scientific research programme should be published for various reasons. As O'Connor (1995) observed it is essential for the progress of the science. It is essential for the ideal and legal protection of the intellectual property [3] and equally important for the gain of reputation by getting more citations of the published paper.

A scientific paper consists of well-organized components. The most common components are introduction, methods, results and discussion (IMRaD). However a paper can be divided into the following seven parts [4].

1. The Title and the Abstract that encapsulate the paper.
2. The Introduction which describes where the paper's research question fits into current science.
3. The Materials and Methods translate the research question into a detailed recipe of operations.
4. The Results which is an orderly compilation of the data observed after following the research recipe.
5. The Discussion consolidates the data and connects it to the data of other researchers.
6. The Conclusion gives the one or two scientific points to which the entire paper leads.
7. The References which lists the many links that tie the paper to the rest of science.

This format can be followed by experimental areas of science; but cannot be strictly followed by descriptive areas of science. Various organizations are followed in such areas. Usually, a journal issues instructions to the authors where detailed descriptions of the format of the paper, method of submission, etc., are made clear.

7.1.1 Title

Title of a research article has an important role in gaining the attention of the readers, because researchers usually search for the literature and find relevant papers based on the title of the papers. Title reflects the theme of the research paper or in other words what a researcher want to communicate to the scientific community. Title should be of fewest words that should accurately describe the content of the paper. The title should concisely state the topic of the paper and the variables or theoretical issues that are being explored in relation to that topic. Very often the decision of the review committee whether the paper is going to be accepted for publication or not, depends on the title itself.

Robert A Day defined a good title "as the fewest possible words that adequately describe the contents of the paper".

The title should not be too long that contains unwanted words like '*investigations on…….*". However if the title is too short it may not convey the content of the paper effectively. According to Peat J et al. (2002) the following are the characteristics of an effective title [5].

- It identifies the main issue of the paper,
- It begins with the subject of the paper,
- It should be accurate, unambiguous, specific, and complete,
- Title should not contain abbreviations.
- It should attract readers.

Titles of research papers can be classified as (i) indicative and (ii) informative titles. Indicative titles reveal the area of investigation, and not answers the paper might offer, whereas informative titles convey messages of the paper on all its relevant elements [6].

Many journals limit length of the title to 10 to 12 words (about 100 characters). The titles should be written with distinct letters, not underlined and without a full stop at the end. Avoid subtitles and supertitles. 'Titles should never contain abbreviations, chemical formulas, trade names and jargon' [7].

Not much guidelines are available regarding presentation of the title of a research paper. Perry et al. (2002) offered the following recommendations regarding the title [8].

- A title should attract reader's attention.
- Journal editors prefer formal titles that are not too clever or cute although it grabs the attention.
- The title should clearly reflect the main theme, issue or position discussed in the article. Because it creates the expectation about the contents of the article, the title should accurately reflect the nature and focus of the study and not create false expectations (Feldman, 2004).
- The title should be as specific as possible given the restrictions on length.
- Some of the key words listed after the Abstract should appear in the title.
- A title preferably answers the following questions.
  - ♦ What will be researched?
  - ♦ How will the topic be researched?

- With whom? (Describes the research population and units of measurement).
- Where/in what context the study will be conducted?

In addition to these the title should be in a form suitable for computerised indexing systems. According to Robert A Day (p. 18) the title is a label of the paper, not a sentence, but the order of the words very important. The considerations mentioned above are important for all users including those who access the paper through secondary sources.

To make the title more effective Jean Luc Lebrun (2007) proposed the following six techniques [9].

1. *Placement of contribution upfront in a title:* In a full sentence (containing a verb), new information usually appears at the end (stress position) and old information at the beginning (topic position).
2. *Addition of verbal forms:* The gerundive and infinitive verbal forms add energy to a title.
3. *Adjectives and numbers to describe the strong point of a contribution:* Besides specific keywords, adjectives and adverbs are often used to describe the key aspect of a contribution. Adjectives like fast, highly efficient and robust can be used.
4. *Clear and specific keywords:* The specificity of a paper is proportional to the number of specific keywords in its title.
5. *Smart choice of keyword coverage:* Readers find new articles through online keyword searches. Therefore choosing effective keywords is vital.
6. *Catchy acronyms and titles:* Acronyms provide a shortcut to help other writers refer to your work succinctly.

7.1.2 Abstract

Abstract is a short summary of a research article. An abstract helps the reader to understand the essentials of the research paper in a few words; typically 100 to 200 words. After reading the abstract one can decide whether to read the complete the paper or not. Abstract highlights the major points and describes the content and scope of writing. 'It needs to be dense with information but also readable, well-organized, brief, and self-contained'[10]. Houghton (1975) defined an abstract as a summary of information in a document.

The following points should be kept in mind while writing an abstract.

- *Accuracy:* An abstract should include only the information presented in the original document.
- *Brevity:* A good abstract gets straight to the point, contains precise language, and does not include superfluous adjectives.
- *Clarity:* A good abstract does not contain jargon or colloquialisms and always explains any acronyms.

The importance of abstract can be summarised as follows:

- It helps the readers decide if they should read the entire article.
- It helps the readers and researchers remember key findings on a topic.
- It helps the readers to understand the text by outlining key points prior to reading the full document.
- It indexes articles for quick recovery and cross-referencing.

Abstracts are generally classified into two:

(i) informative abstract and

(ii) indicative abstract.

Informative abstract is the most common type of abstract. It is also known as the structured abstract. An abstract written on an experiment, investigation or a survey is called an *informative abstract.* The four components of an informative abstract are (i) objective, (ii) methodology, (iii) results and, (iv) conclusion.

1. *Objective section:* The objective section of the abstract states the primary objectives of the experiment or investigation. The purpose section of an informative abstract might also contain the hypothesis of the experiment.
2. *Methodology section:* The methodology section of an informative abstract should describe the techniques used in conducting the experiment. This section should give only as much detail as is necessary to understand the experiment; the abstract should not focus entirely on research methods.
3. *Result section:* The results section of an informative abstract should relate the observations and data collected during the experiment. This section should be concise and informative, and only the most important results need be included.
4. *Conclusion section:* The conclusion section of an informative abstract should state the evaluation or analysis of the experiment results. It should also briefly state the implications of these results. This conclusion section might also state whether the driving hypothesis of the experiment was correct.

Informative or structured abstracts have got the following advantages [11]:

- They contain the most significant data of the paper.
- They are more explicit and it is easier to understand the text written in shorter paragraphs, repeating the format of the whole paper.
- Structuring helps authors not to omit relevant data.
- The structured abstract is intended to make it easier for the reader to select important details and to assess the value and applicability of the study.
- Structured abstracts remind readers about the whole paper, and reveal errors regarding used methodology.
- Structured abstracts in English are convenient to non-English authors.

Indicative abstracts are also called *descriptive abstracts*. An abstract written for a less-structured document like an essay, editorial, or a book, is an indicative abstract. An indicative abstract deals with the content of the paper. The main components of an indicative abstract are (i) scope, (ii) arguments used and (iii) conclusion.

- *Scope:* The scope section of an indicative abstract should state the range of the material dealt within the original document as well as the starting premise of the document.
- *Arguments:* The arguments used section of an indicative abstract should state the main arguments and counterarguments used in the original document. These arguments should be stated in the same progression in which they appear in the document.
- *Conclusion:* The conclusion section of an indicative abstract should state the document's main closing argument and its implications as suggested by the original author. This section might also state plot resolution when the original document is a work of fiction.

Abstract is the most read part of a research paper. A well prepared and well organized abstract enables the reader to identify the basic content of the paper quickly and accurately and determine its relevance to the present topic of research. Since the abstract attracts the primary attention of a reader it should be prepared carefully. The following guidelines will be helpful in preparing an effective abstract [12].

Write the abstract last: An abstract should be written as the final stage of a complete paper. Otherwise it will tend to be vague or incomplete.

Follow any guidelines you have been given: There are no strict rules or exact format for writing an abstract. However, if the abstract is being submitted to a conference, journal, grant agency, or is part of a thesis or dissertation, the organization or department may issue guidelines for abstracts. These rules are to be followed while preparing the abstract.

Be accurate: The term accuracy is used in a broad context. Make sure that it uses the same language as the paper, especially key words and concepts. Abstract should include only information that actually appears in the paper. Make sure that the abstract correctly reflects the purpose and content of your paper. It is to be ensured that the abstract reports what it is in the body of the paper but doesn't comment on it or make claims.

Be self-contained: The reader should understand what the researcher wants to say. Therefore define all acronyms and abbreviations, except standard units of measurement. Also define all unique terms.

Be clear, concise and specific: Make each sentence as informative as possible. Only the most important concepts, findings, or implications should be included in the abstract.

Emphasize points in proportion to the emphasis they receive in the paper: For a research paper the amount of space devoted to results should reflect their importance and level of complexity. For a research proposal devote fairly equal sections to background, literature review, and the proposed method in the abstract.

Use of correct language and tense: Usage of adequate language and words are necessary for preparing an effective abstract. The following points should be remembered.

- Use present tense for the topic, problem and question.
- Use past tense to describe your method, results and analysis.
- Use a cautious present tense for implications and recommendations.
- Use simple future tense in a proposal.

Select key indexing terms: Choose keywords and phrases that will make the research paper readily and accurately searchable in databases and include these words or phrases in the abstract.

7.1.3 Keywords

Keywords are included in a research paper after the abstract. They describe the concepts of the whole paper. 'Keywords should reflect the discipline, sub-discipline, theme, research design and context of the study' [13]. The number of keywords in a paper can vary from 5 to 10 and should be typed in sentence case and in italics.

Easy and fast access to precise information is possible using keywords assigned to each document. 'They are of great help to those creating databases, to indexers in cross-indexing scientific articles and to users providing easy access to scientific sources' [14]. Indexing is necessary due to information explosion, huge quantities of knowledge and need for its methodical organization.

Including keywords in a research paper is very important because using the Internet, abstract databases, full text databases and electronic publications it is possible to search literature using any word from the text. Such searches provide a huge quantity of information, but only a small number is relevant. Therefore, the selection of right information from such data bases heavily depends upon the use of correct keywords. If adequate keywords that reflect the theme of the paper are not used in a research paper, then that paper may not be included in the search result even though a reader may highly be benefited by that paper. He will not get the opportunity to read that paper.

Keywords cannot be picked up at random. According to Ann M Korner (2004), "your choice of keywords should reflect the most important aspects of your paper. You should choose them in such a way that people interested in your field will find a reference to your paper when they refer to certain words in the journal's annual index" [15]. Therefore important words from the title as well as from the content of the paper can be used as keywords.

7.1.4 Introduction

The introduction of a scientific paper leads the reader from a general subject area to a particular field of research. Perry et al. (2003) described introduction as "an executive summary that gives the reader an enticing glimpse of what is to come". Introduction to a research article should be written in such a way that it grabs the attention of the readers and should stimulate curiosity, interest for further reading of the article. Many of the authors opinioned it is the most difficult part of an article to write.

The introduction is sometimes called the background and is usually written in two or three paragraphs. The first paragraph is to provide background information where some important references are mentioned. The statement of the research problem is also made in the first paragraph. The second paragraph usually elaborates the importance of the problem and lists unresolved issues. The final paragraph describes the rationale for the current study and should contain the research question and the hypothesis [16]. The introduction should not be too long; but can include more information if the writer wants to inform the reader with sufficient background knowledge. Therefore; when writing the introduction the following information should be included for making the introduction effective [17, 18].

- Historical background.
- Research question and the working hypothesis. Mention about three to five most important references.
- The relevance of the problem or issue. A few recent references to demonstrate that research is active on the subject can be included here.
- Methodology, instrumentation, materials and analytical tools.
- Inferences from the results of the study.

According to Perry et al. (2003) and Summers (2001) an introduction should include the following six elements [19, 20].

1. The writer has to state the broad theme or the topic of the study.
2. The academic and practical importance of the theme has to be explained.
3. The author summarises the available literature and cites the most important studies that are relevant to the current research. If an existing study is replicated it should be clearly stated here.
4. The author indicates the most important gaps and inconsistencies in the literature that the current study will address.

5. The introduction must give a clear indication of the core research problem, the specific research objectives, the context in which the study will be conducted and the units of analysis of the study.
6. In the introduction the writer has to provide an outline of the structure of the rest of the article.

Ann M Korner (2004) identified the following purposes of introduction [21]:

- to place your research in the context of earlier relevant work by others;
- to explain your reasons for performing your study;
- to mention the methods that you used in your study; and
- to provide an indication of the conclusions that you will draw from your results in the Discussion at the end of your paper.

As mentioned above the introduction introduces the research question, summarise and cite the research done to date, and identify a question that has not yet been answered; which is the research problem. At the end of the introduction, the hypotheses that the researcher tested is stated. The introduction is to be written with care. It should not be too long or too short. It should fulfill the purpose for which it is written. Some guidelines are available for writing the introduction of a research paper. Psychology Writing Center, University of Washington suggests the following steps/guidelines for writing an introduction to a research paper [22].

- Find a review article in the related field. This will provide background knowledge for the researcher. Also look for later studies by the authors cited in the review article and find other studies that cite the authors cited in the review article.
- Make an outline that shows the progression of research that has led to the formation of the hypotheses.
- For each main point; start by citing non-controversial assumptions of findings. Then discuss areas in which conflicting results, if any, have emerged. Try to explain the source of the disagreement.
- At the end of the introduction, identify questions that have not been addressed that led to your hypotheses. If there are more than one or two hypotheses, list them.
- Avoid plagiarism, give credit where credit is due. Whenever you cite someone else's ideas or use their language, give the name of the author and the year of publication.
- In scientific writing, it is much more common to paraphrase an author's ideas than to use direct quotes. If you use direct quotes, however, also cite the page number.
- Use specific language and support your arguments with concrete examples.

- Don't hesitate to evaluate and critique what you have read. Evaluation requires more work and entails more risk, but without it, your paper lacks original synthesis, which falls short of the goal of the paper: to make an original contribution to a research area.

These general instructions can be followed by researchers in all areas of research. However, there may be variations depending upon the discipline and also on the journal. Start the introduction with sentences that are adequately general, and simple enough to understand even for those who are not experts in the same topic. This is then proceeded onto narrower details. This is often referred to as *logical funnel*, which is the most practical way of building up an effective introduction. While writing the introduction use present tense except for what the researcher did or found; the findings should be presented in the past tense.

7.1.5 Materials and Methods

Never direct your readers to look in the methods section for something that is critical to understanding the paper; present everything the reader needs to understand your claims directly in the results section [23].

—Kevin W. Plaxco

In a research paper the introduction is followed by the Method section where the researcher explains the methods adopted and experiments performed to prove the validity of the hypotheses under test. The purpose of Method section can be identified as 'it provides the information needed for another competent scientist to repeat the work' [24].This is important because the replicability is considered to be the distinguishing characteristic of scientific methods. The method section provides an evidence for the answer to the research question. It should be organized under meaningful subheadings. The number of subheadings depends upon the type of paper. The experimental techniques used should be described in sufficient detail so that others can replicate the study. It is to be noted that for published and well known methods brief description and references should be given. Any novel method should be described in full [25].

The Method section should help the readers to evaluate the results. Therefore the Method and Result sections should be related. M. Cargill and P. O'Connor (2009) pointed out the following two strategies for establishing a relation between Method and Result sections [26].

1. Use identical or similar subheadings in the Methods and the Results sections.
2. Use introductory phrases or sentences in the Methods that relate to the aims.

A *structured method* section usually has the following components: subjects, treatment protocol, equipment, study procedures and statistical methods [27].

However, there may be variations depending upon the discipline, study methods and journal types, etc.

Subjects: This part describes the subjects under study, i.e., animals, patients, plants, chemical compound, etc. If there is a control group it should be described here. Describe how the subject is selected by explaining the selection criteria. For example, if a researcher wants to study the effect of annealing on a particular thin film, he should select a control group (without annealing) and experimental group (subjected for annealing); both deposited under identical conditions.

Treatment protocol: If the study involved human subjects and animals, then a statement of approval from the institutional review board must be included, and should describe how informed consent from the study subjects was obtained. Indicate who signed the consent form; the subjects or their legal representatives (for human subjects).

Equipment: A brief description of the equipment used for the study is given here. Also explain how the equipment is calibrated. If the study is an equipment evaluation then a detailed description of the devices is to be included.

Study procedure: This section is important for descriptive studies as well as experimental studies. This section describes what procedures were followed and what data were collected. The procedure or methods falls under four groups [28].

1. Methods with which everyone in your field and related fields is familiar.
2. Methods that are in less common use but have been well documented elsewhere.
3. Methods that are relatively uncommon or that require specification of experimental conditions for each application.
4. Novel methods that you developed for the research, describe in your paper.

Methods in the first group need not be mentioned; but methods in the second group are to be described briefly with appropriate references. Methods in the third group should be described in sufficient detail. This is for someone who wants to repeat the experiment by referring to the original description of the method and using the specific conditions that the author used. Methods in the fourth group should be described in detail, with all reagents, conditions, and equipment carefully specified. Since the researcher describes a novel method, he should also indicate when and how special care should be taken to ensure success.

Data analysis: The data analysis section should describe how data were handled, what statistical tests were done. The relevance of the statistical test used should also be explained. References can be cited for support.

The following guidelines will be useful in writing the Material and Method part of a scientific research paper.

- Give a brief account of the materials used including their source.
- Describe the specific experimental method only.
- Need not include the equipment list.
- The method section should be written in a paragraph form not as a series of steps.
- Reasons for certain measurements and the use of certain equations should be included.
- Do not write the result in this section. It should be written in the Result section only.
- Use sub-sections if required.
- Include visuals that are labelled and referenced in the text. Visuals should be large enough to read the units.
- Decimal quantities include a 0 before the decimal point. (For example; 0.0028).
- Mathematical expressions must be in the prescribed format and equations must be spaced apart from the text.

While preparing the Method section of a research paper for publishing in a journal, it should be in the particular format suggested by the journal. Different journals accept different formats. For example, the heading "Materials and Methods" is used in the same format in some journals whereas in some other journals the format "Materials and methods" is accepted. Again some journals prefer the central alignment of the headings and some others prefer left alignment. Further when we use chemicals for our study give the full name of each item and if the name is complicated appropriate abbreviations can be introduced. For each item the name and location of the manufacturer is to be specified. When using units for measurements SI system is preferred by many of the journals. The writing should be direct, precise and should use past tense. Description of unimportant details and compound sentence structures must be avoided.

7.1.6 Results and Discussion

'The Result section leads the reader along a path that follows the steps in your reasoning and the sequence of experiments from the working to a conclusion that corresponds exactly to the title of your paper' [29]. The results of a scientific research are presented in this section. The result section should state the findings without bias and interpretation. It should follow the same sequence as the Method section. Some journals prefer the results and discussion section separate whereas some others combine these two under the heading *Results and Discussion.*

One important point to remember is that a researcher should publish only those results that are reproducible. He should ensure the reproducibility of the result by performing the experiment a several times. Interestingly, there is a place for irreproducible results in a journal named, *The Journal of Irreproducible Results*, (http://www.jir.com/home.html).

'The results of a study do not prove anything' [30]. Research results can only confirm or reject a hypothesis. 'Each individual study adds to the collective understanding of the problem and adds evidence to support or refute a given interpretation' [31]. The result should directly relate to the title of the paper. Negative results, if any may be mentioned with the phrase *results not shown.* References to any unpublished results should be indicated clearly. Since the results of the study are presented here, it is common to include a large number of data and information. However, it is advised to use graphs and tables whenever large volumes of data are to be presented. Moreover the information represented in a graph or a table should not be included in the paragraph. The results are to be organized under subheadings and the format of the subheadings should be exactly as specified by the Instructions to Authors of the target journal. However, if the paper is a short communication or a letter to the editor, such subheadings are not possible.

In the case of journals that allow a single section Results and Discussion, the researcher can discuss the importance of each result after presenting the relevant data.

Psychology Writing Center, University of Washington, suggested the following to remember when preparing the result section [32].

- Keep your hypotheses in mind while you write. Each result must refer to a stated hypothesis.
- Describe all results that are directly related to your research questions or hypotheses. Start with hypotheses you were able to support with significant statistics before reporting non-significant trends. Then describe any additional results that are more indirectly relevant to your questions.
- If you present many results (i.e., many variables or variables with many levels), write a brief summary, then discuss each variable in separate subsections.
- Report main effects before reporting contrasts or interactions. Briefly mention problems such as reasons for missing data, but save discussion of the problems for the discussion section.
- Use tables and figures to summarise data. Include descriptive statistics (such as means and standard deviations or standard errors), and give significance levels of any inferential statistics.

An equally important thing is the usage of correct tense in writing the result section. Use past tense (either active or passive voice) when the sentence focuses on the completed study: what was done and found. Present tense is used to describe an *always true* situation or to represent a document which is always there. Model verbs; may could, etc. can be used in comments.

The discussion of the results is aimed at relating them to the hypotheses under test. In this section the researcher interprets the results against a background of existing knowledge. The researcher explains the implications of the research and makes some suggestions for future research. The discussion should not be elaborated much; 'it should be as short as possible but at the same time it should clearly and fully state, support, explain, and defend the answers and discuss other important and directly relevant issues' [33]. 'The discussion section should be constructed like a (normal, upright) pyramid, stating the most specific research conclusions at the top before broadening out to encompass wider and wider ideas' [34].

According to Dean R Hess (2004) and M Cargill and Patrick O'Conner (2009) the following elements should be included in the discussion [35, 36].

- A reference to the main purpose or hypothesis of the study, or a summary of the main activity of the study.
- State the major findings of the study, generally in order of their significance, including;
  - ♦ whether they support the original hypothesis, or how they contribute to the main activity of the study, to answering the research questions, or to meeting the research objectives; and
  - ♦ whether they agree with the findings of other researchers.
- Explain the meaning of the findings and why the findings are important.
- Relate the findings to those of similar studies consider alternative explanations of the findings.
- State the relevance of the findings.
- Acknowledge the study's limitations. Limitations of the study that restrict the extent to which the findings can be generalized beyond the study conditions.
- Make suggestions for further research.

Apart from these any unexpected finding can also be discussed. When discussing an unexpected finding, begin a new paragraph with the finding and then describe it.

Dean R Hess further pointed out that the following things should be avoided while writing the Discussion [35].

- Overpresentation of the results: The interpretation should not go beyond what is supported by data.
- Unwarranted speculation: The discussion should be focused on the data or device of the study.
- Inflation of the importance of the findings: Should not attribute unwanted importance to study findings.

- Tangential issues: The author should remain focused on the hypothesis and study results. Injecting tangential issues into the discussion section distracts and confuses the reader.
- The "bully pulpit": Do not use the discussion section to criticize other studies.
- Conclusions that are not supported by the data: Avoid the temptation to allow your biases to enter into the conclusions.
- Inclusion of the "take-home message": Give the "*take-home message*" in the form of a conclusion. (What does the researcher want the reader to remember from the study is called the take-home-message).

It should be ensured that the discussion is connected with the issues raised in the introduction section. After preparing a draft discussion make a visit to introduction and make a check for a close fit. Any deviations noted should be rectified properly. A statement on the significance of the result should be included at the end of the discussion.

7.1.7 Conclusion

The Result and Discussion should be followed by a Conclusion. It should be a paragraph or two in length. It summarises the idea discussed in the paper. According to Byne (1998) the body of a good paper is a "*thunderbolt in reverse*": it begins with thunder (Introduction) and ends with lightning (Conclusion). 'Conclusion should be linked with the goals of the study, and should be limited to the boundaries of the study' [37]. Conclusion that is not supported by data should not be included. Authors should note that the negative findings could be as important as positive findings and therefore it should be included both in the Discussion and Conclusion. Conclusion should not present a totally new result which was not included in the Result section.

Many readers have the habit of jumping from abstract to the conclusion skipping the remaining section of the paper. They may find same sentences repeated in both sections. In order to retain the attention or interest of the reader the author has to differentiate the Conclusion from the Abstract. Conclusion is different from Abstract through the following points [38].

- The tense used in the abstract is present tense whereas the tense used in the conclusion is past tense.
- Conclusion has to be more detailed than the abstract.
- The Abstract briefly mentions the impact of the Contribution, the conclusion dwells on this aspect to energize the reader.
- The Abstract adopts a factual, neutral tone. The Conclusion keeps the reader in a positive state of mind.
- Everything in an Abstract is new to the reader. When reached the Conclusion nothing is new. It allows readers to understand the contribution

better and in greater detail than in the abstract so as to evaluate its usefulness to them.

According to Jean Luc Lebrun (2007) Conclusion has the following qualities [39].

- Conclusion is positively charged. It maintains the excitement created in the introduction.
- Conclusion has *predictable content.* There are no surprises. Everything has been stated in the other parts of the paper.
- Conclusion is *concise.* Restate the contribution.
- Conclusion is *coherent* with the title, abstract, and introduction.

While writing Conclusion it is better to avoid downplaying negative results and deeming them significant when there is no proof, and making statements based on personal opinion without scientific support.

7.1.8 Acknowledgements

Acknowledgments usually appears between the Discussion and Reference sections. The author uses one or more sentences to specify contributions that need acknowledgements. 'It is the section where the author thank all those people who are not listed as authors but whose efforts contributed to the research described in the paper and to the preparation of the paper itself' (Ann M Korner, 2004, p. 64). Any substantial help received from organizations or individuals, whether they provided grants, materials, technical assistance, or advice, etc., can be included in acknowledgements. Persons who have contributed intellectually to the paper but whose contributions do not justify authorship may be named and their function or contribution described. Do not list people who did not contribute directly to the reported work.

The Acknowledgements section should start as simply as "The author(s) thank......." and any complex sentence must be avoided. Some authors use "We thank.......", however "The author(s) thank....." is considered to be the better usage. The following order of preference can be followed in the acknowledgements (Ann M Korner, 2004, p. 65).

- People who provided intellectual input that formed part of the basis for your study or who encouraged the authors in their work.
- People who assisted with some specialized measurements or allowed the use of specialized equipment.
- People who provided specific materials.
- People who provided technical assistance.
- People with whom the authors discussed their results during or after the study.

- People who commented on the original or revised manuscript.
- The person who prepared the manuscript.

Source of funding, with the initials of individual authors included, in parentheses, when funds were allotted to a specific author; and fellowships awarded to specific authors, with the initials of the respective authors, again in parentheses.

7.1.9 References

Reference citation is an integral and important part of scientific articles. It includes a list of all sources cited in the paper, using the appropriate bibliographic format. 'It provides communication among the authors and binds scientific papers as well as whole scientific knowledge in certain fields. Citing scientific sources means quoting results, ideas, considerations, definitions, tables, figures, schemes, texts and so on, published in scientific papers or available in some other way' [40]. References are also known as *literature* or *bibliography*. Reference citation is very important for good scientific practice, intellectual integrity and observation of ethical principles.

There are different types of references and all these are commonly used in scientific writings. They are *footnote*, *endnote* and *bibliography*.

- *Footnote:* Footnote provides details about the source of an in-text citation and placed at the bottom of the page where the citation appears.
- *Endnote:* Endnote provides details about the source of an in-text citation and are placed at the end of the article or paper, on a separate sheet of paper.
- *Bibliography:* It is a complete list of reference books, articles and other source materials cited in your work, in alphabetical order by author. Bibliography is placed at the end of the article or paper, on a separate sheet of paper.

The reference list should be useful to the readers; therefore the writer should be selective. The references included in an article should have direct bearing on the described work. It should be valid, important and most recent. The number of references should be restricted. Halsey (1998) observed it is rarely necessary to have more than 40 references in the longest paper. For brief communications it is restricted to a maximum of 10. The references should be verified against the original documents and it should be carefully organized. This helps to avoid the misquoting of references which is a serious problem in scientific writing.

There are a number of variations of literature reference style. For example *Harvard system* uses *name-and-year* style and *Vancouver system* uses a *consecutive number* style for reference citations. Different journals prefer different styles for reference citations. Some bibliographic programmes are available for effective organization of the references.

The following points will be helpful in selecting appropriate references.

- For links to information about directly related research studies, cite the original papers.
- Avoid citing review articles and general books. Summaries and reviews do not have the space to include all the original data with its full range.
- Do not cite incomplete sources. Do not cite abstracts and personal communications; because these often contain preliminary information and early speculation based on incomplete data.

While writing the references the usage of appropriate citation format is necessary. The complete bibliographic format is used in the References section at the end of the paper. A brief citation format is used in the body of the text. Some example of bibliographic format is give below.

- Gopen GD, Swan JA (1990) "*The science of scientific writing*", Am Sci 78:550–558.
- G. Fowles, "*Introduction to Modern Optics*", 2nd Edition, Dover Publications, New York, 1989.
- R.B.H. Tahar, T. Ban, Y. Ohya, and Y. Takahashi, "*Tin doped indium oxide thin films: electrical properties*," J. Appl. Phys. 83, 2631 (1998).
- W. Cheng, E. Wang, J. Phys. Chem. B 108 (2004) 24.
- M. Mandal, S.K. Ghosh, S. Kundu, K. Esumi, T. Pal, Langmuir 18 (2002) 7792–7797.

However, in the text the complete bibliographic format cannot be used. In text the references can be cited by including name of the author(s) and year in parenthesis. For example:

- "Highly efficient photovoltaic conversions, combined with ease of manufacturing and low production costs, make the 'Dye Sensitized Solar Cell' technology an attractive approach for large-scale solar energy conversion" (G Smestad, 1994).

Majority of the journals number all the bibliographic entries in the References section, and the text citations are simply the entry numbers. In such cases the text citation would look like;

- "Highly efficient photovoltaic conversions, combined with ease of manufacturing and low production costs, make the 'Dye Sensitized Solar Cell' technology an attractive approach for large-scale solar energy conversion" [3]; and the reference section would include the complete bibliographic format with the same number.
- [3] G. Smestad, Sol. Energy Mater. Sol. Cells 32 (1994) 259.

Most of the journals do not allow citation of personal communications, unpublished works or conference abstracts. If a journal allows citation of personal communications, then the author need to obtain written permission from the person cited and give their title, position, and affiliations. Authors can also cite information from websites. The correct format for citing information from a website is as follows:

- "Using Citation Indices", University of Texas Libraries, http://www.lib.utexas.edu/chem/info/citations.html (accessed on 25-05-2013).

An important point to remember is that never enter citations that you have copied from another paper. In such cases chances for citing wrong reference is more and therefore, always get a copy of the original article and verify the correctness of the citation.

7.2 WRITING A SCIENTIFIC PAPER

"Work, Finish, Publish"

—**Michael Faraday**

In the previous section we have seen the different components of a scientific paper. A scientific paper reports an original research result. It should communicate and record information as accurately and concisely as possible. The author should be careful in writing the paper. It needs an expertise which can be gained through experience. 'Excess wordage, tortuous construction, unnecessary detail, duplication, repetition, third-person passive, pseudo-objectivism, etc., obstruct rather than facilitate communication' [41]. Precise words and expressions of unmistakable meanings should be used while writing a research paper. Ambiguous, vague and unnecessarily complex sentences must be avoided.

While writing a paper the author should keep the readers in mind. The readers may be at different level of knowledge in a particular area. Booth et.al. (1995) suggested that an author should keep the following questions in mind while writing the paper [42].

- Who are the readers?
- What do they expect the author to do for them? The author tries to inform them, not entertain them.
- How much do they know?
- Do they already understand the research problem/question? The author should provide some background and explanation of what have studied.
- How will they respond to the solution/answer? The author must try to anticipate questions or misunderstandings that they might have and provide enough explanation to avoid them.
- In what forum will they encounter your report? The author should write technical research paper as if it might be published in a journal, professional society magazine, or conference proceedings. Because of this, there are some standards for formatting that you should use, and are discussed later.

Through a research paper the author communicates his findings to the scientific community. The acceptance by the readers mainly depends how clearly the author presents the results. The solution to the research problem should be empirical and systematic. The assessment of a research paper is mainly based on the following points.

- Is there a theory or model that the research is designed to explore?
- Is there a specific question or questions that the research will attempt to answer?
- Does the research cite related research that seems appropriate to the study?
- Are the techniques for data collection specified? These may be quantitative or qualitative in nature. Do these techniques seem appropriate to the questions asked?
- If statistics are used, are the underlying assumptions for the statistical tests met?
- Do the conclusions logically follow from the evidence collected?

From the beginning of the identification of the research problem each component is important. Equally important is that all these components should be clearly explained to the readers. Any ambiguity in any part of the research paper reduces the acceptability of the paper. Many guidelines are available in selecting the prose, tense, abbreviations, legends, footnotes, etc., in a scientific paper.

7.2.1 Importance of Scientific Writing

"What is written without effort is in general read without pleasure".

— **Samuel Johnson**

The publication of a scientific paper has various objectives such as to present new, original results or methods, to rationalize published results and to present a review of the field or to summarise a particular topic. The editors of journals take every measure to maintain high standards of the journals. A quality scientific paper should have the following features.

- Originality
- Significant advances in field
- Appropriate methods, discussion and conclusions
- Readability
- Studies that meet ethical standards.

Also a scientific paper should not include duplication, reports of no scientific interest, out of date work, inappropriate or incomplete methods or conclusions and studies with insufficient data. Therefore, the work that is intended to publish should meet some basic standards. The research work has to be assessed constantly in order to be able to decide whether the work is suitable for submission. "A paper that records significant experimental, theoretical or observational extensions of knowledge, or advances in the practical application of known principles is worth publishing" [43].

The scientific writing is important because, apart from its general purpose to communicate the new findings to the scientific community it has got various benefits. W C G Peh (2007), categorised the benefits into groups such as *career benefits, professional benefits, institutional benefits* and *practical benefits* [44].

Career benefits: Major reason for many people to start scientific writing is to fulfill some job requirements by employers. This includes the initial appointment, confirmation of the appointment and promotion.

Professional benefits: It is a well-known fact that for young scientists, the publication of articles in journals of international reputation will be very helpful in applying for fellowships. Such publications enable them to gain recognition and acknowledgement as experts in a particular field at national and international levels. For scientists who have good publications in international journals, it is comparatively easy to get research funding from various agencies.

Institutional benefits: Publication in peer-reviewed journals is very important to get international recognition for individuals, institutions or universities. If the topic has some major importance the author's country or his institution will get benefit out of it. The academic productivity of an institution is usually counted in terms of the number of research papers published in peer-reviewed journals.

Practical benefits: During the manuscript preparation the scientist is training himself in the process of scientific writing. He performs a thorough literature search, gathering and analyzing information and writing and revising the manuscript a number of times before preparing the final paper.

The Manual of Scientific Style: A Guide for Authors, Editors and Researchers (2008) emphasizes the importance of scientific writing through the following characteristics of science [45].

Science is a social enterprise: The development of science is through the contribution made by many people throughout the world and through their interaction. Science can grow only if it is communicated to the entire humanity. Collaborative research projects are essential for the exchange of scientific knowledge and the creation of new knowledge.

Science is a political enterprise: The impact of science on politics can never be neglected. In many countries government is the main funding agency for research programmes. Government usually allocates fund to carres out research projects that has a common national interest. For smooth functioning of the government, difficult problems faced by the humanity must be solved and only science could help in seeking a solution for such problems.

Science is an educational enterprise: To become an efficient and effective scientist one should educate and practise throughout his life. Science education is a never ending process. The researchers must convey the value and importance of their findings to the scientific community. For this the researcher should write effectively. 'The most successful scientists have almost universally been recognized as much for the clarity and effectiveness of their prose as the constructs and consequences of their theories', [46].

Science is a cultural enterprise: Science can very well influence the beliefs of humanity. Humanity seeks solution for almost all confounding issues through science, whether it is practical or metaphysical. Moreover the society greatly depends on the technology which is considered the practical side of scientific knowledge. The social as well as cultural development of any nation greatly depends on how effectively the scientific knowledge and technologies are being communicated to the humanity. All these demand publication of research findings through various media.

The skill of scientific writing is essential for a successful scientist. The quality of a scientific manuscript is usually evaluated based on the following standards [47].

- *Purpose:* The purpose of the manuscript must be obvious and unambiguous.
- *Conformity:* Text has to conform to the given formats, e.g., for journals, books, etc.
- *Accuracy:* The language must be grammatically correct, concise, accurate, and precise.
- *Consistency:* Terminology should be consistent and appropriate.
- *Logic and flow:* The manuscript should be a "story" with a clear message based on a logical train of thought.
- *Context:* The work must be presented in the context of established literature or other reported work, and must be congruent with accepted institutional or regulatory values.
- *Structure:* A logical structure like headings and subheadings, paragraphs, figures, and tables should be chosen.
- *Data presentation:* High-quality data should be presented clearly, using tables and figures as appropriate.

A scientific writing has various purposes to fulfill. Major one is 'to impart thoughts or ideas and their bases and implications in such a manner that the audience, with at least a moderate knowledge of science, can understand the material presented within a paper' [48]. Therefore, the manuscript should be written from the perspective of strength rather than weakness. As a result, appropriate word usage, syntax and punctuation are very important in scientific writing.

7.2.2 Characteristics of Scientific Writing

"Science is nothing but trained and organized common sense".

— **Thomas Henry Huxley**

No doubt; scientific writing is entirely different from writing a novel or a poem. Authors should never use unnecessary adjectives and elaborations while writing a scientific paper. The words should be precise, sharp and adequate. Anybody who wishes to write scientific articles should have knowledge about the characteristics of scientific writing. Robert Barrass (2002) speaks about the following commonly accepted characteristics of scientific writing [49].

Explanation: In a scientific paper the author explains the discoveries to the readers. Therefore, they must have his readers in mind while writing the paper. The group of readers he wants to address, their knowledge level, etc., is important in deciding the extent to which the explanation is to be provided.

Clarity: Clarity is required in every phase of scientific research. Clear statement of the problem and selection of appropriate method will lead to a successful research work. The clarity should also be maintained in presenting the results of the research. Appropriate illustrations must be provided wherever necessary.

Completeness: Every statement used in a scientific paper must be complete. There should not be any ambiguity in any of the statements. Vague and erroneous sentences distort the attention of the reader. There is nothing wrong in seeking help from experts while preparing the manuscripts.

Impartiality: Scientists must be free from biases. Every statement must be supported by underlying principles which has got wide acceptance. Every observation may have some limitations. Such limitations and other possible sources of errors must be indicated in the writing. Any evidence that is against the proposed hypotheses must not be omitted. Also, it is very important that one should never undervalue the findings of other scientists. "Any assumption, extrapolation, or generalization, should be based on sufficient evidence, and should be in accordance with all that is known on the subject" (*Robert Barrass,* 2002, p. 31).

Order: A scientific paper should be presented in an appropriate order. It is common to arrange the steps in the order of performance. This helps the reader to follow what the researcher did exactly, what result he obtained and how he interpreted the result.

Accuracy: Accuracy is one of the key features of a scientific method. In all phases of the research work like observation, recording of data, data analysis etc., the researcher has to maintain accuracy. Report writing should also be accurate. Appropriate words and tenses must be used while writing the article.

Objectivity: Objectivity is a main feature of science. All scientific knowledge is objective. It is based on evidence and not on opinion or imagination. "In scientific writing nothing should be implied or left to the reader's imagination" (Robert Barrass 2002, p. 33). Words like *possible, probably* and *it is likely that*; definitely imply that the author is not sure about the fact.

Simplicity: Simple writing is a sign of clarity of thought. While writing a scientific report the author should present the simplest explanation in accordance with the supporting evidence. "Scientists should write direct, straightforward prose, free from jargon, verbosity and other distracting elaborations", (Robert Barrass, 2002, p. 34). William of Occam (a theologian of fourteenth century) suggested that the entities must not be unnecessarily multiplied. This is known as *Occam's razor.*

7.2.3 Rules for Scientific Writing

Scientific writing should be simple and direct. It should be easily understood by the readers. Even though there is no strict rule for writing scientific papers, observation of the following rules will be useful [50].

- Interest, inform, and persuade the reader
- Write for your reader and write clearly
- Eliminate unnecessary redundancy
- Avoid digressions
- Don't overexplain and avoid overstatement
- Avoid unnecessary qualifiers
- Use consistent tenses
- Use precise words
- Simpler words are preferred over complex words and use concrete words and examples
- Simpler sentences are preferred over more complicated sentences
- Use the active voice (except generally, in methods)
- Make sure the subject and verb agree
- Use affirmative rather than negative constructions
- Avoid use of the indefinite "this"
- Use transitions
- Cite sources as well as findings
- Proofread your paper carefully.

These rules are generally accepted by the scientific community for effective prose writing. At the beginning everybody faces difficulties in observing these rules. However through continuous reading of scientific literature, one can understand the technique and acquire the skills of scientific writing.

For the communication to be effective, the writing should be interesting to the readers. The following rules can be useful for making communication effective [51].

- Before starting to write, decide whom you hope to interest, why you wish to interest them, what must be said, and how you should say it. Readers are most likely to understand grammatically correct English, and are interested in evidence.
- Write about things you know, if you have something interesting to say.
- Plan your work so that information and ideas can be presented in an appropriate order, so that the whole composition has the qualities of balance and unity.
- Write for easy reading. Begin well. Keep to the point. Be clear, direct and forceful. Maintain the momentum of your writing, if possible by writing at one sitting.
- Check your work, and revise if necessary.

7.2.4 Effective Scientific Prose: Gopen and Swan's Guidelines

"Writing is easy. All you do is staring at a blank sheet of paper until drops of blood form on your forehead".

—**Gene Fowler**

Basic purpose of scientific writing is communication of scientific knowledge. As already mentioned, while writing, the author must have his readers' in his mind. The writing should be according to the readers' expectations. This is important because the readers usually interpret what they read and their interpretation is based on the information or tips that they receive from the structure of the writing. Therefore, quality of writing should be improved which is actually a sign of improved thought.

George D Gopen and Judith A Swan (1990) studied the requirements of an effective scientific literature and formulated a series of guidelines. These are the guidelines on the structure of scientific prose and the designing of a sentence or paragraph so that the reader clearly understands what the author intends to say. 'The essence of Gopen and Swan's guidelines is to ask what readers expect when approaching any body of text, and what reading habits guide them, even if unconsciously, as they make their way through any piece of prose' (Harold Rabinowitz and Suzanne Vogel, p. 11). 'Information is interpreted more easily and more uniformly if it is placed where most readers expect to find it' (Gopen and Swan, 1990).

Some of the Gopen and Swan's guidelines are given below [52, 53].

Verb position: According to Gopen and Swan 'the reader's expectation stems from a pressing need for syntactic resolution, fulfilled only by the arrival of the verb'. Without verb one cannot understand the purpose of a sentence. As a result, the reader expects the arrival of a verb and resists recognizing anything in the interrupting material as being of primary importance. Readers expect the verb that informs what the subject of the sentence did to come soon after the subject is identified. Therefore the verb of a sentence is to be placed as close as possible to the grammatical subject. Anything of length that separates subject and verb is regarded as an interruption and leaves the reader with a sense of unfulfilled expectations. The reader may forget just what the subject of the sentence is by the time of the verb appears.

Stress points: "Point" here refers to a piece of new information. Such information should be placed in the latter portion of the sentence or paragraph. This is because the readers expect that this is the place where a new idea is to be revealed. Gopen and Swan referred this location as "stress points". Readers instinctively sharpen their attention and prepare for the climax or the point that the writer wishes to convey as they sense that they are nearing the end of the sentence or paragraph.

If the author places the new information in any place other than the stress position; according to Gopen and Swan two things can happen. First, the reader might find the stress position occupied by material that clearly is not worthy of emphasis. Second, the reader may find the stress position occupied by something that does appear capable of receiving emphasis, even though the writer did not intend to give it any stress. Both are undesirable. In the first case the reader recognises that what else in the sentence may be the most likely candidate for emphasis without any additional structural clue. In the second case the reader is highly likely to emphasize this imposter material, and the writer will have lost an important opportunity to influence the reader's interpretive process.

Topic position: Topic position refers to the place where the reader expects the subject of the communication to appear. The place of the subject of the sentence or paragraph is the early portion of the sentence or paragraph because readers expect the subject of the writing to appear early and perceive this positioning as a prompt to prepare themselves for information or observations later in the sentence or paragraph. Gopen and Swan demonstrated that this expectation is so strong that tables that fail to place the subject material on the left and the findings or conclusions on the right become virtually indecipherable.

Gopen and Swan provided a narrative example: "Bees disperse pollen" and "Pollen is dispersed by bees" are two different sentences about the same fact. The first tells us something about bees; the second tells us something about pollen. If what follows the first sentence is about pollen, or if what follows the second is about bees, readers are certain to be confused and will miss the point that the author wishes to convey. In other words, "Bees disperse pollen" will be superior if it appears in a paragraph that tells about bees.

Also the readers expect the material occupying the topic position to provide them with *linkage* and *context*. Gopen and Swan describe them as the *old information*. The information in the topic position prepares the reader for upcoming material by connecting it backward to the previous discussion. The old information that links backward is to be placed in the topic position and the new information that the author wants the reader to emphasize is to be placed in the stress position.

Perceiving logical gaps: Always place the context before or near the topic position. The author should include both background information supplementary to the subject of the paragraph, and a contextual connection to the points in a paragraph. When the context does not appear at all in a sentence, whether in the topic position or elsewhere, readers are left to construct the logical linkage by themselves. This often happens because the connections are so clear in the author's mind that they seem unnecessary to state. The authors often underestimate the difficulties and ambiguities inherent in the reading process. Readers should get proper connection with the context and linkage. Each sentence must proceed logically from its predecessor.

Verbs and action: It is a common expectation on the part of the reader that the action that is attributed to the subject of a sentence or paragraph is going to be described by a verb. This means connection between the subject and the verb must be clear. Complex sentences used by the author usually fail in creating a connection between the subject and the verb in reader's mind. Thus the author fails to convey what he wants to. The author should remember the following points.

- The verb should be appropriate to the subject.
- The verb should apply only to the subject and not to another element of the text.
- The intervening material should not divert the reader's attention from the connection between the subject and the action verb.
- The verbs and actions in the text should be limited, so that the readers can understand them easily.

Relative placement of context: While writing a scientific report the context and background information should be presented before presenting the new information. This is again a consequence of what readers expect when they read. Scientific writing is not just an accumulation of information; but it is an instrument that brings the author and the reader closer and into a relationship that makes communication possible. If the reader is not equipped with the necessary background knowledge, he may not understand the ideas presented in the report and he will not be able to evaluate it. Therefore, an author should never share the new findings and conclusions before presenting the context.

The conclusion drawn by Gopen and Swan is important in this context. "It may seem obvious that a scientific document is incomplete without the interpretation of the writer; it may not be so obvious that the document cannot *exist* without the interpretation of each reader."

Emphasis and structure: The structure of the writing will create an expectation in reader's mind regarding the relative emphasis of the substance of the sentence, paragraph or chapter. An effectively constructed paragraph has the ability to deliver the information uniformly and coherently. The structure of the prose can create understanding, confusion, enlightenment, etc., in reader's mind. A good prose should follow the structural directives provided above. That is, following the basic structural directives of subject placement, verb proximity, reader preparation, point placement, and overall paragraph design, the author can communicate the intention contained in the text to the reader. Gopen and Swan summarized the structural principles as follows.

- Follow a grammatical subject as soon as possible with its verb.
- Place in the stress position the "new information" you want the reader to emphasize.
- Place the person or thing whose "story" a sentence is telling at the beginning of the sentence, in the topic position.
- Place appropriate "old information" (material already stated in the discourse) in the topic position for linkage backward and contextualization forward.
- Articulate the action of every clause or sentence in its verb.
- In general, provide context for your reader before asking that reader to consider anything new.
- In general, try to ensure that the relative emphasis of the substance coincide with the relative expectations for emphasis raised by the structure.

Gopen and Swan derived their guidelines based on readers' expectations. The basic idea is that the resulting structure of the prose significantly influences the reader during the act of interpretation even more than the meanings of individual words. None of these principles are considered *rules*. In other words, there is no fixed algorithm for scientific writing. However, these principles revolutionized scientific writing.

7.2.5 Effective Word Selection

"The difference between the right word and the almost-right word is the difference between lightning and the lightning bug".

— **Mark Twain**

We have understood the importance of constructing adequate sentences or paragraphs for effective communication of scientific findings. Another equally important matter is the selection of appropriate words or phrases. A good scientist must be able to select right words or phrases and to eliminate words or phrases that confuse the reader. It is the duty of the author to convince the reader on the correctness and importance of his writings. Therefore, anything that creates ambiguity or distortion must be avoided.

Scientific research demands precision in all its phases; right from the statement of the research problem to the publication of research findings. In the instructions to the authors the journal *Ecology* says that "write with precision, clarity and economy; every sentence should convey the exact truth as simply as possible". Many guidelines are available for the selection of appropriate words while writing a scientific article. Some of them are discussed below (Harold Rabinowitz, Suzanne Vogel, 2008, pp. 17-33):

Be clear: Clarity is essential in reporting the research findings. It is the supreme criterion of scientific writing. The clarity has two phases: (i) simplicity and (ii) unambiguity. Many authors make sentences complex using complicated words and phrases. On most of the cases such words and phrases can be substituted with simple words that have same meaning. Simple words convey ideas more directly. For example, the word *end* can be used instead of *finalize*, *use* instead of *utilize*, and *previously* instead of *hitherto*. The usage of complex words creates barrier between the author and the readers thereby limiting the effectiveness of communication. Also there are unnecessary words or phrases such as *unless*, that are added for formality and will not add any information. The following points should be kept in mind (Harold Rabinowitz, Suzanne Vogel, 2008, p. 19).

- Writers should avoid the use of a string of modifiers (adjectives and nouns that serve the same purpose).
- Writes should not include prepositional phrases and subordinate clauses in a sentence.
- A single sentence should not be used when too many ideas are to be communicated effectively.

The author should also avoid ambiguity in any word or phrase. The ambiguity in a word or phrase refers to a situation where the reader understands the word or phrase in many ways because of other problems with the prose. The main reason is the incorrect syntax. For example the sentence 'I studied *only* quantum mechanics for examination' has a different meaning from the sentence 'I studied quantum mechanics *only* for examination'. Therefore each sentence is to be constructed with proper positioning of right words to convey the same idea that the author wants to communicate.

Be precise: Precision is highly appreciated in scientific writing. The words should be used carefully. The author should not confuse the readers. For example, the words *mass* and *weight* should be used only in correct situations. The value of precision in science may be the highest value that a science writer should espouse. He should not compromise on precision. A momentary lapse in precision may win a point, but will weaken the purpose of communication.

One of the major problems faced by writers is belief that the repetition of words is a sign of poor writing. As a result they go for alternative words that could serve only as synonyms. If a word or phrase perfectly describes a situation, writers should not hesitate to use it repeatedly.

Writers must also be careful about the connotation of the words, because words or phrases might have acquired connotations in the course of common usage. This may lead the reader to understand the exact opposite of what the author intends. For example, according to *The American Heritage Dictionary* the word *adequate* has got two definitions: (i) able to satisfy a requirement and (ii) barely satisfactory or sufficient. Also there are words like *aggravate* that has a negative connotation and such descriptive words should not be used if they convey the opposite of what the author wishes to convey.

Jargon also presents the same sort of problems when they are used in scientific writings. Connotations and associations with words and phrases introduce meanings that are irrelevant or contrary to the intent of the writer. Many readers fail to understand the right meaning of such words and phrases. Therefore it is very important that the use of jargon and inappropriate language in scientific writing must be avoided.

Be direct: One of the distinguishing characteristics of scientific writing is that it is *direct.* The writer uses no flowering language; he uses precise and right words to communicate the idea. Writers should avoid the use of *pretentious* and *arrogant* words, and *clichéd* language. Pretentious words and phrases are those that are needlessly long, but which express simple ideas. Writers must use simple words. For example, instead of the word *facilitate* use the word *cause*; similarly, the word *implement* can be replaced by the word *use*. The phrase, *in close proximity to,* can be substituted by the simple word *near*.

Arrogant language creates a feeling of insult in reader's mind. The usages like, *it is obvious to any reader*, convey to the reader the sense that only a fool would dispute the statement made or the conclusion drawn. In such cases the readers will develop a negative attitude towards the author and the article. In a similar way clichés and colloquial languages also create a gap between the reader and the author.

The reader has an important role in scientific communication and he should understand what the writer says. The writer should use direct and *strong* nouns and simple, active verbs to convey the information. For example, it is suggested that the writers should always use active voice instead of passive voice. However, there are instances where the passive voice is more effective than the active voice. So the writes should have the ability to use active and passive voices wherever necessary. Always direct, active and concrete writing attracts the attention of the reader more easily than indirect, passive and abstract writing.

Pronouns and tenses also must be used carefully in describing the work and the conclusion. Authors usually use passive voice to increase the objectivity of their work. It is interesting to notice that the elimination of the author as a participant from scientific writing creates the impression that the effects and experiments happened by themselves without any human effort. This misleads the reader by suggesting indirectly that the experiments *had* to happen just as they did. Such misleading situations can also arise from inappropriate use of tenses. Correct usage of tenses in scientific writing will be discussed in the following section.

One of the main challenges faced by an author is how to introduce a new terminology to the reader. Every technical term is actually a jargon which is familiar to a limited number of people in the related field. The right way to introduce a technical term is by giving a definition or an explanation of the term. The writers should avoid the use of abbreviations, as far as possible, unless it has universal acceptance; for example the SI units. Every abbreviation used in the writing should be defined properly. Also care should be taken for not to use abbreviation in the title. Writers can use unfamiliar words and phrases in explaining phenomena that are familiar to the reader. This helps the reader to extract the contextual meaning of those words and phrases. Use examples, analogies and comparisons wherever necessary. A concrete and specific example can do much in making the reader to understand the concept. On the other hand analogy and comparison creates a picture in reader's mind. However, these should be clear and simple for effective understanding of the situation drawing of inferences.

Be concise: The scientific writing should present only the information that is important and relevant. This is an essential skill that a good scientist should have. Concise writing can be achieved through revision and editing process. The author has to prepare a draft first and then he has to work on it to meet all the requirements of a scientific prose. At this stage the author checks thoroughly the language, words and phrases and makes necessary corrections. The following are some points to consider (Harold Rabinowitz, Suzanne Vogel, 2008, p. 30).

- *Redundancy:* Phrases that are filled with words that add nothing to the basic idea being conveyed are annoying and should be eliminated. For example the usage *at present* is more appropriate than *at the present time*; for the latter is a dated usage.
- *Deadwood:* This refers to empty verbiage and unnecessarily ornate language; words and phrases that offer no information to the reader.

Such things are often used to create an impression about the writer and not convey any information. Examples are: *as a matter of fact, it is significant to note that; it should be pointed out that,* etc. Certainly such phrases add no information and should not be used in scientific writing.

- *Fat:* These are phrases that use many words to say what can be said in a word or two. The *phrase at present time* can be replaced by *at present* and even this can be written in a better way as *now*. All these usages bring about the same meaning.

Be fluid: Scientific writing should be able to maintain a flow while reading. Even though the construction of the sentences and choice of right words and phrases are made with utmost care, the reader will feel difficulty if the writing is dull. The writing should be interesting to the reader. Scientists can develop the skill of fluid style of writing through practice. However, there are a number of practices and habits that a writer can apply to make the writing more fluid. They are (Harold Rabinowitz, Suzanne Vogel, 2008, pp. 31-33):

- *Vary sentence rhythm and length:* 'Sentences of nearly the same length in a writing have a numbing effect on the reader and make it impossible for most readers to stay focused and engrossed in the writing'. Variety can be introduced by simply varying the length of the sentences, or by varying their rhythm. Using an introductory close or introducing prepositional phrases of varying length and using variety in subject placement are some methods to make the writing more fluid and interesting.
- *Vary sentence style:* Every sentence should not be simple declarative sentence. Parenthetical remark or rhetorical questions can be introduced in writing.
- *Vary opening sentences of paragraphs:* The opening sentence of a paragraph is very important and should be set with proper tone. It should announce the importance of the argument and that of the new information that is going to be presented soon.
- *Clear up the "logjams":* There are some sentences that divert the attention of the reader from the main argument. Such sentences may be useful in providing background information but should not be placed in the vicinity where the new information is going to be placed.
- *Use surprise and the unexpected:* A writer should not hesitate to use words or a sentence that seems to convey what is meant better than the conventional, rule-obeying writing that are trying to compose.
- *Follow correct usage:* 'Communicating difficult, complex, and unfamiliar ideas to a wide variety of readers with great variation in their backgrounds demands that the writing be carefully wrought and precisely worded' (Harold Rabinowitz, Suzanne Vogel, 2008, p. 32). The writer must always be conscious about their readers and their background knowledge.

He must follow the standard rules of style in preparing the reports so that readers at various levels of background knowledge will be able to understand the piece of information.

7.2.6 Scientific Usage of Tenses

As Robert A Day (2006) pointed out, in scientific writing the choice of verb tenses are quite limited. Only simple present, simple past and present perfect tenses are mainly used in scientific writing. Other tense forms are rarely used.

Present tense: Present tense is used to state an established knowledge; fact or a principle. It is mainly used in the introductory section of a paper. In scientific writing present tense is used in the following situations.

- When writing about the topic or background.
- When explaining the purpose of the study.
- When presenting the results; if the findings are general facts.
- When making general statements especially in conclusion.

Some examples are given below:

- Titanium dioxide (TiO_2) *is* known as the best photo-catalytic material. (about the topic)
- Here we *investigate* the photo-catalytic properties of titanium dioxide. (purpose of the study)
- ......the refractive index of samples *is decreased* and then *is increased* with their thickness when the photon energy is smaller than the band gap energy. (result of a study)
- The refractive index of the film in this study *is* constant in the long wavelength range and increases in short wavelength that *shows* good agreement with Cauchy dispersion model. (Conclusion).

Also the *active* and *passive* forms are used in different situations. Present tense in passive form is used when describing a process or procedure. The active form is used in physical descriptions, such as describing a piece of apparatus or equipment.

Past tense: Past tense is used when referring to specific tasks carried out, such as taking measurements, conducting experiments, describing methods actually used, or presenting the results and conclusions from the particular piece of work. Therefore the abstract, method and result sections are mainly written in past tense. Some other situations where the past tense is used are given below.

- Simple past in passive form is mainly used when we report a particular procedure related to only one particular occasion in the past.

  Example: The electrical measurements *were performed* by using the Van der Pauw method.

- The past tense is used in the introduction to give historical background or development in the field of interest.

  Example: Recently, Taga et al. successfully *introduced* a new approach to broaden the photo-response of TiO_2 by doping with nitrogen [54].
- Past tense is commonly used in acknowledgements.

  Example: This work *was supported* by the research for the Future Programme of the Japan Society.

Present perfect tense: Present perfect tense is commonly used in the introduction. It is used when the author wants to recapitulate the state of the work and to say what work has been done and is still in progress.

- Photo-catalysis, through the use of various powdered semiconductors, *has received* much attention for its potential in the conversion of light energy into useful chemical energy.
- 'In the present work, we *have taken* the concept of nitrogen doping of TiO_2........'

Present continuous: This tense form is mainly used in the introduction. It is used when the author wants to stress that something is currently in progress or is only temporary.

- "Dye sensitized solar cells *are gaining* attention from both academic and industrial groups........."
- "We *have been developing* ways to apply TiO_2 coating on various materials such as glass.................."

Past perfect tense: In scientific writing this tense form is very rarely used. It is used to indicate that one action preceded another action in the past.

- "Titanium dioxide that *had been doped* with nitrogen showed enhanced photo-response".

Future tense: In scientific writing future tense is used to explain how the author intends to present information. It is also used in giving examples and describing visuals. It can be also used to indicate future events or the part of the work that come later.

- "We *will* consider the effect of annealing................."
- "Increasing the O_2 flow-rate leads to a significant decrease in the sputtering voltage and it *will* reduce the dissociation of gas followed by the plasma density and deposition rate".
- "The effect of pressure on the thickness of the film *will be examined*......"

It should be noted that the future tense should not be used in subordinate clause in time or in conditional sentences (when, if, etc.). In scientific writing conditional sentences are mainly used in acknowledgements.

- "The author(s) *would like* to acknowledge ....

For a scientific paper to be effective appropriate tense form should be used in writing. The following points serve as a guideline in using the tense forms properly [55].

- Within one paragraph, do not shift tenses unless there is a good reason to do so.
- When all verbs describe a sequence of actions or states, their tenses should be the same.
- Summarise the purpose of the paper in the past tense.
- State the current state of development in the area of interest in present perfect tense.
- Indicate the aim of the paper in future tense.
- Report the results of the study performed by the authors in past tense.
- Variation between passive and active constructions is strongly recommended to keep the reader interested.
- Check the tense used in subordinate clauses or in conditional clauses.

7.3 OCCAM'S RAZOR

Any discussion on scientific writing will be incomplete without mentioning Occam's razor. William of Occam (1285-1349), proposed a general principle regarding explanatory adequacy. It is a principle of economy in scientific explanation. Occam's razor is "*entities should not be multiplied unnecessarily*." According to Occam's razor whenever we have two competing hypotheses that explain the data equally well; choose the simpler one. Occam's razor serves as fundamental principle of simplicity. It can be well extended to the entire process of scientific writing. In general Occam's razor says: 'if two words or phrases can say the same thing with equal clarity, use the simpler one; if the simpler wording is considerably longer, then use the one most appropriate for the intended audience' (Jeffrey A Lee, 2000, p. 88).

Occam's razor is not a theory, on the other hand it serves as a maxim for choosing among such theories or explanations. Occam's razor is a kind of *principle of parsimony*. Such principles apply parsimony considerations to a number of different ontological items. "They have the general form: all other things being equal, where there are rival theories one of which postulates m items and the other postulates n items and $m < n$, then one ought to choose the theory with the lesser number of items, m" [56]. One of the best examples is the principle of least action. The classical mechanics can be presented in different ways; some of which employ a principle of least action and others that do not, the former often being taken to provide a more economic set of fundamental laws than the latter.

In many disciplines scientists postulate new entities in order to account for new empirical discoveries. Occam's razor prevents such multiple entities and suggests accommodation of other theories compatible with the data in the question. 'Occam used simplicity as a criterion of concept formation and theory construction. He suggested that superfluous concepts are to be eliminated; so that the simpler of two theories that account for a type of phenomena is to be preferred'[57]. In other words simplest or most obvious explanation of several competing ones is the one that should be preferred until it is proven wrong.

As an example Occam applied his razor on the nature of projectile motion. During medieval period it was believed that motion of a projectile is caused by an acquired *impetus* which resides somehow in the projectile as long as it is in motion. Occam argued that the concept of *impetus* is superfluous. A statement about the *motion of a body* is shorthand for a series of statements that attribute to the body various positions at various times. However, motion is not a property of a body, but is a relation to other bodies and to time. According to Occam, since change of position is not a *property* of a body, there is no need to assign an efficient cause to this relative displacement. Therefore, Occam maintained that to say *a body moves because of an acquired impetus* is to say no more than *a body moves*, and he recommended elimination from physics of the concept of impetus.

Encyclopedia Britannica provides another example. Consider the following two theories about the planetary motion around the sun.

- The planets move around the sun in ellipses because there is a force between any of them and the sun which decreases as the square of the distance.
- The planets move around the sun in ellipses because there is a force between any of them and the sun which decreases as the square of the distance. This force is generated by the will of some powerful aliens.

In these statements the second one has an additional baggage which is not necessary to explain the motion of planets around the sun. Then Occam's razor would clearly reject the second theory.

There are several arguments for and against Occam's razor. People who appreciate Occam's razor tries to show that 'other things being equal, the simplest hypothesis proposed as an explanation of phenomena is more likely to be the true one than is any other available hypothesis, that its predictions are more likely to be true than are those of any other available hypothesis, and that it is an ultimate a priori epistemic principle that simplicity is evidence of truth' [58]. Thus, the principle of simplicity has got a simple statement that if the former yields the data just as well as the latter does, it is more probably true; which is a consequence of Occam's razor. The argument against Occam's razor is that 'even though simple explanations are more likely to be true than complicated ones; every now and again the complicated one will be closer to the truth' [59]. We can imagine the example of biological system, say the working of eye. We can explain the working of eye in simple terms that we understand; however such an explanation is definitely incomplete. This is because the biological system is so complex and an explanation based on the principle of natural selection (which is often used by the biologists) may go wrong. In the concluding remark of the article "Occam's razor - essentials and applications", Alberto Murta wrote that "Occam's razor has not been proven to be valid, neither theoretically nor experimentally, and there is nothing mystical in simplicity that makes a simpler theory to be more likely to be true than a more complex one. A simpler theory is just more comprehensible, hence its advantage is of a practical nature."

An interpretation of Occam's razor was presented by Domingos (1999) [60]. Domingos considered the two versions of Occam's razor. Let a model (theoretical or experimental) have two types of errors: (i) *generalization error* and (ii) *training-set error*. Generalization error of a model is its error rate on unseen examples and training-set error is its error on the examples it was learned from. Then the two versions of Occam's razor are as follows.

First razor: Given two models with the same generalization error, the simpler one should be preferred because simplicity is desirable in itself.

Second razor: Given two models with the same training-set error, the simpler one should be preferred because it is likely to have lower generalization errors.

First razor stands closer to Occam's original razor and there is no controversy; while the second one is false literally. Domingos reviewed evidence and theoretical arguments that were put forward to support the second razor. He also reviewed several arguments against the second razor and showed that contrary to the second razor's claim, greater simplicity does not lead to greater accuracy. Domingos interpreted Occam's razor in two ways: 'as favoring the simpler of two models with the same generalization error because simplicity is a goal in itself, or as favoring the simpler of two models with the same training-set error because this leads to lower generalization error'. In spite of all arguments for and against Occam's razor one can conclude that 'for knowledge discovery research and applications, we should prefer simpler models only when we honestly believe the target phenomenon to be simple' [61].

7.4 WRITER'S BLOCK

"People have writer's block not because they can't write, but because they despair of writing eloquently."

—**Anna Quindlen**

We have understood that scientific writing is very important for communication of knowledge. However, some scientists often feel difficulty to start the writing of research findings even though they appreciate the value of good writing in advancing science. This *inability to write, despite the desire and ability to do so is known as writer's block*. Almost every one faces writer's block at one time or another. It can be experienced either as a sustained difficulty in producing text or as an ongoing dissatisfaction with the text produced. The origin of the term *writer's block* is generally attributed to Edmund Bergler, an American psychiatrist. Writer's block has many manifestations and many causes. In general, all blocked writers do not write despite being intellectually capable of doing so, and they suffer because they are not writing.

The specific reason for writer's block may vary from person to person. However, there are some common contributing factors. They include:

- *Uninteresting topic:* This situation occurs particularly when a topic was assigned by a senior or a supervising guide. In such cases one cannot pick the topic of interest or even deviate from it or narrow it.
- *Difficulty in understanding the topic:* This often happens when the topic is difficult or new. One may feel difficulty in understanding the topic properly. This creates a feeling of anxiety and a sense of failure in the mind.
- *Feeling of stress and anxiety:* The deadline placed over a writer to finish his writing will definitely create stress in his mind; for every person there are other obligations too. This finally leads to anxiety. Stress and anxiety is often projected as main reasons for the writer's block.
- *Feeling self-conscious:* Many writers feel self-conscious about writing. Sometimes even after writing the entire day they feel that their writing is inadequate.

There are many suggestions to overcome writer's block. In *The Manual of Scientific Style: A Guide for Authors, Editors and Researchers* Harold Rabinowitz, Suzanne Vogel (2008, pp. 33-35) provides the following guidelines to overcome writer's block. (These guidelines are derived from James G. Paradis and Muriel L. "Zimmerman; *Avoiding Writer's Block*;" *The MIT Guide to Science and Engineering Communication*, MIT Press, 1998, pp. 11-13).

Gather sufficient data: It is common to observe that in many cases, an author will not have conducted sufficient research. In some cases the author will not have enough confidence to start writing even if he has conducted sufficient research. Such situations create fear and anxiety of failure in author's mind. It is recommended that the author should set a definable and proscribed programme of research and begin writing as soon as it is completed.

Define the task of writing specifically: Many authors are confused about the exact nature of their writing aims. They may often fail to distinguish for whom are they writing; at what level should they be writing; how comprehensively should their writing be; and at what level of technical sophistication should their writing aim? The author should identify the purpose of writing, the audience, their knowledge level, etc., before start writing. This will help the author to overcome such difficulties.

Organize the material: In many discipline the complexity and the scope of the subject may be unrealistically large. Such large area of content definitely reduces the confidence level of the author. In such a situation the author must create a detailed outline that organizes the main points of the work. This will help the writer to advance to the next stages of the work.

Discuss the work: Many authors believe that discussing a work in progress impedes the work rather than helps it. However, if an author is facing difficulty in going forward, there is nothing wrong in making a discussion about the topic with a colleague or a supervisor. This discussion can be either a face-to-face conversation, or in the form of a preliminary draft or notes. This will help to overcome the difficulties and can lead to productive work on the project.

Sketch the graphic components of the work: 'Some people organize material best orally; some in written form; and some in graphic form. To exploit the graphic proclivity that a writer might have, preparing sketches and "story boards" of the graphics program of the work may stimulate and crystallize thought to the extent that the writing will become easier and more forthcoming' (Harold Rabinowitz, Suzanne Vogel, 2008, p. 34).

Create a conducive environment: The writing environment has an important role in successful writing. Writing requires sufficient time, an environment free of distractions, and the materials necessary for writing. The authors should set a time of day for writing when there is least disturbance. Also, a comfortable place with no distractions is to be arranged; and ensure the availability of sufficient writing material on hand.

Don't insist on writing a perfect first draft: According to Justice Brande: "there is no great writing; only great rewriting." The revision is critical in the creation of effective scientific prose. While rewriting, a writer should be prepared for the following (Harold Rabinowitz, Suzanne Vogel, 2008, p. 35):

- The ideas (sometimes central to the thesis and argument of the work) may change in the course of, possibly even as a result of, writing.
- Research conducted during the course of writing (or previous research first understood during the course of writing) may alter elements of the work.
- The process of articulating ideas and arguments will, by itself and by its very nature, crystallize and clarify ideas and connections in the writer's mind that would otherwise have gone unnoticed.

"Get thee an editor": Editing is a critical and integral part of scientific writing. Without editing one cannot create an excellent scientific prose. The art of editing involves the ability to read a piece of writing with sufficient detachment to identify the areas of confusion and miscommunication and to suggest means of improving the writing. Every author should acquire this ability to edit a piece of writing without imposing the editor's own style on the writing.

Every author should understand the fact that the writing skills required for effective communication at an advanced level is developed through years of practice in writing at all levels. It is evident that the clarity and effectiveness of a scientific paper heavily depends on the experience and skill of the author.

7.5 DUTIES AND RESPONSIBILITIES OF AUTHORS

Scientific writing is intended to present new findings before the scientific community. The findings will be discussed by a variety of groups: students, researchers, scientists, professionals, etc. Therefore; any finding that is going to be reported should be genuine and should have relevance. These facts assume that the writer should have some responsibilities (both professional and ethical) towards the scientific community. American Nuclear Society (ANS) *Ethical Standards on Publishing* pointed out the following responsibilities of the authors.

- Assure that the work submitted is new and of significance.
- Accurately report methods and results so that the work can be replicated.
- Retain and provide access to data.
- Faithfully reference and attribute sources of information and factors that have influenced the work.
- Clearly distinguish between actual results of work reported in the submitted paper and author's own interpretation of these results.
- Assure that co-authors have made substantive contribution to the results being reported or its documentation.
- Assure that co-authors have had access to the manuscript and have agreed to its submission for publication.
- Ensure that the publication is complete and is not divided into a series of lowest publishable units.
- Guard against plagiarism.
- Disclose conflicts of interest and explain why the work should not be affected by such conflicts of interest.
- Correct inadvertent, essential errors that may be discovered after a paper has been submitted for publication.
- Assure that the work has not been published or submitted for publication elsewhere and that if it is accepted for publication, the same or similar material shall not be submitted for publication elsewhere.
- Recognize that peer review by volunteers is undertaken to improve the work by noting omissions, alternative interpretations, and inadvertent errors, and to raise questions likely to arise among readers of the work.
- Recognize that editor's decisions are made in the best interests of the journal, its contributors, and the larger technical community.
- Do not seek identities of reviewers, and even if known, do not disclose the reviewer's identity.
- Make no substantive changes to a manuscript after it has been accepted for publication except to fix essential errors.

7.6 COMMUNICATING TO A JOURNAL

After completing a research paper with possible corrections made; next aim of an author is to get it published in a peer-reviewed journal. There are hundreds of journals almost in every discipline. Therefore, the author must be careful in considering the target journal. The selection of right journal will increase the chance for getting the article published. Before completing the paper the author must have a list of three to five most appropriate journals. While reading the literature the author might have noticed the journals that publish articles similar to the one in hand. Selection of such a journal from the list is more appropriate. After selecting the target journal the author should ensure that the prepared article is in the same format and style as prescribed by the target journal. This is because different journals allow different formats, style and organization.

The choice of right journal optimizes the speed and ease of publication. There are many factors to be considered for the selection of an appropriate journal for publishing the research article. The following factors are important [62].

Is it a peer-reviewed journal? The peer-review process is important for establishing the quality of the research work. In order to develop a good research profile it is essential that the articles are to be published in peer-reviewed journals. If the first journal rejects the article then the author can select the next journal from the list.

The scope and aims of the journal: In the introduction section of an article we usually cite journals where related works are already published. Those journals are the right choice for publishing the manuscript. Then check the websites or issues of these journals to identify those with scope and aims most appropriate for the prepared manuscript.

The audience for the journal: 'The audience for a journal is largely determined by the scope and aims of the journal, the journal's reputation and history of publishing in the field, and the accessibility of the journal to researchers'(M. Cargil and Patrick O' Connor, 2009, p. 69). Majority of the journals allows free online access of the title and abstracts. However, the body of the paper can be accessed by making payment. Now most of the journals can be accessed from university libraries at a reduced price or at free of cost. Journal subscription rate for universities are usually less than that for industries.

Journal impact factor: The quality of a journal is assessed through its impact factor. A higher impact factor often indicates a more critical or more prestigious journal. Publishing an article in a journal with high impact factor is considered a sign of recognition as a good scientist. Also such journals are most appreciated by the scientific community and an article published in such journals gains the attention of the scientific community easily.

Time for publication: Different journals take different time to complete the peer-review process and to publish the article. Every author wants to publish the research quickly to ensure that others do not publish similar work before getting his article published. Report on the average time to publication and can be obtained from the journal website. A high quality journal is more likely to ask for major revisions, and if the article is sent to the best and the toughest journal, the time required for publication will be large.

Page charges: There are some journal charges fees for publishing the article. The charges may be fixed or based on the number of pages. There may be extra charges for colour pages. These are to be considered before submitting the article.

Open access: Every author wants his article to be accessed by a wide range of readers. Some journals allows full open access so that anybody can copy or download the articles published in it. There are some journals that allow open access if the author pays an upfront fee. If an author wants to provide this facility to his readers he can select such a journal. The payment can be made for a affixed period of time. Open access of articles will be very much helpful in popularizing the research findings. For paid versions a subscriber has to pay a relatively large amount. However, for universities and research institutions many journals provide low rate subscription. A directory of open access journals can be found at: http://www.doaj.org/

7.6.1 Electronic Publication

Electronic publications are popularly known as e-publications. With the development of the Internet, communications are mainly through e-mail. Recently, social media networks are also used for communication. Internet is a globally connected system of computer networks. Many journals publish their articles through online mode in addition to print journals. The advantage is that the article can be accessed from any part of the world without any delay. The printed material takes many days to reach the person who purchases it. The open-access journals are readily available on the Internet at free of cost. Also, the articles submitted to e-journals will get published in a short period of time than a normal article publication in a printed journal.

The following advantages of e-publication may be noted.

Cost-Effective Publishing: Expenditure is one of the most important factors that must be considered while publishing an article. The cost of e-publication is very much lower than that of publishing a printed journal. This is because digital publishing eliminates a lot of processes that are involved in traditional publishing.

Provides an Interactive Reading Experience to Users: A digital publishing platform offers an interactive reading experience. Users cannot only turn pages in an article or eBook but can also zoom and change the font size as per their comfort. You can add multiple interactive elements to his publication to make it more appealing to your audience. Videos related to the research work can also be included in the article pages.

Editing is Possible Even After Publishing: e-publications are mostly created on cloud-based digital publishing platforms. So, if one ever wants to update any information or make some other edits, they can do so easily. We cannot say the same for print publishing though. So, digital publishing helps to keep the articles error-free and updated at all times.

Wider Reach: The aim of e-publishing is to reach a large number of readers. With the majority of people owning a mobile device with the Internet, they have instant access to a variety of content online. Online content knows no boundaries and does not charge for shipping and packaging. Anyone can download the article from anywhere in the world.

Publishers can Earn Revenue from Advertising: If the publisher has much commercial interest, they can allow advertisement on their website and revenue can be earned from the advertisements. However, a scientific journal is meant for promoting research activities, and therefore, good journal websites do not have an advertisement space.

Eco-Friendly Publishing: For the manufacturing of paper millions of tonnes of wood are used every year. The electronic publication certainly helps to save trees. Nothing is going to waste in e-publishing.

Less Storage Space: This is another great advantage of e-publications. One can download the article can save it on a computer or on a pen drive. One can imagine the space required for a 1000 pages book on a shelf. If it is in electronic form, hundreds of such books can be saved in a small pen drive that weighs less than 20 grams.

7.6.2 Submission Methods

In earlier days when computer was not popular, the articles were accepted in hard copy; either hand written or typed. Now the scenario is changed and the computer is so popular and efficient in preparing the documents. It has the major advantage that the author can edit the article at any time at no cost, which was not possible in handwritten or typed copies. Therefore, all journals now insist on the electronic submission of articles. In addition to the standard word-processing programs commonly used in computer-based communication, each scientific discipline has developed *authoring tools* that may be used to format papers in a manner that is more likely to present a paper in accordance with rules and strictures of specific journals in that field.

If an author is not using a formatting program such as those available in mathematics, physics or chemistry; it is advised to submit the article as both a word processing document and as PDF. This is because when transferring from one computer to another, any word processed document is subjected to some change in appearance due to variations in font editions and interaction with other programs on the machines. On the other hand PDF can retain its appearance and provides a better reproduction. In recent days since many PDF editors are available, submission of articles as word processed document is not necessary. Author can submit the article only as a PDF document.

Different submission methods are allowed by journals. However, as the technology advanced many of them became outdated as they consume more time in reviewing and publishing.

Handwritten or typed: This was the earliest method of submission of an article. Evidently such a submission consumes more time; in addition to the time for review time is to be allotted for re-typing the document. Now all journals discourage this submission method.

Camera-ready copy: In the past, submission of articles in camera-ready form was accepted by many journals. This was superior to handwritten or typed format as it expedited the composition process and allowed for quick turn-around and publication. Now journals are not accepting this format because advanced and convenient programs are available.

Electronic submission: Submission through electronic media revolutionized the process of publication of an article in a journal not only by reducing the time but also by increasing the efficiency. The author who wishes to publish an article in a journal first sends a query letter to the editor. If the editor is satisfied about the article that the author wises to publish, then an FTP site address, user name, and password will be issued. Recently, some journals allow the creation of an account through direct signup and registering.

Journals usually provide instructions to the authors in the journal website. The author can download the instructions. This includes information about the text font, font size, line separation, alignment of headings and paragraphs, preparation of graphs and tables, etc. The author has to prepare the article in accordance with these instructions. The requirements may be different for different journals. The articles are submited in PDF format. Once an article is submitted it will be subjected to the process of peer review. Based on the report of the reviewers the article may be accepted, asked for a revision and resubmission or rejected.

7.6.3 Peer Review

Peer review is an important process that determines the *fate* of a research paper submitted by an author. 'Scientific peer review is the evaluation of scientific research findings or proposals for competence, significance and originality, by qualified in experts in the same field' [63]. It is used by the editors of scientific journals where the experts provide written opinions about research papers that have been submitted for publication.

Peer review provides a system to select which research findings should be brought to the attention of other researchers. It also gives authors feedback to improve the quality of their research papers before publication. The peer review system judges the originality of the work, the importance of the questions addressed, the appropriateness of the techniques used, the quality of the data and the reliability and significance of the conclusions are the most important criteria. Some journal keep the editorial staff and for many other journals this work is done by professional scientists who act as editors as an additional activity.

Peer review is applied to a number of activities in the research process. They include [64]:

- Evaluation of applications for funding, to determine which applications are successful,
- Review of reports submitted by researchers once their funding award has come to an end, to assess whether the project has been completed satisfactorily,
- Evaluation of draft of conference presentations, journal articles and monographs, before they are published, to assess whether they meet quality standards,
- Evaluation of publications once they have been published, through reviews and review articles.
- Evaluation of the quality of work produced by individuals, teams, departments and institutions to help determine appointments, promotions and levels of funding.
- There are different types of peer review system in practice. They are: *Single-Blind review*, *Double-Blind review* and *Open review* [65].

Single-Blind review: This is the most common system of peer review followed in science disciplines. In this system the reviewers know who the authors are, but the authors do not know who the reviewers are. The advantage of this system is that this allows reviewers to provide honest, critical reviews and opinions without fear of reprisal from the authors. The disadvantages include the lack of accountability; allow unscrupulous reviewers to submit unwarranted negative reviews, delaying the review process and stealing of ideas.

Double-Blind review: This is the main system of peer review used in humanities and social sciences. Here the authors and reviewers do not know each other. This system reduces possible bias resulting from knowing who the authors are or where they come from. Work assessed on its own merits. The disadvantage is the difficulty involved in anonymizing the manuscript. Also there is a possibility that the reviewers can often guess the authors if the authors have cited their own papers.

Open review: This system is open in the sense that the reviewers know who the authors are and the authors know who the reviewers are. This system has greater accountability and reduced opportunity for bias or inappropriate actions. The problem associated with this system is that many reviewers may decline to review because revealing reviewer's identity may lead to animosity from authors, and some other related issues.

The peer reviewing system can be imagined as a quality control system. The importance of peer review is made clear through the following words. 'Researchers may make mistakes that render their conclusions worthless; and even when they conduct their research properly, they are also all too likely to exaggerate its importance. A review by scientists familiar with the subject matter is likely to detect mistakes and the exaggerated claims. Thus, peer review is important because it helps determine whether a study's substantive conclusion follows logically from the procedures used to arrive at it and whether the conclusion makes a significant contribution to our knowledge.' [66].

There are many criteria based on which the review is carried out. The following questions are usually asked by the reviewers [67].

- Does the paper fit the standards and scope of the journal it is being considered for?
- Is the research question clear? Was the approach appropriate?
- Is the study design, methods and analysis appropriate to the question being studied?
- Is the study innovative or original?
- Does the study challenge existing paradigms or add to existing knowledge? Does it develop novel concepts?
- Does it matter?
- Are the methods described clearly enough for other researchers to replicate?
- Are the methods of statistical analysis and level of significance appropriate?
- Could presentation of the results be improved and do they answer the question?
- If humans, human tissues or animals are involved, was ethics approval gained and was the study ethical?
- Are the conclusions appropriate?

After receiving the reviews the editor will decide whether the paper is suitable for publication or not. This decision is made upon the comments made by the reviewers. There are five basic responses that an author receives from the editor. They are:

1. The paper is accepted as it is.

2. The paper is accepted with minor amendments. The paper will not be sent for peer review again.
3. Major revisions are proposed and acceptance is dependent on whether the author can deal with those satisfactorily. The paper may be sent out again for peer review.
4. The paper is rejected but the author is advised to publish it elsewhere.
5. The paper is rejected and further submission of it to any journal is not advised because the work is seriously flawed.

Once a review process is completed, journals often inform reviewers whether the paper is accepted or rejected, and send them a copy of the other reviews. This helps the reviewer to see the assessments and opinion of other experts and gets a chance to examine whether there is anything missed in his own review.

When a paper is rejected the author usually submits it to another journal. The editor of that journal will send the paper for peer review to some other reviewers. Thus, a paper can go through several rounds of peer review. Scientific community often considers this a wastage. This wastage can be reduced through the so called *cascade peer review*. Here, when a paper that has been rejected after peer review, is passed to another journal along with the reviewers' reports. The peer review process of the second journal is now relatively short because the review reports from the first journal are also taken into account.

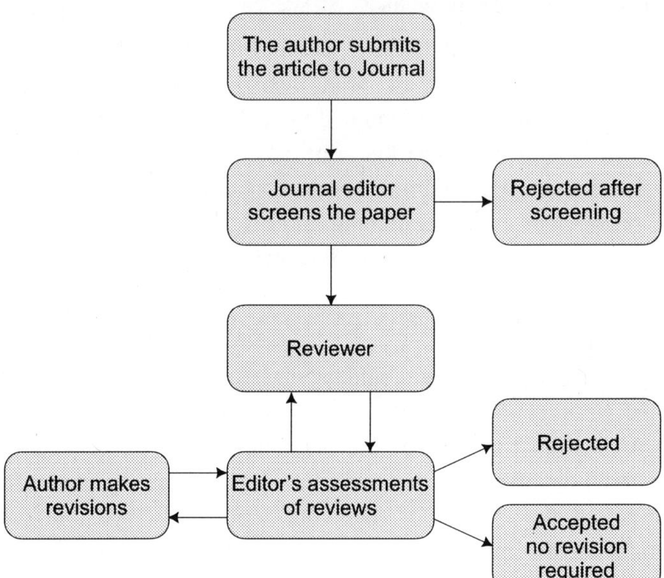

Fig. 7.1 Flowchart showing the various processes involved in a typical peer reviewing system. (Adapted from *The Peer Review: The nuts and bolts - A guide for early career researchers*)

Even though peer review is an integral part of scientific publication it is not a perfect system for a variety of reasons. It heavily depends on *trust*. Since the reviewers are human beings there is possibility of misconduct. According to Collette Teasdale, Development Editor—*Economics Journals,* Routledge Journals, "Reviewers are trusted to deliver an opinion but the editor knows this to be subjective and so will carefully consider this when making a final decision on a paper. And journals rarely accept papers based on only one review" (Cited in *Peer Review: The nuts and bolts,* p. 18).

One main problem associated with the peer review is that the reviewers can unnecessarily delay the publication to enable them to get their paper out first. In most cases the reviewers are given a deadline. If there is a considerable delay the editor will invite an expedited review from a backup reviewer or consider the reviews they already have in-hand at an editors' meeting. The other issues are [68]:

- It *shuts down new ideas* as research that goes against the status quo may be rejected by reviewers.
- New research that goes against current thinking might take longer to pass peer review.
- Most often cases the so called scientific fraud may go undetected.
- The peer review process often doesn't pick up plagiarism.

Reviewers might recommend rejection of a paper because of personal prejudices, rather than because the work is weak or flawed.

As Mammo Muchie (Editor of the *African Journal of Science, Technology, Innovation and Development*) viewed 'regardless of its weaknesses, peer review is something the scientific world cannot do without'. 'Peer review may have its limitations, but it is also a remarkable process which relies on the trust and co-operation of the scientific community and acts as a quality control ensuring that published research is valid, significant and original. The process is essential for the dissemination and advancement of scientific knowledge' [69].

7.7 RESPONSIBILITIES AND ETHICS OF REVIEWERS

The review work is usually unpaid and is undertaken as part of professional contribution of researchers to the development of science. Usually, the reviewers are experts in the field of the paper and have many publications in the related fields. The reviewers are asked to go through the manuscript and write a report about the quality of the work along with possible corrections and recommendations to improve the paper. Peer reviewers play the central and critical role in the peer review process. Therefore, it is essential that everyone involved in the peer review process should behave responsibly and ethically. American Nuclear Society (ANS) *Ethical Standards for Publishing* suggests the following responsibilities of reviewers.

- Respond in a timely manner to the editor's invitation to perform the review after pondering on adequacy of technical expertise, lack of conflict of interest, and ability to abide by the confidentiality guidelines listed below.
- Take steps to facilitate the review process, being careful not to ignore it to complete some other work for a similar or competing publication.
- Perform the review within the time period requested by the editor. If this is not possible, notify the editor promptly.
- Review the manuscript based only on the technical review guidance provided by the editor, making editorial changes only if necessary for clarity.
- Provide comments in a professional and objective manner, based on the technical review of the manuscript.
- Employ no personal biases to influence the review.
- Restrict asking for citations of reviewer's work unless justified and necessary.
- Comment on a manuscript in a way only that is well grounded, as complete as possible, as detailed as possible.
- Be concise and to the point for the benefit of both the author/s and the editor.
- Keep away from performing the review if lacking the appropriate technical expertise. Notify the editor immediately.
- Don't pass on the manuscript to others who may have the appropriate technical expertise for review without written permission by the editor.
- Disclose and discuss all actual or apparent conflicts of interest to the editor before performing the review.
- Notify the editor immediately if for any reason an objective review cannot be performed.
- Consider the manuscript as privileged information that is provided only for review purposes.
- Don't discuss or provide the manuscript to others for any reason.
- Don't cite or use the information contained in the manuscript as part of the reviewer's technical work before actual publication.
- Be aware that all communications with the author/s should be carried out by the editor.
- Do not discuss the manuscript with the author/s or with any other person, except the editor, either during or after the review process, until it is published.
- Indicate clearly any remarks that are intended only for the editor, or include the remarks in the appropriate location on the review form.

In addition to the professional responsibilities every reviewer should observe some ethical standards. *Committee On Publication Ethics* (COPE) suggests the following principles to be observed by everyone who acts as a reviewer (www.publicationethics.org).

- The peer reviewers should only agree to review manuscripts for which they have the subject expertise required to carry out a proper assessment and which they can assess timely.
- They should respect the confidentiality of peer review and not reveal any details of the manuscript or its review, during or after the peer-review process, beyond those that are released by the journal.
- They should not use information obtained during the peer-review process for their own or any other person's or organization's advantage or to disadvantage or discredit others.
- They should declare all potential conflicting interests, seeking advice from the journal if they are not sure whether something constitutes a relevant interest.
- They should not allow their reviews to be influenced by the origins of a manuscript, by the nationality, religious or political beliefs, gender or other characteristics of the authors, or by commercial considerations.
- They should be objective and constructive in their reviews, refraining from being hostile or inflammatory and from making libelous or derogatory remarks.
- They should acknowledge that peer review is largely a reciprocal endeavour and should carry out their fair share of reviewing in time.
- They should provide journals with personal and professional information that is accurate and a true representation of their expertise.
- They should recognize that impersonation of another individual during the review process is considered serious misconduct.

These are only some basic principles. COPE issued much detailed ethical standards to be observed by every reviewer at various stages of the peer review process. A detailed version of these principles can be downloaded from the website www.publicationethics.org. Many other organizations like Council of Science Editors (CSE) also issued a number of guidelines regarding the responsibilities and ethical practices in the peer review process (www.councilscienceeditors.org).

The reviewers should be aware of various ethical issues when deciding whether to review a paper. Journals rely on their reviewers to evaluate the quality, importance, and novelty of the science presented in the manuscript. It is not easy to write a review. They must be clear, concise and accurate. It helps in the revision of the paper and suggests the ways to improve the paper by including additional data or experiments. Even after the publication of the paper the review process should remain confidential. Also the reviewer should show interest to read the final published paper, because the conclusions in the final paper may be altered or may be restated more precisely.

REVIEW QUESTIONS

1. What is a scientific paper?
2. What are the features of a scientific paper?
3. What are the various components of a scientific paper?
4. Discuss the various components of a scientific paper in detail.
5. How the title of a research paper should be written?
6. Give some guidelines for writing the title of a research paper.
7. Give the six techniques for making the title of a research paper effective.
8. What is abstract? Explain the features of an abstract.
9. Why an abstract is important?
10. What are the advantages of a good abstract?
11. What is an indicative abstract?
12. What are the main components of an indicative abstract?
13. Give the guidelines for preparing effective abstract.
14. What are keywords and why they are important?
15. What things must be included in the introduction part of a research paper?
16. What should not be included in the introduction part of a research paper?
17. What are the purposes of the introduction?
18. Give the guidelines for writing a good introduction to a research paper.
19. Explain the contents of the materials and methods section of a research paper.
20. Give the guidelines for preparing the materials and methods section of a research paper.
21. What is the importance of the result and discussion section?
22. Give the guidelines for preparing the result and discussion section of a research paper.
23. What are the elements that should be included in the result and discussion section of a research paper?
24. What should be avoided in the result and discussion part of a research paper?
25. Give the guidelines for writing an effective conclusion for a research paper.

26. What are the qualities of a conclusion?
27. What is acknowledgements in a research paper? Give the order of preference in placing the acknowledgements.
28. Why reference is important in a research paper?
29. What are the different types of references used in scientific writing?
30. What points the author must keep in mind while writing a scientific paper?
31. What are the points based on which the assessment of a scientific paper is made?
32. Explain the importance of scientific writing.
33. What are the general features of a good scientific paper?
34. Discuss the various benefits of scientific writing.
35. Explain how the quality of scientific writing is assessed?
36. Discuss the characteristics of scientific writing.
37. Give the rules for scientific writing.
38. Give the five rules that make scientific communication effective.
39. Explain Gopen and Swan's guidelines for writing effective scientific prose.
40. What are the structural principles of scientific writing?
41. Discuss the importance and various features of effective word selection in scientific writing.
42. Give the guidelines for making scientific writing more fluid.
43. Discuss the scientific usage of tenses.
44. What is Occam's Razor? Explain its importance in scientific writing.
45. Give Occam's first and second Razor.
46. Explain writer's block.
47. What are the contributing factors of writer's block?
48. Give some suggestions to overcome writer's block.
49. What are the duties and responsibilities of an author?
50. How will you select an appropriate journal for publishing your research paper?
51. What are the factors that must be considered for selecting a journal for research publication?
52. What are the features of electronic publication?
53. What are the advantages of e-publication?
54. Discuss various methods for the submission of a research paper.

55. What is peer review? What are the different types of peer review?
56. Gives the points based on which the peer review is conducted.
57. What are the possible responses in a peer review process?
58. Draw the flowchart of a peer review system.
59. What are the responsibilities of a reviewer?
60. Give the principles to be followed by a reviewer.

REFERENCES

[1] Robert A Day, "*How to Write and Publish a Scientific Paper*". 1983, Second edition. ISI Press, Philadelphia, p. ix.

[2] O'Conner, M and Woodford, F.P., "*Writing scientific papers in English*", 1976, Elsevier, Amsterdam, p. 108.

[3] O'Connor, M, "*Writing Successfully in Science*". 1995, Chapman & Hall, London, p. 1.

[4] Michael Jay Katz, "*From Research to Manuscript: A Guide to Scientific Writing*", 2006, Springer, p. 53.

[5] Peat J., Elliott E., Baur L., Keena V, "*Scientific Writing - Easy when you know how*", 2002, BMJ Books, London, p. 94.

[6] Silvija Brkic et al., "*Title, abstract, keywords and references in biomedical articles*", 2003, Archive of Oncology;11(3): 207-9.

[7] Ibid.

[8] Perry C, Cavaye A, and Coote L, "*Technical and Social Bonds with Business-to-Business Relationships*", 2002, Journal of Business and Industrial Marketing, 17(1): 75-88.

[9] Jean Luc Lebrun, "*Scientific Writing: A Reader and Writer's Guide*", 2007, World Scientific Publishing Co. Pte. Ltd. pp. 110-113.

[10] Dena Bain Taylor, "*A Guide to the Abstract*", 2007, University of Toronto.

[11] Silvija Brkic et al.

[12] Dena Bain Taylor.

[13] Theuns Kotze, "*Guideline on Writing a First Quantitative Academic Article*", 2007, p. 6.

[14] Silvija Brkic et al.

[15] Ann M Korner, "*Guide to Publishing a Scientific Paper*", 2008, Routledge, pp. 32-33.

[16] Richard D Branson, "*Anatomy of a Research Paper*", Respiratory Care, October 2004, Vol. 49, No. 10, 1222-1228.

[17] Ibid.

[18] Ann M Korner, pp. 43-44.

[19] Perry C, Carson D and Gilmore A, "*Joining a conversation: Writing for EJM's editors, reviewers and readers requires planning, care and persistence*", 2003, European Journal of Marketing, 37(5/6): 652-657.

[20] Summers J O, "*Guidelines for conducting research and publishing in marketing: From conceptualization through review process*", 2001, Journal of the Academy of Marketing Science, 29(4): 405-415.

[21] Ann M Korner, p. 41.

[22] "*Writing an Empirical Paper in APA Style*", Psychology Writing Center, University of Washington, http://www.psych.uw.edu/psych.php#p=339.

[23] Kevin W. Plaxco, "*The art of writing science*", Protein Science, 2010, Vol.19: 2261-2266.

[24] Margaret Cargill and Patrick O'Connor, "*Writing Scientific Research Articles: Strategy and Steps*", 2009, Wiley-Blackwell, p. 35.

[25] Ibid.

[26] Margaret Cargill and Patrick O'Connor, p. 36.

[27] Richard D Branson.

[28] Ann M Korner, pp. 49-50.

[29] Ibid., p. 56.

[30] Chatburn, RL. "*Respiratory care research*". Cleveland: Cat in the Dark Productions; 2002, 254–255.

[31] Richard D Branson.

[32] "*Writing an Empirical Paper in APA Style*", Psychology Writing Center, University of Washington.

[33] "*Fourteen Steps to Writing an Effective Discussion Section*", San Francisco Edit, www.sfedit.net

[34] Kevin W. Plaxco.

[35] Dean R Hess, "*How to Write an Effective Discussion*", Respiratory Care, October 2004, Vol. 49, No. 10, 1238-1241.

[36] Margaret Cargill and Patrick O'Connor, p. 56.

[37] Mahmoud F. Fathalla, "*A Practical Guide for Health Researchers*", 2004, World Health Organization, p. 138.

[38] Jean Luc Lebrun, pp. 199-200.

[39] Ibid., pp. 205-206.

[40] Silvija Brkic et al.

[41] Marc E. Tischler, "*Scientific Writing Booklet*", p. 9.

[42] Booth, W. C., Colomb, G. C., and Williams, J. M., "*The Craft of Research*". 1995, The University of Chicago Press.

[43] O'Connor M, p .3.

[44] WCG Peh, "*Scientific writing and publishing: its importance to radiologists*", Biomed Imaging Interv J, 2007; 3(3): e55.

[45] Harold Rabinowitz and Suzanne Vogel, "*The Manual of Scientific Style: A Guide for Authors, Editors and Researchers*", 2008, Elsevier, pp. 6-7.

[46] Ibid., p. 6.

[47] Rogers S M, "*Mastering Scientific and Medical Writing: A self help guide*", 2007, Springer, p. 4.

[48] Leslie N Carraway, "*Improve Scientific Writing and Avoid Perishing*", Am. Midl. Nat. 155:383–394.

[49] Robert Barrass, "*Scientists must write*", 2002, Routledge, pp. 30-35.

[50] Marc E. Tischler, p. 2.

[51] Robert Barrass, p. 88.

[52] Harold Rabinowitz and Suzanne Vogel, pp. 12-17.

[53] George D. Gopen and Judith A. Swan, "*The Science of Scientific Writing*", American Scientist (Nov-Dec 1990), Volume 78, 550-558.

[54] Torbjorn Lindgren et al., "*Photo-electrochemical and Optical Properties of Nitrogen Doped Titaniumn Dioxide Films Prepared by Reactive DC Magnetron Sputtering*", J. Phys. Chem. B 2003, 107, 5709-5716.

[55] Zuzana Svobodova, "*Writing in English: A Practical Handbook for Scientific and Technical Writers*", 2000, Leonardo da Vinci programme, European Commission, p. 64.

[56] Robert Nola and Howard Sankey, "*Theories of Scientific Method*", 2007, Acumen Publishing Limited, p. 41.

[57] Anderson D, "*Occam's razor: Simplicity, complexity, and global geodynamics*", 2002, Proceedings of the American Philosophical Society. 146, 56-76.

[58] Alberto Murta, "*Occam's razor - essentials and applications*", http://www.complexsystemsstudies.eu

[59] Ibid.

[60] Domingos, P, "*The role of Occam's razor in knowledge discovery*", 1999, Data Mining and Knowledge Discovery. 3, 408-425.

[61] Alberto Murta.

[62] Margaret Cargill and Patrick O'Connor, pp. 69-72.

[63] Tracey Brown, "*Peer Review and the acceptance of new scientific ideas*", 2004, http://www.senseaboutscience.org,

[64] "*Peer Review: A Guide for Researchers*", www.rin.ac.uk/peer-review-guide, p. 4.

[65] "*Peer Review: The Nuts and Bolts*", 2013, pp. 13-14, www.senseaboutscience.org

[66] Murray, D., Schwartz, J., Lichter, S.R., "*It ain't necessarily so: how the media remake our picture of reality*", 2002, New York: Penguin Books, pp. 148-149.

[67] "*Peer Review: The Nuts and Bolts*", p. 8.

[68] Ibid., pp. 17-20.

[69] Ibid., p. 23.

Suggested Readings

1. Bjorn Gustavii, "*How to Write and Illustrate a Scientific Paper*", 2008, Cambridge University Press.
2. Page G, Campbell R and Meadows A J, "*Journal Publishing*", 1997, Cambridge University Press.

3. W I B Beveridge, "*The Art of Scientific Investigation*", 1957, W.W.Norton and Company Inc. New York.
4. Robert Barrass, "*Scientists Must Write*", 2002, Routledge.
5. Meadows A J, "*Communicating Research*", 1998, Academic Press, San Diego.
6. Frederickson, E.H. (ed.) "*A Century of Scientific Publishing*", 2001, IOS Publishing, Amsterdam.
7. Straw, Jane Anne, "*Unstuck: A Supportive and Practical Guide to Working Through Writer's Block*", 2003, St Martin's Press, Boston.
8. Kerry L. Dearfield, and A. Robert Flaak, "*Peer Review Handbook*", 2000, U.S. Environmental Protection Agency.
9. Elizabeth Wager, Fiona Godlee and Tom Jefferson, "*How to Survive Peer Review*", 2002, BMJ Books.

Chapter 8

Preparation of Research Thesis

"As for the search for truth, I know from my own painful searching, with its many blind alleys, how hard it is to take a reliable step, be it ever so small, towards the understanding of that which is truly significant".

— **Albert Einstein**

"Nothing in life is to be feared, it is only to be understood".

— **Marie Curie**

8.1 THESIS AND DISSERTATION

Many of us are usually confused between a thesis and a dissertation. Many consider both as the same. But there is a large number of differences between a thesis and a dissertation. During a Ph.D. programme, a researcher learns and understands many things about his topic of research, and in the end, he has to submit a thesis for evaluation purpose. A thesis is a lengthy piece of written work that includes original research or expanded research on a new or existing topic. The thesis should be written as per the guidelines set by the university or the research institute. It should contain the details of the study made by the research during the Ph.D. programme. On the other hand, the dissertation is a report that is to be submitted at the end of a course such as an M.Phil. Some universities demand the submission of a dissertation which is usually a report of the project work carried out during the course, for the fulfilment of the award of master's degree. A dissertation is also a scholarly piece of writing. A thesis allows students to record their knowledge and expertise within the subject matter they have been studying.

It is to be noted that different countries define the words thesis and dissertation differently, i.e., in some countries they are used interchangeably, while in some countries thesis is related to bachelor's or master's degree course and dissertation is used in the context of a doctorate degree, whereas in some countries the reverse is true. In countries like India and UK, Ph.D. scholars have to submit a thesis, while M.Phil. students submit a dissertation. Since the Ph.D. is a 3–5-year programme in India, the thesis contains a greater number of pages and elaborate discussions than an M.Phil. dissertation.

A Ph.D. programme is designed such that the researcher studies a topic in detail and he has to publish the possible number of research articles in internationally reputed journals. Publications in international journals are mandatory for the award of a doctoral degree. In the next section, we will see the components of a thesis in detail. The following differences between a thesis and a dissertation may be noted.

- The primary difference between a thesis and a dissertation is the time when they are completed. As mentioned earlier, a thesis is presented at the culmination of a master's programme, whereas, a dissertation is presented to earn a Ph.D.
- A dissertation is a compilation of research ensuring that the researcher is well informed and has knowledge about the research topic learnt in the study programme. On the other hand, a thesis provides an opportunity for the researcher to contribute new theories and information to the existing literature in the research field.
- A dissertation is the presentation of learnt and existing information, while the purpose of a thesis is to develop a unique concept and defend it with the basis of theoretical and practical results.
- A master's or M.Phil. dissertation is approximately 100 pages in length. However, a Ph.D. thesis should be much longer than a thesis and must include background and research information. A thesis must include your research proposal, grant proposal, literature review, ideation of research topic, and every other minute detail about your research. Consequently, the length of the thesis is approximately three times of a dissertation.

Further, thesis and dissertation share some common features. Some of them are [1]:

- Both these papers have the same purpose: commonly, in many institutions, these two terms are used interchangeably.
- A dissertation and a thesis have certain terms, they should be done in time without delays. If a student fails to defend his / her document, they could try to do it many times.
- Both dissertation and thesis require students to choose a topic for research and create complex work to demonstrate the level of their skills and knowledge they have got for years of studying various programmes on their faculty.
- Both dissertation and thesis have a similar structure and format.
- If students are going to get their M.Phil. / Ph.D. degree, they should create a master thesis proposal/dissertation thesis proposal before writing a final paper. The main goal of this work is to introduce to readers the main aim of writing the future document.
- A student has to defend both a completed thesis and a dissertation to get a certain degree.

- If students are writing a dissertation/thesis, they should avoid copyright infringement: this term means they couldn't copy works from other authors because their rights are reserved. That is why it is important to be careful with every word during creating a thesis/dissertation if high quality is the aim.

8.2 COMPONENTS OF A RESEARCH THESIS

The contents of a thesis can be divided into three sections; the title pages, the subject proper, and the references. [2]. Various components of the title pages and the subject proper are given below.

1. Preliminary Pages

- Title page
- Signature page
- Declaration by the student
- Certificate from the guide
- Acknowledgements
- Table of contents
- List of figures
- List of tables
- List of symbols, abbreviations, or nomenclature.

2. Subject Proper

- Introduction
- Review of literature
- Materials and methods
- Results
- Discussion
- Summary and conclusion.

A discussion of important components is given below.

8.2.1 Preliminary Pages

Preliminary pages include the details such as the title, certificates, acknowledgements, table of contents, etc. These pages must be written neatly and proper typesetting must be done.

Title Page: For a Ph.D. thesis, the title of the thesis is already fixed while submitting the synopsis at the time of registration. However, if the researcher feels to make some modifications in the title, usually he can give an application to the authority and get approval of the modified title. "For a thesis, two title pages are usually given; the first is the cover page with the title of the thesis, author's name, and department and college name. On the second title page, title name, author's name, mandatory name of the thesis (whether the degree is Master of Science or Doctor of Philosophy), faculty name, name of the department, and name of the institute or college are given".

Certificate Page: The certificate page contains the declaration by the researcher and a certificate from the supervising guide. The main objective of these certificates is to declare that the work done by the researcher is bona fide and unethical practices like plagiarism are not employed while preparing the thesis. Usually, a certificate after plagiarism checking is also to be included while submitting the thesis.

Acknowledgements: The acknowledgements is included on a separate page before the table of contents. The researcher acknowledges the help from the authorities, research supervisor, friends, institutions, teachers, various departments, funding agencies, etc., in this section. Acknowledgements should be written briefly and any fancy sentences must be avoided. Help received from friends and relatives may be acknowledged, but without specifying their names.

Table of Contents: The table of contents is included after the acknowledgements page. All major divisions of the thesis are included on the content page. Title, certificates, and acknowledgements are not usually listed. Chapter headings and subheadings in the order of page number are given in the content list.

List of Tables and Figures: A separate list of tables and figures included in the thesis is to be given after the table of contents. Separate pages must be used for the list of tables and the list of figures. The list of tables must include the table number, details of the table, and the corresponding page number. Similarly, the list of figures must also include the figure number, figure caption, and page. All the details must be given in order. Therefore, for including figures or tables in a thesis, there are four considerations to take into account [3]:

1. The graphical display (does it look good and show what it ought to?)
2. Annotation (to explain features in the figure or table)
3. The figure or table title
4. The figure or table legend (a short piece of text that tells the reader what they are looking at).

The easiest way to number your figures and tables is by giving the chapter number and then an individual figure or table number; for example, in Chapter 1, the figures should be numbered like Figure 1.1, Figure 1.2, Figure 1.3, and so on. Similarly, the tables are to be numbered as Table 1.1, Table 1.2, Table 1.3, and so on.

8.2.2 Subject Proper

This section is the main part of the thesis. All the research work done by the researcher is presented in this section in an organized manner. The usual format for preparing this section is the same as that for preparing a scientific paper. For writing a scientific paper we usually follow the IMRAD pattern [4] and the same can be used for writing the thesis also. However, separate chapters can be used for writing the 'Introduction', 'Review of literature', 'Materials and methods, 'Results and Discussion', and 'Conclusion'. A brief discussion of these is given below.

Introduction: Apart from a research paper, in a thesis, the researcher gets enough space to describe the background knowledge and motivation before starting the research. The main objective of the introduction part is to inform the reader about the justification and rationale in taking up the research work. The research problem, its importance, and practical implications and applications can also be included in the introduction part. In the introduction, the researcher must justify the reason for undertaking the research work. Some of the relevant previous works can also be indicated in the introduction. Do not include a large number of previous works as the researcher gets enough space to discuss all those in the section of the literature review. The following points may be noted while writing the introduction of a thesis [5].

- State the general topic and give some background
- Provide a brief review of the most relevant literature related to the topic
- Define the terms and scope of the topic
- Outline the current situation
- Evaluate the current situation (advantages/disadvantages) and identify the gap
- Identify the importance of the proposed research
- State the research problem/questions
- State the research aims and/or research objectives
- State the hypotheses
- Outline the order of information in the thesis
- Outline the methodology

In brief, after reading the introduction of a thesis, one must get an idea about the research topic, its importance, and practical implications.

Literature Review: In IMRAD structure, the literature review is dedicated to a review of the previous works related to the research topic. The literature review is included in a separate chapter. Through the literature review, one could be able to understand the current level of knowledge related to the research topic. Also, he will be getting the details of researchers working in the same field. This will be helpful in the future if he wants to make some collaborations with them. Moreover, the topics can be discussed with eminent researchers. In a thesis, the review is meant to present not only up-to-date knowledge of the field but also the intellectual ability of the student to synthesize and put together information from different sources. Proper guidance from the research guide is necessary to have a good review of the literature. For a literature review, "there should be a clear structure and there should be an explanation for that structure. The literature should be presented in a planned order, for which there is a clear rationale" [6]. A literature review helps readers to appreciate something of the sequence and growth of knowledge.

There is a misunderstanding that the literature review is to include as many research papers as possible. Actually, the number does not matter. One should include only those publications which are most relevant to the problem under study. The researcher needs to identify the key trends in the research area of his interests and need to formulate the research problem accordingly. Too old and outdated ideas and problems need not be considered. Finally, it is very important to select only the original and primary research articles published by genuine researchers with good research experience. When we search over the Internet, we get a number of related information from various websites. However, very often one cannot verify the authenticity of that information.

Materials and Methods: The methods and materials used for the research is discussed in this section. The researcher presents a detailed description of the experimental procedure. For theoretical research, this section is usually represented as the 'theory' or the 'model'. In a research paper, only a brief account of the methods or theory is given. In a research thesis, the researcher has the freedom to give a detailed experimental description of the methods used in the laboratory or theoretical derivations.It is vital that this section is clear, comprehensive, and sufficiently detailed so that other researchers can adequately judge the results obtained by the researcher and can validly replicate the study. The search process will only be convincing if the researcher explains the following questions [7].

- Why a particular method has been chosen?
- What are the characteristics of the population, sample or experiment?
- Why some possibilities have been rejected?
- What instruments will be used?

Therefore, the description of the materials and methods should be given in simple language and unnecessary complexity is to be avoided. For a particular problem, there may be different methods to solve it. The researcher should not switch to one another and he should concentrate on the selected method and explore its potential advantages and disadvantages.

Results: There are two possibilities; either one can separate these into two chapters or write as a single chapter. If a separate chapter is used, then the data collected during the research is compiled into the chapter 'Results'. "The 'Results,' therefore, should communicate the facts, measurements, and observations gathered by the student supported by data in tables, diagrams, graphs, and photographs" [2, p. 309].

During the research, the researcher may collect a large amount of data. However, in the "Results" chapter, only the most relevant data related to the hypothesis must be selected after a critical evaluation. It is to be noted that in the "Results" chapter no discussion is to be included. Most relevant observations are to be pointed out. A more comprehensive method to represent data is the inclusion of graphs (line or bar) and charts. Care is to be taken in the positioning and placing of the plots. Tables should be placed vertically rather than placing it horizontally.

Discussion: Discussion is the most important part of a research thesis. The researcher presents and interprets the findings of the research. At the end of the discussion, the researcher must add something new to the existing knowledge. Positive and negative results are to be discussed in detail. The conclusion must be justified with proper evidence. In the discussion, the results need not be written again, but a discussion of the results given in the previous chapter is to be presented. The following points may be noted [2, p. 310]:

- Some students, due to lack of experience or proper guidance, present a superficial interpretation, sometimes, simply re-stating the results without bothering to explain underlying mechanisms.
- Any attempt to prove or substantiate a point should take into account the findings of earlier research workers. State clearly how your results and interpretations agree or contrast with the published works of others.
- Remember that in life sciences, as phenomena are so complex, it is possible to have more than one interpretation.
- While interpreting the results, including previous results or comments to buttress the arguments. However, no need to mention all the previous works, but choose the most relevant works and include them based on their relative importance to the present study.
- Do not encourage the tendency of some students to conceal negative results.
- Any deviation from the expected observation must be interpreted properly.

Summary: Note that the summary is different from an abstract. The "summary" contains a summary of all the findings of the research. A thesis summary must include an introduction, materials and methods, in brief, main findings, and conclusion. This can be organized as:

- A brief statement of the problem statement
- A brief description of the context of the thesis work
- The importance and significance of the work
- Objectives of the study

- Statement of the hypothesis
- Summary of the materials and methods
- Results or contributions.

In the conclusion, the research outcome that contributes to the scientific knowledge must be included. An indication about the related study in the future can also be included at the end of the summary.

References: The Bibliography or List of References appears after the summary chapter. It is a complete listing of all cited resources used to develop the research thesis. This complete reference list of all citations must appear at the end of the thesis. In science disciplines, the *Author-Date* system of referencing is usually followed. *Harvard* system of referencing is an *Author-Date* system. In this system, sources are briefly cited in the text by the surname or last name of the author closely followed by the year of publication, often in parentheses. Other referencing systems like CSE Style Manual [8], APA Style Manual [9], and Chicago Style Manual [10], recommend the Harvard system of referencing with some minor variations.

A detailed description of all these components can be seen in *Research Methodology and Scientific Writing* by C George Thomas, (2021, Springer).

8.3 THE THESIS ABSTRACT

The abstract is an important component of a research thesis. Presented at the beginning of the thesis, it is likely the first substantive description of the research work read by an external reader or a reviewer. The abstract is a summary of the whole thesis. It presents all the major elements of the research work in a highly condensed form. An abstract often functions, together with the thesis title, as a stand-alone text. The abstract will help the reader to get an overall idea about the research work presented in the thesis. A good abstract should answer the following questions.

- What is the reason for writing the thesis?
- What are the current approaches and gaps in the literature?
- What is the research problem and what are the aims?
- Which methodology is used?
- What are the main findings?
- What are the main conclusions and implications?

The abstract is limited in words. For a Ph.D. thesis, the abstract may have a light of around 500 words. The primary function of the abstract is not to tell readers what the researcher did during his Ph.D. work, it is to tell them what he has discovered. Therefore, it is important to include the result of the study in the abstract. A typical thesis abstract should have information on four aspects [2, p. 312]

- the rationale and main objectives of the study,
- the materials and methods used,
- the summary of the results obtained, and
- the conclusion of the entire thesis.

It is to be noted that citations or references to tables or figures given in the main part of the thesis should not be included in the abstract. In the abstract, the results must be stated in the past tense. Also, abbreviations, acronyms, fancy words, etc., are to be avoided in the abstract. The following things must be taken into account while writing a good abstract [11].

- Give a good first impression by writing in short clear sentences
- Don't repeat the title in the abstract
- Don't cite references
- Use keywords from the document
- Respect the word limit
- Don't be vague. The abstract should be a self-contained summary of the research, so don't introduce ambiguous words or complex terms
- Focus on just four or five essential points, concepts, or findings. Don't, for example, try to explain the entire theoretical framework
- Edit it carefully. Make sure every word is relevant (you haven't got room for wasted words) and that each sentence has maximum impact
- Avoid lengthy background information
- Don't mention anything that isn't discussed in the thesis
- Avoid overstatements
- Don't spin the research findings, contribution or significance to make your research sound grander or more

8.4 FORMATTING THE THESIS

Before printing, the thesis is to be formatted as per the guidelines issued by the authorities such as universities. Usually, the thesis is printed on A4 size paper. Therefore, while formatting the thesis, the page size must be selected as A4 from the page setting. Further, the thesis must be printed on high-quality bond paper. Good quality bond paper lasts long without considerable fading, discolouration, or deterioration. The thesis is printed only on one side of the paper. The line spacing is to be adjusted to 1.5 points. The double spacing is also desirable, however, it consumes more pages and if the thesis is long, it becomes bulky. Recently, in Kerala universities like Calicut University and Kannur University allows book-sized theses which are very easy to handle. The template for formatting the thesis is available on the university website. George Thomas pointed out the following various steps involved in formatting a thesis [2, pp. 313-316].

Text Spacing: The thesis is usually prepared in A4 size bond paper and must be word-processed with a good quality printer. "Most universities now permit one-and-a-half spacing for regular texts, and single space for block quotes, footnotes, headings, tables, table and figure captions, reference entries, and for the 'Table of contents". Between paragraphs and between each reference, one space is to be left if the single spacing is used. For spacing of 1.5 points, no need to leave any space between paragraphs.

Margins: Proper margin is to be provided on all sides of the paper. Usually, on the top and left sides, a margin of 1.5 inches is recommended. A margin of 1.5 inches on the leftside is required for the correct visibility of the text when binding the thesis. On the right side and bottom, a margin of 1 inch is usually recommended. "One can use a larger top margin on main heading pages. Similarly, smaller margins may be used with appendix material, but the page numbers must appear in the same location as on the text pages".

Page Numbering: All the pages in the thesis must be properly numbered. The preliminary pages are usually numbered using lower case roman letters. After the preliminary pages, Arabic numerals are used for page numbering. The page number of the title page is "i", however, the page number is not shown in the title page. From the next preliminary pages onwards, the pages are numbered from "ii". Similarly, for the content page also the page number is not shown. Page numbers are to be put on top of the pages in the same typeface and size as the text. Also, the position of the page number must be the same throughout the thesis.

Design and Formatting Chapters: The average thickness of a thesis is assumed to be 250 pages. The thesis is typed with 1.5 spacing with a font size of 12 points. A thesis normally contains six chapters; Introduction, Review of literature, Materials and Methods, Results, Discussion, and Summary. It is to be noted that while placing the figures, graphs, tables, etc., it should come close to the relevant text where the reference about such figures, graphs, and tables are made. Each chapter is to be started from a new page with a chapter number and chapter title. A chapter may contain various sections and all sections should be properly numbered. In other chapters also, the sections should be numbered in the same manner. The guidelines regarding the font size of the chapter heading and its position, etc., will be given in the guidelines provided by the universities. In many cases, 16-point letters are used for typing the chapter heading. It is important to follow the guidelines issued by the universities while preparing the thesis. "Indicate the major divisions of a chapter by centre headings. There must be at least two major divisions of a chapter if centre headings are to be used. For centre headings, capitalize all the words in boldface. If no centre headings are possible as noted above to indicate major chapter divisions, major side headings can be used."

Numbering the Sections: A chapter usually contains different sections. Each section is given a heading. For easy access to such sections, they should be numbered properly and is to be indicated in the content pages. For the first chapter, the heading of major sections is to be given as 1.1, 1.2, 1.3, etc. The sub-heading below the major heading 1.1 can be numbered as 1.1.1, 1.1.2, 1.1.3, etc. The same method is to be used for numbering the major headings and sub-headings in all chapters.

Layout of Tables: In a research thesis or in a research paper, the tables are used for presenting a large amount of data in minimal space. It is to be noted that, the data explained in the text part should not be repeated in the table. The table should be self-explanatory and self-contained. Simple data can be explained in text form and if the data are more complex and detailed, then tabulation is more appropriate. Usually, in a thesis, the tables are placed in portrait style and not in landscape style. The portrait is more convenient for a reader. More details about the tables will be given in the next section.

Language and Style: For writing a scientific paper we use technical language, The same is to be used in writing the research thesis also. "If the language is apt and expressive, the results of research become comprehensible and appreciable". A researcher can get help from his supervisor, other experienced faculties or an English teacher for structuring and correcting the sentences in a more appropriate manner. All relevant points should be explained clearly which requires good linguistic skills. "Long words and phrases, clichés, redundant words, and repetitive usages" are to be avoided.

Typeface and Fonts: Typeface is the overall visual appearance or style of lettering. The font is the appearance of a letter like roman, boldface, or italic fonts. There are two categories of typefaces: *serif* and *sans serif.* "Serif typefaces give a visual illusion of connecting the letters in a word. Therefore, it is generally agreed that a serif typeface is easier to read than a sans serif typeface in the print form".In a thesis, one should use the same typeface throughout the thesis. If the guidelines of the university insist on any typeface, then the same must be used for the thesis. However, for figure and table captions, different typefaces can be used.

8.5 TABLES, GRAPHS, AND OTHER ILLUSTRATIONS

Tables and graphs are used to represent more data in a minimum space. When the amount of data is large and cannot be included in the text, one uses the table to show the relationship between variables. The reader can easily understand the relationship between different quantities through a table. Similarly, graphs are also used to show the relationship between different variables included in the study. A brief discussion about how to use tables and various types of graphs is given below.

8.5.1 Tables

Tables are used to display the research data in a compact form. In all research thesis, tables became an essential part. The data included in the table must be easily comprehensible and the table should look attractive. Therefore, it is very important to give proper attention while preparing the tables. Earlier, researchers used to draw tables on paper and put data in the table. Then, a photograph is taken, which is included in the thesis. Now, with the development of computer technologies, the preparation of attractive tables became a very easy process. One can use Microsoft Word or Excel programs to prepare a table.

A table contains five main parts: the title, the stub, the box headings, the field, and the supplements [2, p. 320]. A brief explanation of these is given below.

The title: The title of the table contains details of various relationships included in the table. It is written on the top of the table. The title normally includes the table number, heading, and a sub-heading (if necessary) of the table. The table heading is also known as the table caption. The caption is written in 'sentence case', with only the first word and proper nouns in capital initials. If any headnote is needed, it can be given below the title. All the tables in the thesis must be properly numbered.

The box headings: A table contains vertical columns and horizontal rows. The first row of a table is usually reserved for including the details of various quantities included in each column of the table. These details are known as the box headings. The box heading should be very brief and it must clearly explain the quantity included in the corresponding column. Any units of measurement should be written in abbreviated form in the box heading.

The stub: The first column in a table is known as the stub. It lists the items about which information is provided in the table in horizontal rows to the right. This depends on the type of data included in the table. Always, the first column need not represent such details.

The field: This is the main part of the table. It is also known as the body of the table. The data are included in the field. Usually, the data are numerical values. If the value of the data is large, the approximation is to be used. For example, if the value is 2564.68, in the table one should use 2564. On the other hand, if the value of the data is small one can use the decimal points. For example, if the data is 6.42, the same value is to be entered in the field. All the boxes on the table must be filled. If there is no data to enter in a particular box in the field, the "NA" is to be written in the table. Also, a footnote NA-Not available is to be included below the table.

Supplements: The supplements are given below the table as footnotes. The supplement usually includes footnotes about the sources of the data given in the table, special features of the table, and symbols used in the table. Four types of supplements are generally used: "source notes, general notes, notes on specific parts of the table, and notes on the general level of probability" [2, p. 321]. If the data from a different source is included in the table, it is to be acknowledged below the table.

8.5.2 Arrangement of Data in Tables

As already mentioned, the data must be included in the table comprehensively. The following must be remembered while arranging data in a table.

- The most appropriate form of table is in the portrait form. Landscape form is used only if it is unavoidable. Since the thesis is typed in A4 size paper, the table should be within the margins of the paper. If needed, the font size of the data in the table can be reduced maximum by 2 points. It is very important that all the columns of the table must be on the same page. If there are more rows that cannot be included on a single page, the table can be continued on the next page. In such cases, on the top of the table on the next page, table number and the word "continued" must be given. If the columns cannot be included in a single page, then the table must be arranged in a landscape form.
- Overcrowding of data in the table must be avoided. This means one should not include too much information in a single table. A maximum of five related parameters can be included in a table. This has the advantage that this will reduce the number of columns required and the table can be placed in the portrait format.
- If the value of the data is very large, it must be approximated. If the data contains more than 4 digits, then it can be expressed in powers of 10 as exponents. However, the exponent must be written in the heading box and the numerical value in the table. For example, if the data is 2564358.86, even if we neglect the value after the decimal point, the data is big. Then it can be written as 2.56×10^6. Then, in the field box, we write 2.56 and the multiplication factor 10^6 is included in the column heading.
- The data must be arranged logically. If the data collected during different years are presented in a table, it must be included in chronological order. If we are studying the variation of the resistance of a material with temperature, the data in the table must be presented in the increasing order of the temperature.
- The table must be placed immediately after the mention of the table in the text. If the table is not possible to include on the same page, it must be placed at the beginning of the next page.
- If an explanation about a certain data is required, then with the data, a symbol like *, #, †, etc. can be used as a superscript, and with the same symbol, the explanation can be given as a footnote.

8.6 ILLUSTRATIONS

Different types of illustrations are used to represent data in the most compatible form. Illustrations are usually used when a large number of data are to be presented. In such cases, these data cannot be included in a table. The commonly used illustrations are photographs, line drawings, charts, graphs, and tracings. Illustrations make the data readily comprehensible. The comparison between different variables is very easy in illustrations like graphs and charts. It should be comprehensive as well as self-explanatory. In the text, the term 'figure' is used to represent avariety of graphs, charts, maps, sketches, diagrams, and drawings. While writing the caption under the figure we use "Fig." followed by the number of the figure and then the caption. It is to be noted that the photographs, diagrams, and drawings are less explanatory than graphs. Therefore, the most important illustration in the thesis as well as in research papers is the graphs. The advantage is that many hundreds of data can be included in a graph and the comparison between various parameters can be done in a single graph. A brief discussion of various illustrations is given below.

8.6.1 Figures

The term "figure" is misused in a broad sense. It includes various illustrations like photographs graphs, charts, maps, sketches, diagrams, and drawings. While preparing a figure for a research thesis, the following points must be taken care of [2, p. 324].

- *Create Simple Figures:* It is very important that the figures used in a thesis should be simple and must convey a clear idea. The reader must be able to understand the details of the figure without giving much description. "Present the data carefully and accurately to avoid oversimplification, misrepresentation, or distortion of facts. Ensure that numerical data upon which the figure is based are presented in the text or in an accompanying table unless they are included in the figure itself".
- *Use Figures Sparingly:* It is not good to use too many figures in a research paper or in a thesis. Even though the figures are an easy way to communicate with the readers, if too many figures are used, they may only detract the readers from the paper instead of elucidating it.
- *The Caption to Figures:* To explain the content in a figure, the figure caption is used at the bottom of the figure. Before caption, the figure number is to be added. For example, for a figure showing the photograph of a telescope, below the figure it should be written "Fig. 1. Telescope" as the figure caption. For a table, the caption must be given above the table, but for a figure, it must be below the figure only.
- *Numbering of Figures:* All figures in a thesis should be numbered in consecutive series. The pages containing figures should also be numbered in the same series as those of the text.

- *Explanation of Symbols:* Symbols used in the figure must be explained as a footnote. Also, if the author is using a figure from another source, it must be mentioned in the footnote.
- *Selection of Most Appropriate Charts or Graphs:* To present data comprehensively, there are a wide variety of graphs and charts are available. The selection of a particular form of the graph depends on the type and amount of data to be represented. The various forms of graphs include bar graphs, pie graphs, line graphs, area graphs, etc. A detailed discussion of the graphs is given in the following sections.

8.6.2 Line Graph

Line graphs are usually used for plotting the relationship between continuous variables. The independent variable is usually taken along the x-axis and the dependent variable along the y-axis. Since we make scientific interpretations based on the plotted graph, the accuracy of the graph is very important. The relation between a dependent and an independent variable is easily observed from the line graph. Any abrupt change is easily visible. Therefore, it will communicate with the reader quite easily and effectively. In a single graph, we can represent the variation of different parameters with respect to the independent variable. The line graph must be observed to understand in the following sequence.

- First, take a look at the two axes and try to understand what these axes denote.
- Then look at the graph and examine the values of the points situated in the lines of the graph.
- Follow the lines and figure out if there is any kind of rise or fall. Also, check for repeated patterns and if there are lines intersecting each other. This way, one will get to know the purpose of the line graph.
- There is also a possibility that one will end up seeing emerging patterns that will help you guess the trend.

The line graphs can be of the following types:

- *Simple line graph:* Only one line is plotted on the graph.
- *Multiple line graph:* More than one line is plotted on the same set of axes. A multiple line graph can effectively compare similar items over the same period of time.
- *Compound line graph:* If information can be subdivided into two or more types of data. This type of line graph is called a compound line graph. Lines are drawn to show the component part of a total. The top line shows the total and the line below shows part of the total. The distance between every two lines shows the size of each part.

The graph must include a title that explains what is plotted in the graph. For example, "The plot of resistance versus the temperature" can be used as the title of a graph that shows the variation of resistance with temperature. Then, in the graph, the scale used to plot the graph must be indicated. The scale is the number that explains the units utilized on the linear graph. Further, on both the side and the bottom of the linear graph there must be a label that indicates what kind of data is represented in the graph.

The following are the essential elements of a good graph [12].

- A good title that describes the relationship between the variables.
- The graph should fill the space allotted for the graph. The graph should be as large as the paper will allow. In order to do this, the graph must be properly scaled.
- Each axis should be labelled with the quantity being measured and the units of measurement. The independent variable is usually plotted on the x-axis and the dependent variable is usually plotted on the y-axis
- Each data point should be plotted in the proper position. Use a tiny dot to mark your point, and circle the dot with a point protector so that it doesn't get covered by your line of best fit.
- A line of best fit. This line should show the overall tendency of the data. If the trend is linear, one should draw a straight line that shows that trend using a straight edge. If the trend is a curve, one should sketch a curve which is the best guess as to the tendency of the data. This line (whether straight or curved) does not have to go through all of the data points and it may, in some cases, not go through any of them. Never connect the data points dot to dot.

8.6.3 Bar Graph

Bar graphs are used when the data is not continuous and only a smaller number of data is to be plotted. "These are generally used when a striking difference is to be stressed, and when it is not important to maintain a high degree of precision of the individual mean values". There are different types of bar graphs. A brief discussion of some of them is given below.

Simple bar graph: A simple bar graph is used to compare some discrete data. They are rectangles of equal width arranged either horizontally or vertically. The length of the bar is proportional to the quantity to be plotted. Discrete data like the population in a country in different years, consumption of fuel in different years, monthly recovery rate from a pandemic disease, etc., can be illustrated using a simple bar graph.

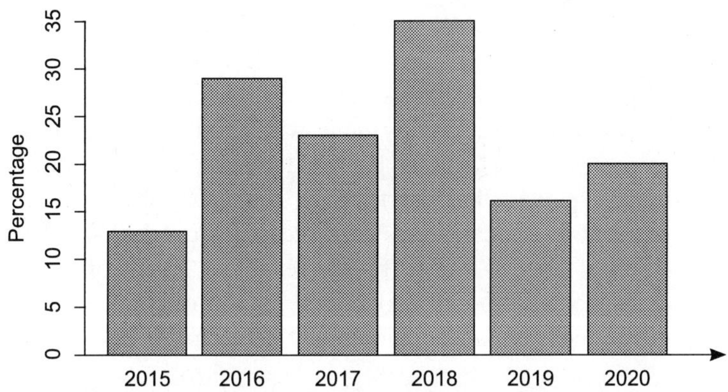

Fig. 8.1 Simple vertical bar graph

Multiple bar graph: In certain cases, the researcher needs to represent more than one characteristic of a single item in a single plot. Then, multiple bar graphs can be used. For example, if we want to represent the vaccination rate and recovery rate for comparison, then we can use a multiple bar graph.

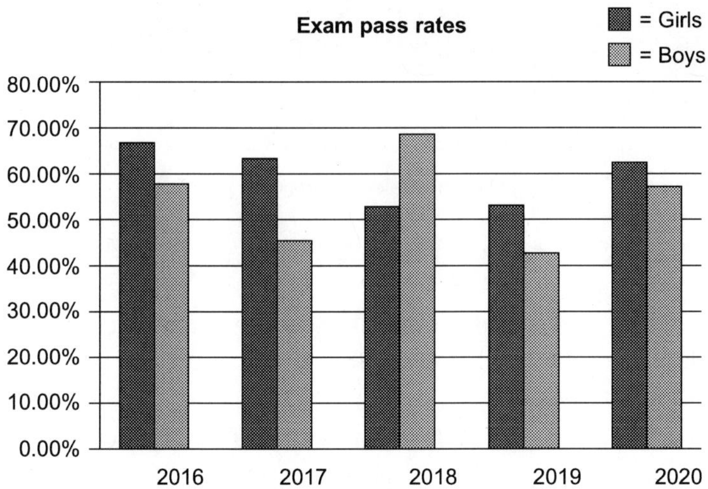

Fig. 8.2 Multiple bar graph

Figure 8.2 shows the exam pass percentage of boys and girls in the certain entrance examination. Easy comparison of pass percentage of boys and girls is possible from the graph. Multiple bar graphs can also be drawn horizontally.

Stacked bar graphs: Stacked bar graphs show the quantitative relationship that exists between the main category and its subcategories. Each bar represents a principal category and it is divided into segments representing subcategories of a second categorical variable. The chart shows not only the quantitative relationship between the different subcategories with each other but also with the main category as a whole. They are also used to show how the composition of the subcategories changes over time. These kinds of charts enable us to show more complex relationships than those usually done with standard bar graphs. SBG is represented by means of rectangular bars that can be oriented horizontally or vertically just like standard bar charts. Each principal category is divided into segments representing subcategories of a second categorical variable. The quantity of each subcategory is shown by the length or height of rectangular segments that are stacked end to end horizontally or vertically. The final height or length of each bar represents the total amount of each principal category.

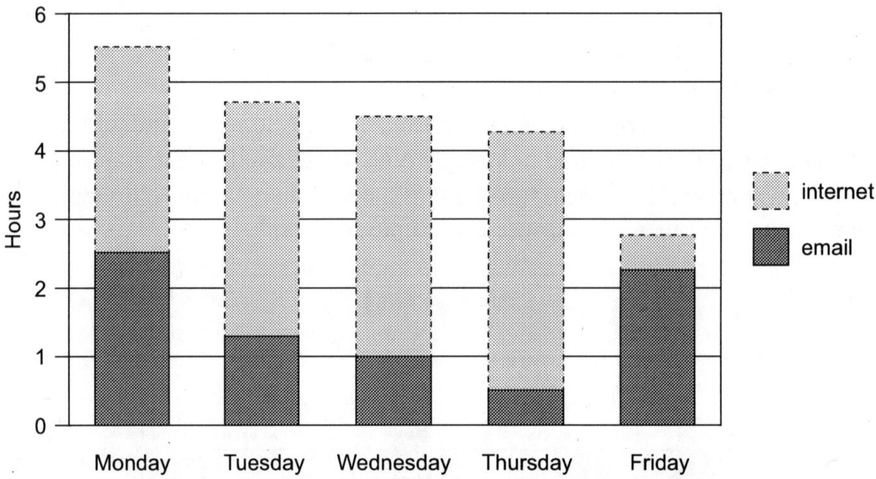

Fig. 8.3 Stacked bar graphs

Figure 8.3 shows the stacked bar graphs showing the time spent by a person in working days over the Internet and checking email. In this case, also, the plot can be drawn horizontally.

100% stacked bar graphs: A 100% stacked bar graph shows the relative percentage of multiple data series in stacked bars. The total of each stacked bar always equals 100%. Like a pie chart, a 100% stacked bar chart shows a part-to-whole relationship. However, unlike a pie chart, a 100% stacked bar chart can show how proportions change over time, like the changes in population in a country with year.

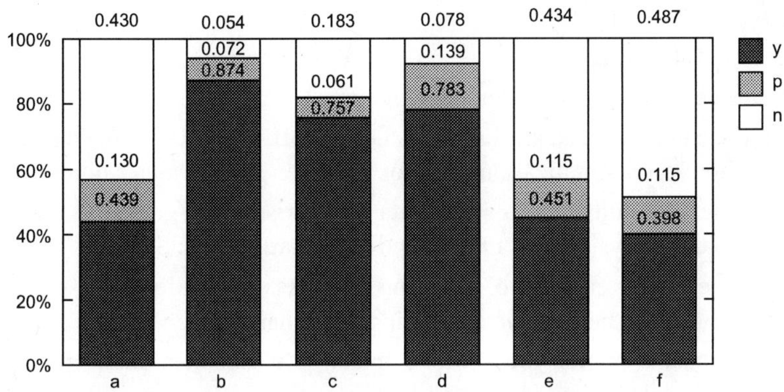

Fig. 8.4 Illustration of 100% stacked bar graphs

Bar graphs show the frequency of each data category. Also, it summarizes large dataset in a visual form. It readily shows that how the variable or group of variables change over time. The relationships between different variables can be interpreted quite easily.

8.6.4 Pie Charts

We have already seen the pie chart and its application in chapter 5. However, we will see some more details of the pie chart in this section. Pie charts are also known as sector charts, circle charts, or circular graphs. Usually, the pi diagram is used to represent the various components of the same item. These are represented inside a circle. The circle is divided according to the percentage of each component. An illustration is given in Figure 8.5.

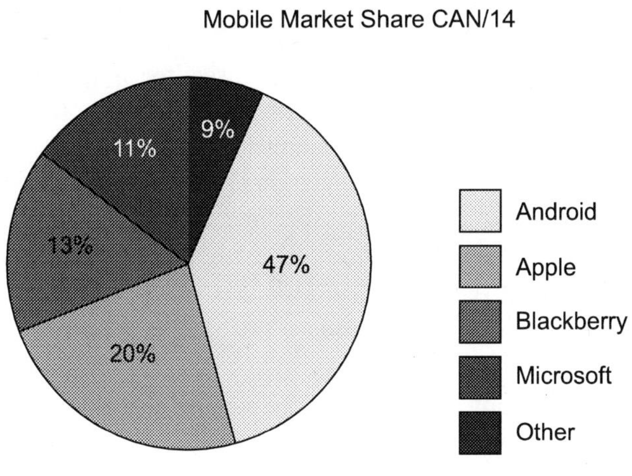

Fig. 8.5 Pie chart representing the mobile market share

There are some advantages and disadvantages to the pie charts. Some of them are given below [13].

Advantages of a Pie Chart

- A simple and easy-to-understand picture.
- It represents data visually as a fractional part of a whole, which can be an effective communication tool for the even uninformed audience.
- It enables the audience to see a data comparison at a glance to make an immediate analysis or to understand information quickly.
- The need for readers to examine or measure underlying numbers themselves can be removed by using this chart.
- To emphasize points you want to make, you can manipulate pieces of data in the pie chart.

Disadvantages of a Pie Chart

- If too many pieces of data are used, the pie chart becomes less effective.
- They themselves may become crowded and hard to read if there are too many pieces of data, and even if you add data labels and numbers may not help here.
- You need a series to compare multiple sets as this chart only represents one dataset.
- To analyze and assimilate information quickly, may make it more difficult for readers.
- As the reader has to factor in angles and compare non-adjacent slices, it has its problems in comparing the data slices.
- Making decisions based on visual impact rather than data analysis leads readers to draw inaccurate conclusions.

8.6.5 Photographs

In research thesis and papers, photographs are used for illustration of some subjects or experimental setups. In recent years, characterization techniques like scanning electron microscopy (SEM), scanning tunnelling electron microscopy (STEM), etc., in materials science and simulation techniques provide their output in picture format. Therefore, photographs became a common element of any research thesis. There are many advantages of using photographs in a research thesis. Some of them are given below [14, 15].

- They are self-explanatory. They reduce the number of words required in the description.
- Normally, they cannot deceive you. Remember the old dictum 'seeing is believing'. However, bear in mind that in the era of advanced digital technologies, photographs can also be 'made'.

- Photographs show a universal language, and therefore, irrespective of the language barriers, everybody can appreciate them.
- Multiple copies can be made, and the message may reach many hands quickly.
- Photographs simplify interpretation.

In selecting photographs for publication, therefore, "look for relevance, scientific interest, sharpness of focus, and effective lighting and contrast, and then consider whether or not a good line drawing or diagram would serve the purpose better" [15].

Any illustrations, tables, photographs, etc., used in a research thesis must be properly checked before finalization. The following points must be noted while verifying the illustrations, tables, photographs, etc. [16].

- Check every drawing or diagram against your original artwork.
- Check for clarity, accuracy, and neatness of line.
- Does the numbering of the figures correspond with the numbers in the text, and are other things referred to in the text included in the artwork?
- Are the letters, words, and abbreviations on the figures consistent with those used in the text?
- Is the labelling clear and are the labelling lines acceptable?
- Is any figure cluttered with too much information?
- If the figure is drawn to scale, is the scale marked on the illustration in metric units?
- Are any diagrams or drawings that are to be compared arranged side by side and drawn to the same scale?
- Are numbers and symbols for SI units of measurement marked clearly on all axes and scales?
- Are all symbols sufficiently explained?
- Should any photograph be replaced by a drawing?
- If the illustration is to be published, check that the information required by the printer is written in the margin or lightly on the reverse in soft black or blue pencil (author's name, title of document, number of figures, reduction required).
- Check that the information in a table is not duplicated in an illustration.
- Check that each table and illustration fits upright on the page (portrait) in the space available, and check that there will be space below the figure, on the same page for the legend.

A thorough checking will help the researcher to minimize the errors and the illustrations, tables, photographs, etc., will convey the correct information to the reader that the researcher wanted to communicate.

8.7 SOFTWARE FOR REFERENCE MANAGEMENT

Proper references are to be placed at the end of a research paper or a thesis. All publishers insist on the citation of references in particular formats. In the paper, the references are to be included as per the requirement of the journal. Therefore, the proper arrangement of references needs a lot of time. With the development of technology, reference management became easier since a large number of reference management software are available free of cost. Some of the referencing software are commercial like EndNote. EndNote is the most popular commercial reference management software. Some of the most common software management software are given below.

- EndNote (https://endnote.com/)
- Mendeley (https://www.mendeley.com)
- Zotero (https://www.zotero.org/)
- JabRef (http://jabref.sourceforge.net/)
- Qiqqa (http://www.qiqqa.com/)
- RefWorks (https://www.proquest.com/)
- CiteULike (http://www.citeulike.org),
- Bibus (http://www.sourceforage.net/projects/bibus-biblio)
- Refbase (http://www.refbase.net)
- BibSonomy (https://www.bibsonomy.org/), and
- Wikindx (https://sourceforge.net/ projects/wikindx/)

The advantage of using the reference software is that the software itself provides various reference styles. One can add the reference papers in the software directory and can be selected directly from there and inserted in the place in text where it is needed. The author needs to select the reference format as per the requirement of the journal. The software will build a reference list in the order that we select the reference from the software library.

8.8 HOW TO IMPROVE WRITING SKILLS

The research presents his research findings before the public through the finished research paper or thesis. Therefore, presenting the research finding in a proper manner is very important. Usually, the research paper and thesis are limited in pages and the researcher should convey the message correctly within the available length of the text. This requires good writing skills and a good knowledge of using appropriate tenses, phrases, etc., incorrect places. Since the English language is preferred for writing articles and thesis, the researcher should have good knowledge of the English language. A brief discussion of some principles which will help in writing a good research paper is given below.

Use of Simple and Plane English: A scientific report must be written in a straightforward way. There must be clarity in presenting the information. The use of ambiguous words like "may", "either", etc., must be avoided in scientific writing. Simple sentences are to be used in scientific writing. By eliminating redundant words and phrases, the length of the text can be reduced. If there is a choice between a short word and a long word with the same meaning, choose the short word. The usage of active voice rather than passive voice will help to reduce the length of the sentence. Therefore, the active voice is to be used wherever possible. It is to be remembered that the improper usage of words always creates ambiguity. Any abbreviations used in the text must be expanded in the place where it is used for the first time itself.

Proper Use of Paragraphs: "The ideas presented in a scientific report must be arranged in proper paragraphs. A paragraph is a unit of thought or information. Hence, the idea expressed in the paragraph should be homogenous in relation to the subject matter. It should be homogeneous in content and should treat content in logical and sequential order. All the sentences in the paragraphs must provide information to the topic, and the first and the last sentence should help to link paragraphs so that readers understand instantly how your thoughts about one aspect lead onto others" [2, p. 402]. Then, the research must have a clear idea about the facts that he wanted to present in a paragraph. After completing the facts reserved for a particular paragraph, then the writing is to be continued in the next paragraph. When information, several facts, etc., are included in a single paragraph, the reader may not grasp the information completely and may get confused.

Correct Usage of Tenses/ Voices: In scientific writing, the correct usage of tenses is very important. The most commonly used tense forms are the simple present, present continuous, and past tenses. The scientific usage of various tens forms is discussed in chapter 7. Also, in normal writing, active voice is preferred over passive voice. If one writes in passive voice, the sentence becomes verbose and slow, as it requires more words than the active form. It is clear from the following sentences.

- Marconi invented the radio
- The radio was invented by Marconi

The active voice is straightforward and needs fewer words. Therefore, one needs to use the active voice while preparing a research article unless there is better explanation to use the passive voice.

Use of Proper Article: To make a sentence meaningful, the articles must be used properly. The adjectives 'a', 'an', and 'the' are known as the articles. The indefinite article 'a' precedes a word beginning with a sounded consonant, and 'an' precedes a word beginning with a vowel sound. The article 'the' is a definite article because it points out a definite person or thing. We usually omit the definite article 'the' in many places where it is actually needed. When we use a singular noun, no article is needed. More details about the proper use of articles can be seen in Ref. 2 on page 413.

Proper Use of Prepositions: The use of prepositions is common in the English language. A preposition connects a noun, pronoun, or noun equivalent and another word. It is to be noted that for the same preposition, there may be different uses. Also, different prepositions may have the same uses. Therefore, the selection of correct prepositions is very important.

Use of Most Appropriate Relative Pronouns: A relative pronoun is one that refers to previously mentioned nouns in a sentence. The nouns can either be places, animals, people things, or ideas. The commonly used relative pronouns are given below.

- *That:* It is a relative pronoun used to introduce a defining clause to state a reason or a purpose. It is used to refer to a person, idea, or an object.
- *Whose:* It is a relative pronoun that means associated with or belonging to a person.
- *Which:* It is a relative pronoun that is used to refer to something that was previously mentioned and is used when a clause giving more information was introduced.
- *Who:* Who is a relative pronoun used to introduce a clause which gives more information about people or a person that was previously mentioned.
- *Whom:* Whom is a relative pronoun used in written or formal English instead of "who" and this is when it is the object of the preposition or object of the verb.
- *Whoever:* Whoever is a relative pronoun that is used to refer to a person whose identity is not yet known.
- *Whomever:* Whomever is a relative pronoun that is used instead of the pronoun "whoever" when it is the object of the verb or object of the preposition

If relative pronouns introduce the restrictive relative clauses, no comma is used to separate the main clause from the restrictive pronoun. Relative pronouns usually introduce relative clauses, and they can function as an object, subject, or possessive pronoun

Phrases: Phrases and translational words are used to provide connections between various ideas. "Transitions improve the logical connections between thoughts. In writing, transitions are used with many objectives such as to generalize, to introduce an example, to add something, to compare or contrast something, to demonstrate cause and effect, to summarize, or to conclude" [2 p. 420]. Classification of various types of phrases are given below.

- *Contradictory phrases:* There are many contradictory phrases used in scientific writing. Some of the contradictory words are 'Although', 'Though', 'Even Though', 'But', 'However', 'Nevertheless', 'Nonetheless', 'Even though', 'Still', 'Yet', 'On the Other Hand', 'on the Contrary', 'Whereas', 'Compared to', 'In Contrast to', 'In Spite of', 'Despite', etc., such phrases are to be used to reflect the meaning that the researcher wants to communicate.
- *Introducing examples:* Examples are to be introduced in scientific writing. While introducing an example we use some transition words. The commonly used transition words to introduce examples are: for example, for instance, such as, like, and including.
- *Adding an idea:* "Some transitional words are used to add something to what has previously been indicated, to reinforce ideas, and to show a continuation of an idea noted earlier". For this purpose, in scientific writing transition words like 'furthermore', 'moreover', in addition, and 'besides' are commonly used.
- *Consequence of effects:* To indicate the consequence or effect of certain things or events, transition words are needed. The commonly used transition words that denote a consequence are: 'Because', 'Because of', 'Due to', 'As a Result of', 'Owing to', 'Since','Thereby' and 'Thus', 'Unless', and 'Otherwise'.
- *Adding a comparison:* For a comparison between two data or events, we need to use suitable phrases. The most commonly used comparison phrases in scientific writing are 'Alternatively', 'Likewise', and 'Similarly'.

8.9 GENERAL GUIDELINES FOR IMPROVING THE WRITING SKILLS

Scientific writing skills can be improved by regular practice. The aim of scientific writing is to communicate with society or people about the findings of research work. For effective transfer of the research ideas and information, the researcher must follow the following guidelines for preparing the manuscript [17].

- *Organize your thoughts, ideas, and action in a logical manner:* Begin with sufficient background information to take your reader along the pathway from your observations or understanding to your hypothesis. Describe the context of the background to appeal to a broad group of readers. Provide sufficient context to communicate the significance of your inquiry and experimental findings. Omit extraneous information so that the reader can obtain a clear picture. Group similar ideas together and state your ideas and thoughts concisely. Present ideas in a consistent manner throughout the manuscript. The most common structure of a scientific manuscript is the IMRAD (Introduction, Methods, Results, and Discussion) format.

- *Provide clear descriptions:* Repeating the background or concepts may be necessary when the concept is complicated. You may need to explain the concept from different viewpoints. Start simple and then advance the complexity only as far as necessary to get the concept across. Consider your audience as you write. Are you writing for a general audience or a specialized audience? Will your audience understand the terms of art (i.e., phrases or words common to your particular field of study) or concepts that underlie your field and your research, or is some prior explanation needed? Keep your explanations simple.
- *Simplify your word choices:* Use simple straightforward language. Your manuscript will be read by students and researchers alike – make it easy for them to understand and care about your research even if they are not in your field or are not proficient in English.
- *Write concisely:* Note that "concise writing" is mentioned several times in this article. Science writing must be concise. No one appreciates long and or unnecessary descriptions or paragraphs. Use simple and direct language.
- *Use passive and active voice appropriately:* In science writing, it is important to know when to use passive and active voices. Active voice is more natural, direct, and engaging, and should be used when referring to widely accepted findings. The Introduction section should mainly be written in an active voice because you are telling the story of "what is". When referring to the findings of a specific study, however, passive voice could be used. In the Methods and Results sections, passive voice should be used to discuss what you did and what you found. In the Discussion section, a mixture of passive and active voices is acceptable, but take care not to mix the two together in a single sentence.
- *Select the appropriate words:* Selecting the appropriate words can be challenging. The best words accurately capture what the author is trying to convey. If a word is not sufficiently precise, use a thesaurus to replace the word or phrase with a more appropriate word. Precise words allow for specific, clear, and accurate expression. While science writing differs from literature in that it does not need to be colourful, it should not be boring.
- *Broaden your vocabulary:* Use clear, specific, and exact words. Expand your vocabulary by reading in a broad range of fields and looking up terms you don't know.
- *Avoid filler words:* Filler words are unnecessary words that are vague and meaningless or do not add to the meaning or clarity of the sentence. Consider the following examples: "it is", "it was", "there is", and "there has been", "it is important", "it is hypothesized that", "it was predicted that", "there is evidence suggesting that", "in order to", and "there is a

significant relationship". All of these phrases can be replaced with more direct and clear language. See our list of words and phrases to avoid here.

- *Read what you write:* Make sure to vary sentence length to keep the reader from getting lulled to sleep by a monotonous rhythm. Do not, however, make overly long or complicated sentences that hinder the reader's ability to follow your story. Reading the manuscript, yourself after some time away or having someone else read the manuscript will help you to refine the readability.
- *Optimize paragraph and sentence structure:* Each paragraph should present a single unifying idea or concept. Extremely long paragraphs tend to distract or confuse readers. If longer paragraphs are necessary, alternate them with shorter paragraphs to provide balance and rhythm to your writing. A good sentence allows readers to obtain critical information with the least effort. Poor sentence structure interferes with the flow. Keep modifiers close to the object they are modifying. Consider the following sentence: "Systemic diseases that may affect joint function such as infection should be closely monitored." In this example, "such as infection" is misplaced, as it is not a joint function, but rather a systemic disease. The meaning is clear in the revised sentence: "Systemic diseases, such as infection, that may affect joint function should be closely monitored."
- *Use abbreviations and acronyms to aid the reader:* Only use abbreviations/acronyms to help the reader more easily understand the paper. A general rule of thumb is to use only standard, accepted abbreviations/acronyms that are used at least three times in the main text of the paper. Whenever using an abbreviation/acronym, ask yourself "Does this help me or the reader?" Exceptions may apply for those abbreviations/acronyms that are so commonly used that spelling them out might confuse the reader.
- *Minimize pronoun use for clarity:* Make sure every pronoun is very clear, so the reader knows what it represents. In this case, being redundant may contribute to clarity. Don't refer to this or that, making the reader go back to the previous paragraph to see what this or that means. Also, limit or avoid the use of "former" and latter".

8.10 USE OF LIBRARY IN RESEARCH

When a researcher starts the research work, the available information in his hand is usually very little in amount. To buildup the knowledge, he has to collect information and research papers related to his research area. With the development of technology, information is available at the fingertip. Many of the research publications are available online. However, the journals charge some amount to purchase the research articles. Many of the researchers cannot afford such charges. This difficulty can be overcome by a researcher if there is a good library that can be used by him. In earlier years, a library is meant for keeping books and printed journals. Good libraries have in-depth resources in the form of books, journals, magazines, articles, and bibliographies. Nowadays, many of the libraries subscribe to important journals and the students and researchers can access the content of the journals from the library.

A report commissioned by Research Libraries UK (RLUK) and the Research Information Network (RIN) in 2011, emphasizes the following roles of the library in research activities [18].

- Good libraries help institutions to recruit and retain top researchers.
- Libraries help researchers win research grants and contracts.
- Libraries promote and exploit new technologies and new models of scholarly communications.
- Repositories increase the visibility of the institution and raise its research profile.
- Outward-facing libraries contribute to institution-wide initiatives.
- Specialist staff work in partnership with academic departments.
- Connecting with researchers enhances the value of the library's services.
- Dedicated spaces provide a better work environment for researchers.
- Easy access to high-quality content is a key foundation for good research.
- Libraries are a physical manifestation of the values of the academy and of scholarship.

For a researcher, a good library provides a wide variety of facilities. Some of them are given below.

- Procures research material (journals, books, special collections, etc.,) and facilitates easy access and discovery of research material
- Provides and promotes access to external research material
- Offers a physical space: to work in, to meet in, to access research material, to access specialists
- Provides inductions and training programmes to develop research skills
- Assists with research tasks (particularly in finding hard-to-get resources and information gathering)
- Provides specialists who have subject and information expertise to help with problems
- Provides information and advice (online and face-to-face) on topics such as publishing your work, copyright, open access, citations
- Provides a named point of contact to prevent Ph.D. students from feeling isolated
- Assist in the grant proposal process: help with citations, resourcing, specialist expertise, and project resources

- Helps are provided for the submissions of their work validation, and checking of metadata, finding full-text articles
- Helps with submitting articles to the open access repository
- Digitizes important research collections

Some good libraries like university libraries own plagiarism checking software. Such software can be used by the research students who are members of the library free of cost or by giving some nominal charges set by the authority.

REVIEW QUESTIONS

1. Differentiate between thesis and dissertation.
2. What are the common features of a thesis and dissertation?
3. Discuss the various components of a research thesis.
4. What are the components of the preliminary pages of a research thesis? Discuss each of them.
5. What are the components of the subject proper? Discuss each of them.
6. What are the general considerations for including figures and tables in a thesis?
7. Write a note on the literature review.
8. Write a note on the thesis abstract.
9. What are the characteristics of an abstract?
10. What are the characteristics of a good thesis abstract?
11. Write a detailed discussion on the formatting of a research thesis.
12. Discuss the various components of a table.
13. Explain the various considerations for arranging data in tables.
14. What are illustrations? What is its importance in a thesis?
15. Discuss the main features of a figure to include in a research thesis.
16. Explain the classification of graphs.
17. What are the essential elements of a good graph?
18. Discuss the characteristics and features of various types of graphs in a research thesis.
19. What is a pie chart? What are its advantages and disadvantages?
20. What are the advantages of using graphs in a research thesis?
21. Write a note on the reference management software.
22. Discuss the various elements that must be practised for improving scientific writing skills.

23. Discuss the general guidelines for improving scientific writing skills.
24. Explain the role of a library in research?
25. How does a good library facilitate research activities?
26. How does a researcher is benefitted from a good library?

REFERENCES

[1] https://dissertationauthors.com/blog/dissertation-vs-thesis

[2] Thomas, p. 304

[3] Fisher, E. and Thompson, R. 2014. *Enjoy Writing Your Science Thesis or Dissertation.*, (2nd Edn.). World Scientific Publishing Company, p. 174.

[4] Day, R.A. 1989. The origins of the scientific paper: The IMRAD format. *Am. Med. Writers Assoc. J.* 4(2): 16–18.

[5] https://www.student.unsw.edu.au/introductions

[6] Paul Oliver,"*Succeeding with Your Literature Review: A Handbook for Students*", Open University Press (2012).

[7] Derek Swetnam, "*Writing Your Dissertation: How to Plan, Prepare and Present Successful Work*" HowToBooks, UK, (2004).

[8] https://www.scientificstyleandformat.org/Home.html

[9] https://apastyle.apa.org/products/publication-manual-7th-edition

[10] https://www.chicagomanualofstyle.org/home.html

[11] https://www.thephdproofreaders.com/writing/how-to-write-an-abstract-for-your-phd-thesis/

[12] https://view.officeapps.live.com/op/

[13] https://www.excelr.com/advantages-and-disadvantages-of-pie-charts

[14] Gustavii, B. *How to Write and Illustrate a Scientific Paper* (2nd Ed.). Cambridge University Press, Cambridge, (2008).

[15] Barrass, R., *Scientists Must Write: A Guide to Better Writing for Scientists, Engineers and Students* (2nd Edn.). Routledge, London (2002).

[16] Barrass R, p. 135

[17] https://www.scitechedit.com/en-gb/helpful-resources/newsletters/502-improve-your-scientific-writing

[18] https://www.rluk.ac.uk/wp-content/uploads/2014/02/Value-of-Libraries-report.pdf

Suggested Readings

1. Elizabeth Fisher and Richard Thompson "*Enjoy Writing Your Science Thesis or Dissertation*", Imperial College Press (2014).
2. Derek Swetnam, "*Writing Your Dissertation*", (3rd Edn.), How-to-Books, UK (2004).

3. John Biggam, "*Succeeding with Your Master's Dissertation*", Open University Press, UK (2008).
4. Felix R Librero, "*Writing Your Thesis*", Universisty of Philippines (2012).
5. Paul Oliver, "*The Students's Guide to Research Ethics*" Open University Press, Philadelphia (2003).

Chapter 9

Ethics in Research

"Ethics is a system of public, general rules for guiding human conduct".

— **Gert. B**

"Science is both contained and constrained by Science Ethics and the primary source of knowledge and understanding in the development of practical ethical norms".

— **John Coggon**

9.1 RESEARCH ETHICS

"Scientific ethics is an institutional code of conduct that reflects the chief concerns and goals of science".

— **David B Resnik**

In addition to individual responsibility, researchers have many responsibilities towards the community. The researcher is expected to devote all his abilities to conducting high-quality research. He should be free from outside influence and manipulation, and should also not act based on personal motives or those of interested agencies. The success and future of researchers and of research are greatly dependent on the faith that the community has in the researcher. The importance of research ethics can be read in this context. It has become the cornerstone of effective and meaningful research.

The term *ethics* has a philosophical foundation and it controls the activities of every human being. The majority of the ethical principles are formulated by the society with various purposes such 'as standards of conduct that distinguish between right and wrong, good and bad, and so on; as an academic discipline that studies standards of conduct; as an approach to decision making; and as a state of character' [1].

Research ethics is defined as the standards of *conduct* for scientists in their professional endeavours. These diverse sets of values, norms, and institutional regulations help to plan, imply and regulate scientific activity. In earlier days the research ethics were mainly concerned with the treatment of the research subjects, especially humans and animals. Now the definition of research ethics is not limited to a particular situation, but covers the entire research process. A scientist has to observe these codes of conduct in every phase of his research. 'The term *conduct* refers to all aspects of scientific activity, including experimentation, testing, education, data analysis, data storage, data sharing, peer review, government funding, the staffing of research teams, etc....' [2]. A researcher should understand the values and principles of ethics that he should observe at various stages of his research. This is because 'the community of scientists is bound by a set of values, traditions, and standards that embody honesty, integrity, objectivity, and collegiality. These values are reflected in the particular principles and practices characteristic of specific scientific disciplines' [3].

Research ethics emphasizes that the researcher has more responsibility towards society. 'Research ethics also concerns the internal relationship among researchers, as well as the relationship between researchers and other people [4]. The reliability and quality of research work are based on good scientific practice based on ethical principles. When even one part of research work is conducted unethically, the integrity of the entire project is called into question.

The observation of the ethical standards is not the duty of an individual researcher alone. There are many other agencies involved in the research process; they include universities, scientific institutions, academies, and industries, etc. Therefore, research ethics has got both individual and communal perspectives and there should be equilibrium between these two. This is evident in the words of Mark Frankel (1989): "promoting ethical conduct does not, and should not, have to be solely the responsibility of the individual. The professional group, as a more visible, more stable, and more enduring entity, has a collective moral responsibility that is non-distributive; that is, a responsibility borne by the profession as a whole independent of the ethical posture of its individual members" [5]. The ethical standards that scientific institutions develop are important both to the institutions themselves and to the individual scientists operating in them. This is because the community places trust in such institutions as well. Also; as Carolyn Tuohy and Alan Wolfson (1977) observed individual scientists are trusted largely 'because the exercise of professional discretion at the individual level is governed by rules which are prescribed and enforced by the group' (scientific community) [6].

Since research ethics is an area that is very broad and is not well defined, many articles and guidelines address the responsibility of a researcher towards research and the scientific community. In this context, this can be regarded as *professional ethics*. Issues in connection with the behaviour of the researcher in various roles, responsibility in connection with the publication, and research misconduct, etc., belong to this category.

9.2 IMPORTANCE OF ETHICS IN RESEARCH

Scientific research is based on the following principles:

- *Scientific honesty*: Do not commit scientific fraud. That is, do not fabricate, fudge, trim, destroy, or misrepresent data.
- *Carefulness*: The researcher should strive to avoid careless errors or sloppiness in all aspects of scientific work.
- *Intellectual freedom*: Researchers should be allowed to pursue new ideas and criticize old ones.
- *Openness*: Scientific research should be open to sharing data, results, methods, theories, equipment, and so on. Research work should always be open to criticism
- *The principle of credit*: Avoid plagiarism. Whenever the work of other scientists is used, give them due credit.

All researchers are liable to observe these principles without failure. Among these, the scientific honesty and the principle of credit are matters of ethics. A failure in practising these principles can be considered *scientific misconduct*. The research conducted by a scientist not only affects those directly involved, but also the public's attitude towards science and scientists. Any challenge in the integrity of science destroys confidence of the public and faith in the research findings. Therefore the observation of ethical principles is unavoidable in scientific research.

Ethical principles are to be followed for the good practice of research. It is essential for fostering scientific integrity and for the preservation of privileged information. In the *Guidelines for Research Ethics in Science and Technology*, prepared by The National Committee for Research Ethics in Science and Technology (NENT), Norway, the following good research practices were discussed [7]. The importance of the observation of ethical principles in scientific research will become clear when we go through these points.

- The researcher and the research institution are responsible for exercising honest research practices.
- The researcher is individually responsible for the activities, subject matter and method of his or her research, as well as for the quality of the results.
- The researcher must respect the contributions of other researchers and follow standards for authorship and cooperation.
- When conducting research, the researcher must follow national and international regulations on ethics and safety.
- The researcher must clarify the degree of certainty and precision that characterizes the research results.
- The researcher must respect the demand for informed consent.

- Research must secure the privacy of the research subjects.
- The researcher must whenever natural seek to incorporate and respect alternative sources of knowledge, such as traditional knowledge.
- The researcher should whenever relevant seek to use participatory methods.
- The researcher is responsible for ensuring openness and scientific quality in contract research.
- The researcher is obliged to be open about possible conflicts of interest.

A scientist has responsibility for all his actions. The scientific community notices every action performed by a scientist. The community has a great impact on developing scientific integrity. The performance of a scientist is often controlled by the various *norms* and *values* set by the community. There are many social elements that can control scientific misconduct. They include [8]:

- Scientists themselves, who serve not only as practitioners but also as reviewers, colleagues, consumers of other scientists' work, and members of professional associations (Frankel 1993, cited in Patricia A. Bolton).
- Editors and publishers of scientific articles, who have an interest in being the first to publish ground-breaking science but also, have an interest in enhancing and maintaining the reputation of their publications and institutions.
- Research project managers, who both conduct science and oversee the work of other scientists.
- Institutional research programme officials, who employ the scientists and therefore have direct line responsibility for ensuring compliance with regulatory and contractual requirements and a need to maintain a volume of research that supports those employees and the institutional infrastructure.
- Officials in federal and other research funding agencies that commission the research and have responsibility for ensuring that the funds are used effectively and provide benefit.

In addition to these *self-regulations* also plays an important role in controlling errors and misconduct in scientific research. Self-regulation is a term used to represent the ability of a professional to oversee and correct the behaviour of their members. The profession itself creates self-consciousness and the researcher follows the ethical principles without any particular guidance. This is actually a consequence of the social nature of science.

9.3 ETHICS: VALUES AND PRINCIPLES

Ethics and morality are philosophical concepts that found an important place in scientific research. There are many fundamental values and principles regarding scientific ethics. These principles are intended to address various ethical issues. Through an understanding of these principles, a researcher would be able to consider all potential risks and ethical conflicts when designing and conducting research. *The Committee for Ethics in Science and Intellectual Property*, Tel-Aviv University, Israel suggested the following values and principles in observing ethical practices in scientific research.

- Truth
- Freedom
- Responsibility
- Integrity
- Collaboration
- Professionalism

Truth: The fundamental aim of the research is the generation of knowledge. Every scientist has the responsibility to contribute towards knowledge expansion. Moreover every creation of a scientist should be beneficial to humanity as a whole. To achieve these goals a scientist in every branch of science has to follow the methods of scientific research in each branch, and the rules of conduct in the scientific community in general.

Freedom: In order to achieve the goals of scientific research by scientists the research should be based on the principle of scientific research freedom, which is one of the most prominent expressions of the democratic system. However, some practical restrictions are imposed upon freedom of scientific research by the principles of the democratic system, for the adequate safeguarding of human life, welfare, dignity, and liberty. Within the framework of democratic principles, the scientists undertake research activities in the areas of development and application.

Responsibility: Every scientist has complete responsibility for every scientific research or experiment he conducts. This statement aims, particularly, with regard to its direct effect on human lives and on human physical and mental health, welfare, dignity, and liberty. Scientists are responsible for any direct effect of the research upon those participating as patients or subjects (animals, plants, etc.), in scientific research or experimentation.

Integrity: A scientific method has well-organized components. Scientists perform every scientific research in accordance with all the requirements of the scientific method with an aim to produce a highly accurate result. Scientists collect and analyze data with high precision and accuracy. The conclusions are based on a thorough analysis of data collected through experiments and their generalization.

Collaboration: The scientist acts within a universal framework of scientific collaboration, based on the shared scientific goals. The scientist fosters scientific collaboration by maintaining an atmosphere of openness, mutual assistance, and trust among scientists, their assistants, and students. It is also to be considered that a scientist may possess pursuant rights to intellectual property for scientific achievements to which he has made a unique or significant contribution.

Professionalism: Many consider research as a profession. The scientist engages in his scientific activities in a professional manner, making judicious and continual use of the scientific knowledge. The investigations made by a scientist lead to some development in his area of expertise. Moreover the activities of a scientist are based on the ethics of scientific research. A scientist imparts the values and principles of scientific research to all those conducting research or experimentation under his supervision, particularly to students in every course of study serving to prepare them for professional activity within the scientific research community.

There may be many variations or additions in these principles. For example, David B Resnik and Adil E Shamoo (2009) pointed out the importance of the following ethical principles in scientific research [9].

Honesty: Honesty is expected in every action of a researcher. This includes reporting data, results, methods and procedures, publication status, research contributions, and potential conflicts of interest. A researcher should never fabricate, falsify, or misrepresent data.

Objectivity: It is the duty of a researcher to preserve objectivity in all stages of the research programme. The experimental design, data analysis, data interpretation, peer review, and other aspects of research are expected to be objective.

Openness: Openness is one of the basic principles of scientific research. Researchers should possess open-mindedness to share data, results, ideas, tools, materials, and other resources. They should always be open to criticism and new ideas.

Confidentiality: Even though a researcher should have an open mind to share the ideas and materials related to research, he should protect confidential communications, such as papers submitted for publication, personnel records, business or military secrets, and records that identify individual research subjects or patents.

Carefulness: The ultimate aim of any research is to find the truth. Therefore, a researcher should be very careful in all actions related to the research. Researchers should not commit errors. Careless errors in experimentation, methodology, etc., should be avoided. The researcher should keep good records of all research activities, such as data collection, research design, consent forms, and correspondence with agencies or journals.

Respect for colleagues: Research is not an individual task. Nobody can complete research in isolation. A researcher is always surrounded by colleagues, students, and subordinates. Give due respect to all of them. Do not hurt or discriminate them against on the basis of sex, race, ethnicity, religion, or other characteristics. It is the duty of every researcher to help to educate, train, and advise the next generation of scientists.

Respect for intellectual property: Intellectual property includes patents, copyrights, etc. They are the results of serious researches made by many scientists for years. Therefore, a researcher should always respect such intellectual properties. Never use unpublished data, methods, or results without permission. Proper credit should be given wherever credit is due.

Respect for the law: Researchers are part of certain institutions. They should understand and follow relevant laws and institutional policies.

Respect for research subjects: This is important in a study where an animal or a human being is a subject for study. Proper respect and care should be given to animals when using them in research. The experiments should be designed properly to minimize harms and risks and maximize benefits.

Stewardship: A researcher should have good managing skills. All resources such as human, financial, and technological must be used effectively. Proper care should be given to materials, tools, samples, and research sites.

Social responsibility: Apart from the responsibility of the effect of research on the subject, every researcher has some social responsibilities. He should promote good social consequences and prevent bad ones through research, consulting, public education, and advocacy.

9.4 INTELLECTUAL PROPERTY RIGHTS

Intellectual Property Rights (IPRs) are legal rights that protect creations and/or inventions resulting from intellectual activity in the industrial, scientific, literary, or artistic fields. The World Trade Organization observes that "the intellectual property rights are the rights given to persons over the creations of their minds. They usually give the creator an exclusive right over the use of his/her creation for a certain period of time". The common type of intellectual property rights includes [10]:

- *Copyright*: Copyright protects written or published works such as books, songs, films, web content and artistic works, "paintings, sculptures, computer programs, and films. In the same way, copyright is applicable for performers such as actors, singers, and musicians. Producers of sound recordings and broadcasting organizations are also eligible for copyright protection. The main purpose of copyright laws is to promote and reward original and creative contributions".
- *Patents*: Patents protect commercial inventions, for example, a new business product or process. "It is an exclusive right granted by law to

the patentee for a specific period in exchange for full disclosure of his/her invention; at the same time, preventing all others from making, using, selling, or importing patented product or processes for business or other purposes. The objective of the patent is to encourage inventions by their promotion and utilization, especially for the development of industries. Patents ensure property rights for the invention".

- *Trademarks*: Trademarks protect signs, symbols, logos, words, or sounds that distinguish your products and services from those of your competitors. "A trademark is a sign or mark capable of distinguishing a product or service of one company from those of another company of any given business. The sign or mark accepted as a trademark can be a unique word, letter, numeral, picture, drawing, shape, colour, sound, smell, logo type, or any combination of these".

The intellectual property rights can be either registered or unregistered. With unregistered intellectual property, one automatically has legal rights over his creation. Unregistered forms of intellectual property include copyright, unregistered design rights, common law trademarks and database rights, confidential information, and trade secrets. With registered intellectual property, one will have to apply to an authority, such as the Intellectual Property Office in the UK, to have his rights recognized. If he does not do this, others are free to exploit his creations. Registered forms of intellectual property include patents, registered trademarks, and registered design rights. Copyright is also registerable.

For the international exchange of ideas and products, there exist many agreements between the member nations. India also signed the following agreements [11].

- *The Paris Convention*: Under this, any person from a signatory state can apply for a patent or trademark in any other signatory state and will be given the same enforcement rights and status as a national of that country would be.
- *The Berne Convention*: Under this, each member state recognizes the copyright of authors from other member states in the same way as the copyright of its own nationals.
- *The Madrid Protocol*: Under this, a person can file a single trademark application at their national office that will provide protection in multiple countries.
- *The Patent Cooperation Treaty*: This is a central system for obtaining a 'bundle' of national patent applications in different jurisdictions through a single application.

9.4.1 Royalty

Royalty is a term used to represent the financial earnings of an author from a book published by a publisher. The royalty is usually expressed as a certain percentage of the total book value. The rate of royalty is usually varied from 10% to 15%. The rate is usually specified in the agreement signed between the author and the publisher. For a well established author, the publisher offers more percentage of royalty.

9.5 SOME ETHICAL ISSUES

Stem Cell Case

Dr. Hwang Woo-Suk, a researcher and professor at Seoul National University, rose to fame after claiming a series of remarkable breakthroughs in the field of stem cell research. In 2004 and 2005, Dr. Hwang published two papers in the journal Science that claimed his team had succeeded in creating human embryonic stem cells through cloning. Allegations later followed from a co-worker that this paper was based on fabricated data. The papers were editorially retracted, Dr. Hwang lost his position at Seoul National University, and the South Korean government ended its financial and legal support of his research [12].

The Baltimore Case

In the summer of 1995, Margaret O'Toole, a post-doctorate fellow at MIT, was asked by her supervisor Dr. Imanishi-Kari to do experiments that would extend the work described in a paper that had been published in the journal, Cell. Unable to repeat aspects of the research documented in this paper, O'Toole came across a laboratory notebook that suggested to her that the Cell study was wrong. Before too long, O'Toole came to believe that the errors in the paper were deliberate. She then challenged the authors of the paper, including the Nobel Prize winner David Baltimore. A long investigation followed, during which the National Institutes of Health and members of Congress became involved. During the investigation, Baltimore was forced to resign as president of Rockefeller University, in part because of his spirited defense of Dr. Imanishi-Kari. Dr. Imanishi-Kari was first found guilty of 19 counts of research misconduct but was later cleared of all misconduct charges in 1996 by a Human Health and Services appeals panel [13].

Good science requires the ethical practice of science. A deviation from ethical principles in science is considered *scientific misconduct*. Many researchers are involved in unethical practices, very often unknowingly, because they are not usually trained in research ethics. A person who just entered the field of scientific research may not recognize the unethical practices in scientific research. An understanding of the unethical practices will help the researcher to act according to the values and principles of scientific research by placing their positions into a logical framework. Unethical practices in research can be identified in various forms. "*Code of Practice for Dealing with Allegations of Scientific or Ethical Misconduct in Research*", Manchester Metropolitan University listed the following activities as scientific misconduct.

- Failure to obtain appropriate permission to conduct research.
- Failure to assess the research proposal using the University Ethics framework.
- Unethical behaviour in the conduct of research, for example, in relation to research subjects.
- Plagiarism, or dishonest use of unacknowledged sources.
- Misquotation or misrepresentation of other authors.
- Inappropriate attribution of authorship.
- Attempting, planning, or conspiring to be involved in research misconduct.
- Inciting others to be involved in research misconduct.
- Unauthorized use of information that was acquired confidentially.
- Deviation from good research practice, in particular where this results in risk of harm to humans, animals, or the environment.
- Fabrication, falsification, or corruption of research data.
- Distortion of research outcomes, for example, by the manipulation or omission of data that do not fit expected results.
- Dishonest misinterpretation of results.
- Publication of data known or believed to be false or misleading.
- Fraud or other misuses of research funds or research equipment.
- Collusion in or concealment of research misconduct by others.
- Deception in relation to research proposals.

Scientific misconduct increased considerably during the past few years. A study conducted by Nature Publishing Group (2011) on the raising *retraction rate* (withdrawing articles from publication, known as retractions) from high-impact factor journals such as *Nature*, *Cell*, *Science*, *Proceedings of the National Academy of Science, USA* and *J. Immunology Findings* showed that "44% of the retracted materials were flagged due to research misconduct" [14, 15].

9.5.1 Plagiarism

"*Plagiarism is the act of taking the writings of another person and passing them off as one's own. The fraudulence is closely related to forgery and piracy practices generally in violation of copyright laws.*"

— **Encyclopedia Britannica**

Plagiarism is the most common scientific misconduct. It is an act of scientific fraud. The word *plagiarism* was derived from the Latin word *plagiarius*, meaning kidnapper or abductor. Plagiarism is defined as '*the act of passing off somebody else's ideas, thoughts, pictures, theories, words, or stories as your own*' [16]. In other words, plagiarism is a deliberate or unintentional disguising of someone's claim to intellectual or artistic ownership. It is a violation of someone else's intellectual property rights. Plagiarism is a literary theft that causes the integrity and trustworthiness of scientific research to be loss. It defeats the purpose of honesty and truthfulness in scientific writing. It is an illegal and punishable act.

Plagiarism is a widely discussed topic, especially by many universities. Almost all university websites provide detailed information about plagiarism and other scientific misconducts. Plagiarism can take various forms. The following are some common forms of plagiarism:

- Authors usually copy phrases or sentences from other authors but fail to cite sources. This includes inaccurate and incomplete citing. The copying can be a major portion of text from a single source, sentences, or phrases from multiple sources.
- Sometimes the authors cannot identify the source of cited information with an introductory or signal phrase and therefore it is not possible to present attribution.
- "The author borrows generously from his previous work, thus the final product is completely unoriginal" [17]. This is termed self-plagiarism. The redundant and duplicate publication comes under this category.
- Misrepresenting the true meaning of the original source when paraphrasing is also a kind of plagiarism. (Paraphrasing is a technique to avoid plagiarism which is discussed later in this section).
- Authors sometimes fail to use quotation marks around the phrase or sentence copied from a source.
- The use of words or sentences that are too closely matched with the original documents is a kind of misuse in the paraphrase technique.

An author cannot avoid using ideas from outside sources in scientific writing. Using other's ideas is a kind of appreciation and recognition given to those authors. Very often ideas can only be built based on existing ideas. However to avoid plagiarism 'it is important that the author should properly identify and give appropriate credit to the outside sources through the proper use of quotation marks, internal citations, and references' [18].

There are various reasons for committing plagiarism. It varies from accidental mistakes to deliberate intention to present a result that actually did not obtain. Many mistakes are related to referencing. One example is the *forgotten footnote*. Sometimes the author mentions the original author's name for a source but does not include specific information in the reference. Even if proper citation is given, authors may forget to include the cited text in quotation marks. These are unintentional or by mistake. Some other reasons are:

- Lack of knowledge about what plagiarism is.
- The absence of a complete understanding of the conventions required in scientific writing.

- Ineffective methods of recording the data and other observations.
- Insufficient or poor time management skills.
- Different cultural values and different practicing conventions.
- Lack of confidence in one's own ability and in the quality of research findings.
- The overwhelming pressure to succeed.
- Feeling of pressure from the higher authorities.
- The ease of copy and paste of text from electronic sources.

Plagiarism should be avoided in scientific writing not only because it is unethical but also it degrades the integrity of scientific research and the dignity of the scientific profession. The first step in avoiding plagiarism is to understand the characteristics of scientific writing, ethical principles, and various forms of plagiarism. The methods to avoid plagiarism in scientific writing include:

- Proper note making;
- Paraphrasing;
- Summarizing;
- Quotations.

Proper note making: Keeping proper notes is important in the research process. Researchers usually refer to a number of articles related to the topic of study and usually keep a brief note. These notes are used when the researcher prepares the report of the findings. During note making researcher may use the same sentences as in the article, but after some time he may forget which is his own sentence and which one is the copied one. To avoid this not take notes in the first instance. Instead, read the article first, understand what the author has said and then summarize the work in your own sentences. It will be easy to recognize the copied sentences if one uses a different ink pen for writing those sentences. It is also essential to record all the necessary details about the original article so that it can be used in the reference list when the final paper is prepared.

Paraphrasing: "Paraphrasing is the restatement of information and ideas in one's own words and style" [19]. While paraphrasing the writer should use his own words. The paraphrased text can be shorter or longer than the original text. However, the essence of the paraphrase should not be different from the original idea. Also, it should be noted that changing only a few words or phrases or restructuring the sentences is not considered paraphrasing. It is also important to cite the source of the paraphrase in the text and reference list. An example of paraphrasing is given below:

Original text:

- "However, some writers object to a double-blind review on the grounds that reviewers can almost always determine the authors of various papers based on their citations, their ideas, and other aspects of content." (David B. Resnik, 1998, p.87).

Paraphrase:

- A double-blind review has the limitation that the reviewers can often guess the author/s from the citations and from other features of the text (David B. Resnik, 1998).

Summarizing: "A summary is a shortened version of the original source without changing the meaning" [20]. For writing an effective summary the first step is to identify the main points in the original text. If the text consists of a number of paragraphs understand the main idea of each paragraph. Then write a brief outline of each paragraph using which the final summary can be prepared. While summarizing unnecessary details and examples are to be removed. If there is any direct quote; word or sentence, it should be enclosed in quotation marks. Do not forget to cite the source of the summary.

For writing an effective summary, the University of Idaho's Advanced Technical Writing site (2006) suggested the following tips:

- State the main ideas first
- Focus on details that are really essential
- Use keywords and phrases
- Break down larger ideas
- Write only enough to convey the gist
- Take succinct notes
- Use a lower level of a technicality than used by the authors. The summary should be simple and easy to understand even without reading the original text.
- Do not add a new idea or do not inject your own ideas.
- Use a simple organization: The main point, main result, and conclusion.

Following is an example for summarizing the text:

Original text:

- "There are several reasons why scientists should take responsibility for the social impacts of research. First, although scientists cannot be held responsible for the unanticipated consequences of research, scientists can be held responsible for the consequences that can be anticipated. Second, scientists are also members of society who have moral duties toward other people, such as beneficence, non-maleficence, and utility. Third, scientists have professional duties to promote benefits and avoid harms" (David B. Resnik, 1998, p. 57).

Summary:

- Being a member of society, in addition to the responsibility of "anticipated consequences" of research, a scientist has moral as well as professional duties towards society (David B. Resnik, 1998).

Quotations: A quote is an exact reproduction of spoken or written words. An author can use the same text that appears in the original article if he thinks that it is more suitable in expressing an idea or if the author feels that the words cannot be bettered. However, such phrases or sentences or even words should be enclosed in quotation marks. It should be an exact copy of the text including capitalization and use of punctuation marks. It should be clear where the quotation starts and finishes. Also, it is very important to cite the source in the text and to include the details of the source in the reference list or in the bibliography.

In some cases, an author may use *secondary quotations*. This is the case where an author takes a quotation from a source without studying the original piece of work. To understand the essence of such quotations the original document is to be read. Minimum usages of such quotations are usually suggested. The format of such a citation is given below.

- "Science is facts; just as houses are made of stones, so is science made of facts, but a pile of stones is not a house and a collection of facts is not necessarily science" (Henri Poincare, cited in Finlay MacRitchie, p. 49).

An author should cite all the sources he has used. If there is a considerable similarity with the original source and if the source is not cited then it is considered a case of plagiarism. This is applicable not only to the text but also to data, diagrams, graphs, and tables. Any graph, data, diagram, or table adapted from other sources should also be acknowledged with proper citation. It is always to be remembered that a short list of well used sources is much better than a long list of sources that have never been looked at.

Detection of plagiarism in universities and other research institutions is an academic challenge. Plagiarism detection is important because in modern days enormous information is available on the internet and it is easy to find them using any search engine. Any author copies and uses them as his own without citing the source, thus cheating the scientific community. Plagiarism can be detected either manually or using detection software. Manual detection is not easy as it consumes more time and is not effective when the number of documents is large. In computer-assisted detection, there are many software and tools used in automatic plagiarism detection. Examples are; PlagAware, PlagScan, Check for Plagiarism, iThenticate, Academic Plagiarism, The Plagiarism Checker, Urkund, Docoloc, etc.

The detection of plagiarism is mainly based on the following methods. These are adapted from Asim M. El Tahir Ali et al. (2011) [21].

Grammar-based method: In this method, the detection of plagiarism is based on the grammatical structure of the document. The similarity between the documents is detected and measured using a string-based matching approach. This method is suitable for the detection of an exact copy without any modification.

Semantics based method: This method detects the similarity between the documents using a vector space model. "It also can calculate and count the redundancy of the word in the document, and then they use the fingerprints for each document for matching it with fingerprints in other documents and find out the similarity". This method is suitable for the detection of complete plagiarism. If only partial plagiarism, then it cannot produce a good result as "it is difficult to fix the place of copied text in the original document".

Grammar semantics hybrid method: This is the most important method of plagiarism detection. It is suitable for the detection of copied text including modified text by rewriting or switching some words that have the same meaning. It can also determine the location of the plagiarized parts of the documents.

External plagiarism detection method: "The external plagiarism detection relies on a reference corpus composed of documents from which passages might have been plagiarized". This method checks for duplication or near duplication of the passages in the document under consideration within the *reference corpus*. This method has the limitation that the reference corpus must be large enough to detect plagiarism in all suspicious documents. Also, it requires a large run time to complete the detection process.

9.5.2 Regulations of Plagiarism

Scientific misconduct, scientific fraud, plagiarism, etc., are serious issues in the field of scientific research. Such malpractices are to be prevented for the existence of real scientific discoveries. There are many regulations that help to prevent such malpractices. Such regulations issue many codes that must be followed by a researcher during the research programme and on publishing the research findings. For the good practice of science, the Council of Science Editors (CSE), published a white paper which is available on their website http://www.councilscienceeditors.org. This white paper clearly specifies the role of authors, reviewers, editors, and chief editors in preventing any kind of malpractices in scientific research. The ethical responsibilities of each are also discussed in detail. Here, I would like to quote a few important responsibilities of the author as they are very important for the good practice of research.

- *Confidentiality*: The author-editor relationship is founded on confidentiality. Authors should hold all communication between themselves and the journal in confidence. Authors should designate a specific contact for all communication about the manuscript throughout peer review and the publication process.

- *Originality*: The authors should provide a statement attesting to the originality of the study they have submitted for consideration. Originality is crucial, because many journals have limited space and editors may give a low priority to studies that, regardless of scientific accuracy and validity, do not advance the scientific enterprise.
- *Disclosures*: Authors have a responsibility to be forthright when complying with journal submission requirements. This entails disclosure about the originality of the content, a statement of an author's actual contribution to the study, and financial and conflict of interest disclosures. Some journals also require statements on the regulatory status of any drugs or devices used in the study.
- *Copyright Assignment*: Assignment of copyright is a legal document in which the authors assign certain rights to the publisher. Alternatively, some journals may use a licensing agreement. Although individual arrangements vary, these agreements generally allow the authors to retain certain rights to the material. In either case, the content in question must be original and not otherwise under copyright elsewhere (in whole or in part). Authors should ensure that the study under consideration is original and does not contain plagiarized content.
- *Permissions*: Authors frequently wish to reuse previously published images and other copyrighted material. It is the author's responsibility to follow journal or publisher guidelines to reuse any copyrighted material and provide proper attribution. This includes the author's own work if the copyright was ever transferred to a publisher or journal. Authors should contact the journal or publisher of the source material or consult the "permissions" information that can be found on many of their websites.
- *Multiple Submissions*: It is not acceptable for authors to submit the report of a study to several journals at the same time, including a manuscript undergoing peer review that has not been formally rejected by the original journal to which the manuscript was submitted. Authors who do not follow this standard may find that editors reject their papers as a violation of policy. In addition, this practice can be a violation of copyright. If authors want to submit their article to another journal while it is under consideration elsewhere, then they must send a formal notification to the editor of the journal in which it is under consideration, requesting that their study be withdrawn from further consideration.
- *Data Sharing*: Data sharing is the practice of making data used for scholarly research available to other investigators. Authors should be aware of their data sharing responsibilities imposed by their funding

agencies. The goal of this policy is to promote reproducibility and availability of underlying datasets.

9.5.3 Tools for Plagiarism Checking

With the development of software technologies, plagiarism checking became an easy process in recent years. Plagiarism checking is made mandatory by every university before the final submission of the research thesis. For this, the university libraries usually purchase plagiarism-checking software like TurnItIn, iThenticate, URKUND, etc. These three are the most popular anti-plagiarism software. This software is paid software. There are many free plagiarisms checking software such as Dupli Checker, Copy Leaks, Paper Rater, Plagiarisma, Plagiarism Checker, Plagium, Plag Scan, Plag Tracker, Quetext, and Viper. The use of such free software is limited to the checking of a few thousand words. Plagiarism checking is a kind of matching verification. The software compares the text content in the article with the previously published articles and provides a similarity index. The similarity index gives an indication of similar contents that have been found in other published materials or theses. Usually, a similarity index of less than 0.1 or 10% is allowed by the universities.

The author should take care while preparing the manuscript to minimize the similarity as smaller as possible. This includes the similarity with own published works. However, in India, UGC provided some exemptions while checking plagiarism. They are [22]:

- Information put in quotations with necessary permission or attribution.
- References or bibliography, table of contents, preface, and acknowledgments.
- Generic terms, laws, standard symbols, and standard equations.
- Common knowledge or coincidental terms consisting of less than 14 consecutive words.
- Phrases, jargon, and terminology, which are commonly used in writing, may be matched by the software but can be excluded.

9.5.4 Misuse of Privileged Information

This is rather the most serious scientific misconduct. This includes the misuse of privileged information taken from the manuscript received from the editor of a journal or from a funding agency for peer review, the information taken from the application for a grant, etc. This is a serious matter since it is the real theft of intellectual property. "It not only deprives the original author of appropriate credit by citation but could also preempt priority of first publication or use of the original idea to which the source author is entitled" [23]. Peer reviewers should follow ethical principles and should not breach the confidentiality of the manuscripts. Any unpublished information may be so precious for an author because when it is published in a peer-reviewed journal it may get worldwide attention.

9.5.5 Misuse of Data

Data analysis is a major stage of research work. In some cases, researchers do manipulate data in order to get the anticipated results. Here the researcher is actually fabricating the result. This is a serious form of scientific misconduct. Therefore, every researcher has the responsibility to avoid a false statement and addition or omission of data that distort the result of a study. To avoid this, researchers should maintain proper records of data. The record should be helpful to replicate the research if required. A well-maintained research record confirms the validity of research conclusions. "The intentional destruction of research records or the failure to maintain and produce research records supporting a questioned research publication or report may be considered to be circumstantial evidence of research misconduct" [24].

The integrity of research demands the reporting of all relevant data. If a researcher fails to report any data that contradict or fail to support the reported conclusions, it is considered a breach of research integrity. Researchers should have a good understanding of different methods of data processing including graphical and tabular methods of presentation, and error analysis. Special care should be taken in using photo images as the possibility of misinterpretation of underlying data is more. Misinterpretation of data is usually known as *falsification*. This is especially true in the case of spectral lines; for many spectral lines will not have sufficient clarity due to the lack of intensity. Should not add or remove any band using some image processing software, should not adjust the intensity of one or more bands, and most importantly should not label an image obtained in an experiment as obtained from another experiment.

Many universities and research institutions claim ownership of research data produced by their researchers. The researchers of groups under these institutions can access the data directly. Such data will be made available to other responsible researchers through written request. The responsible and efficient access of data is essential because it:

- reinforces open scientific inquiry;
- encourages diversity of analysis and opinion;
- promotes new research;
- makes possible the testing of new or alternative hypotheses and methods of analysis;
- supports studies on data collection methods and measurement;
- facilitates the education of new researchers;
- enables the exploration of topics not envisioned by the initial investigators; and
- permits the creation of new datasets when data from multiple sources are combined [25].

The data underlying a research publication should be made available to other researchers upon request after the research results have been published or accepted for publication. Therefore, it is essential to store the research data for a considerably longer period even after the publication of the results.

9.5.6 Authorship and Other Publication Issues

Researchers publish the research findings for the scientific community. In fact, every researcher has the ethical obligation to make the research findings accessible in a manner consistent with the relevant standards of publication. The reported data and method should be sufficiently detailed so that other researchers can attempt a replication.

All researchers who participated in the research should be credited with proper authorship when publishing the article. Many journals and professional associations usually specify the criteria for authorship. The common criteria are that each author should have participated in the formulation of the research problem, interpreting the result, and preparing the research paper. It is important that the name of a person should not be listed without his permission. Also one should avoid the listing of *honorary authors* who do not meet the criteria for authorship.

A person's name should be included even in the oral presentation if his contribution meets the requirements of co-authorship. This is especially important if the abstract is included in the proceedings of the conference where the presentation is made. Persons who provided technical and office assistance, supply of research materials, or facilities are not eligible for authorship. Partial fulfilment of the authorship criteria is also not eligible for authorship.

The corresponding or the senior author is responsible for communicating with the publisher or the editor. The corresponding author's name usually appears in the last of the listed names in a multi-authored paper.

The main publication issues are self-citation and duplicate publication. An author must be very careful in using self-citations. When citing an unpublished work, it should not imply an unwarranted state of the manuscript. An unwarranted state represents a submitted paper before receiving a confirmation from the editor or anticipation of expected publication. The author can cite a paper only if it is approved for publication.

Publication of the same paper in two different places is not usually allowed. This is applicable to abstracts also. However, under special circumstances duplicate publication is allowed. In such cases, appropriate citation should be made in the later publication to the earlier one and the editor should be explicitly informed. The unexplained duplication of publication without citation is referred as self-plagiarism.

In addition to the various issues discussed above, in some cases, the *conflict of interest* also affects the integrity of the science. This can lead to scientific misconduct. The main factor is the financial interest of the researcher. A researcher may distort the result of research work for financial benefits. Many universities and other funding agencies insist on the enclosure of a notice of conflicting financial interests in publications, research proposals, and reports.

Every responsible researcher has the obligation to report scientific misconduct. Many universities and other institutions specified the penalties for scientific misconduct. The investigator has the obligation to submit a correction if any error, either accidental, intentional, or of plagiarism, is noticed. The correction should be submitted in a form specified by the publisher or the editor.

9.5.7 Undesirable Authorships

There are many types of authorships that cannot be accepted in research. Some of them are guest authorship, honorary or gift authorship, and ghost authorship. Some of them are given below [26].

- *Guest authorship*: Guest authorship is defined as authorship based exclusively on the expectation on part of the other authors that the inclusion of a particular name as co-author may improve the chances of the work getting accepted for publication or for other personal gains. It is unethical to include such a "guest" author name who made no useful contribution to the study.
- *Honorary or Gift Authorship*: This is again highly unethical. Gift authorship is the authorship based solely on a feeble affiliation with a study, e.g., head of the department/institution, whose name is included solely because of the administrative position. Another example of 'gift' authorship is the inclusion of names of colleagues to increase the number of publications in the publication list of each other.
- *Ghost Authorship*: In this case, although the persons participated in the research, data analysis, and/or writing of a manuscript, they are not listed as authors in the manuscript or while filing a patent application. It is unethical to exclude names of persons who have, by general principles, to be included as authors. However, the name of a professional writer who is involved in writing the manuscript on a payment basis can be excluded from the author's list.
- *Anonymous Authorship*: Since authorship has to be transparent and requires accountability, it is improper to use pseudonyms or to publish scientific articles anonymously.
- *Surrogate Authorship*: There are instances where manuscripts are written or got written by someone else without any original data of one's own. Such papers are often published in 'predatory'/bogus journals. These acts are unethical and illegal.

9.6 CODES OF ETHICS

In research conducted in many fields such as medicine, most of the experiments are done with human participants, that too for many years. Vaccines for many diseases were developed through such research projects. In earlier days human participants for such experiments were selected from psychiatric patients or prisoners. The best examples for such studies are the medical experiments conducted by Nazis. In developed countries like the United States also such experiments were conducted (for example, the Tuskegee Syphilis Study conducted by US Public Health Service in 1932). However, many international medical communities are vigilant about such cruelty against a human being or any animal. Such organizations were issued many guidelines based on ethical principles for conducting such experiments.

9.6.1 The Nuremberg Code

Among the various codes, the first major international attempt is the *Nuremberg code* developed in response to the Nuremberg trials of Nazi doctors who conducted many unethical experiments during the Second World War. The following are the Nuremberg codes [27].

- The voluntary consent of the human subject is absolutely essential.
- The experiment should be such as to yield fruitful results for the good of society, unprocurable by other methods or means of study, and not random and unnecessary in nature.
- The experiment should be so designed and based on the results of animal experimentation and knowledge of the natural history of the disease or other problem under study, that the anticipated results will justify the performance of the experiment.
- The experiment should be so conducted as to avoid all unnecessary physical and mental suffering and injury.
- No experiment should be conducted, where there is an a priori reason to believe that death or disabling injury will occur; except, perhaps, in those experiments where the experimental physicians also serve as subjects.
- The degree of risk to be taken should never exceed that determined by the humanitarian importance of the problem to be solved by the experiment.
- Proper preparations should be made and adequate facilities provided to protect the experimental subject against even remote possibilities of injury, disability, or death.
- The experiment should be conducted only by scientifically qualified persons. The highest degree of skill and care should be required through all stages of the experiment of those who conduct or engage in the experiment.

- During the course of the experiment, the human subject should be at liberty to bring the experiment to an end, if he has reached the physical or mental state, where continuation of the experiment seemed to him to be impossible.
- During the course of the experiment, the scientist in charge must be prepared to terminate the experiment at any stage, if he has probable cause to believe, in the exercise of the good faith, superior skill, and careful judgment required of him, that a continuation of the experiment is likely to result in injury, disability, or death to the experimental subject.

Another major attempt is *The Helsinki Declaration* of 18th World Medical Assembly held in Helsinki in 1964. This declaration contains twelve ethical principles on ethical considerations in biomedical research. These principles have been revised a number of times and necessary modifications were made. For further details of these principles refer to the "*Bulletin of the World Health Organization*", 2001, 79(4), pp. 373-374

9.6.2 The Belmont Principles

The Belmont report was issued by the US National Commission for the Protection of Human Subjects of Biomedical and Behavioural Research in 1979. According to the Belmont report, there are three underlying principles in the ethical conduct of all research conducted with human participants. They are:

1) respect for persons,
2) beneficence, and
3) justice.

Respect for persons: This principle is derived from two basic ethical pre-assumptions. First, the person should be treated as an autonomous individual. This is because individuals are capable of making their own decisions. They have their own personal goals derived from self-determination. Respect for a person insists that we should encourage and accept an individual's opinion and actions. If we deny an individual's freedom to make a decision, it is against the principle.

The second preassumption is that all individuals are not equally capable of making self-determinations. The self-determination capacity of an individual develops during his life. Therefore, very young individuals may not have the ability to make the right decision. Also, a person may lose the ability for self-determination due to various reasons such as illness, mental disabilities, situations that restrict personal freedom like imprisonment and military service. The principle of respect ensures the protection of such immature, incapacitated, or restricted individuals when subjecting to an experimental study.

Beneficence: "Beneficence is the practice of doing good making all efforts to improve an individual's well-being" [28]. It is an act of kindness or charity. Individuals should be protected from any harm. The principle of beneficence is based on Hippocrates's ethical tradition of medicine that declares "as to deceases, make a habit of two things: to help or at least to do no harm". Then two general rules have been formulated. They are:

1) Do not harm
2) Maximize possible benefits and minimize possible harms.

The first one is directly derived from traditional medical ethics. Do no harm has long been a fundamental principle of medical ethics. In order to follow this principle, one should understand what is harmful and unfortunately any action performed on an individual to understand this may expose the individual to the risk of harm. Therefore, it is very crucial to decide "when it is justifiable to seek certain benefits despite the risks involved, and when the benefits should be foregone because of the risks" [29].

The principle of beneficence serves as a justifying rule in many areas of scientific research involving human subjects. In scientific research, the consequences of a research work affect not only the individual but the society as a whole, especially in medical sciences. Therefore, members of the society should recognize the longer term benefits and risks that may result from the improvement of knowledge and from the development of novel psychotherapeutic, medical, and social procedures.

The second rule brings about some difficulties. Even though it helps to avoid the harms that may result from the application of previously accepted routine practices by improving the method of experimentation, very often it places the researcher in a dilemma. For example, many experiments with child participants "presents more than minimal risk without immediate prospect of direct benefit to the children involved" [30]. Some people are against such experiments and some others pointed out that this limit would rule out much research promising great benefit to children in the future.

The Belmont Report suggested the following five rules to determine the risk-to-benefit ratio of a research program [31].

1) Brutal or inhuman treatment of human subjects is never morally justified.
2) Risks should be reduced to those necessary to achieve the research objective. It should be determined whether it is, in fact, necessary to use human subjects at all. Risk can perhaps never be entirely eliminated, but it can often be reduced by careful attention to alternative procedures.
3) When research involves a significant risk of serious impairment, review committees should be extraordinarily insistent on the justification of the risk.

4) When vulnerable populations are involved in research, the appropriateness of involving them should itself be demonstrated. A number of variables go into such judgments, including the nature and degree of risk, the condition of the particular population involved, and the nature and level of the anticipated benefits.
5) Relevant risks and benefits must be thoroughly arrayed in documents and procedures used in the informed consent process.

Justice: Justice is the third principle identified in the Belmont Report. According to this principle, every individual is to be treated fairly, equitably, and given what he deserves. An injustice occurs when some benefit to which a person is entitled is denied without good reason. Injustice occurs when an excessive burden is imposed on an individual. The principle of justice is also relevant to how research participants are treated, or not treated. The principle of justice can be conceived in a way; *equals ought to be treated equally*. This approach requires the explanation of who is equal and who are unequal individuals. Such a distinction is usually made on the basis of age, experience, competence, deprivation, etc. Also, there are several norms based on which the burdens and benefits should be distributed. They include:

- to each person an equal share,
- to each person according to individual need,
- to each person according to individual effort,
- to each person according to societal contribution, and
- to each person according to merit.

The principle of justice also suggests that there should not be any bias or discrimination of any kind in selecting the participant. When we apply these ethical principles to scientific research it demands the following requirements:

- informed consent,
- assessment of risk and benefits, and
- selection of subjects

Informed consent: The autonomy of the subject ensures them the opportunity to decide whether to participate as informed consent. The subjects who are participating in the research should be given sufficient information regarding the various aspects of the research such as the purpose of research, the research procedure, risks associated with the research and anticipated benefits, etc. Subjects should get the opportunity to ask questions regarding the research and they should be able to withdraw at any time from the research. Such information is to be conveyed comprehensively. The subject's ability to understand depends on various factors such as intelligence, rationality, maturity language, etc. Therefore, the investigator should present this information according to the capacities of the subjects. Another important fact is that consent by a subject should be given voluntarily. It should be free from any kind of influence; either coercion or any undue influence. If the subject has any such fear or interest, it would distort the result of the study.

Assessment of risk and benefits: The assessment refers to the process of gathering systematic and comprehensive information about the proposed research. In the ethical context, more focus is given to the consequences of the research. It became the responsibility of the researcher to justify the research work on the basis of a favourable risk or benefit assessment. The term risk refers to the possibility for harm to occur and the term benefit refers to something of positive value. The assessment of risks and benefits associated with research is important because it not only affects the individual but also the society. It should be noted that the risks involved in research are sometimes justified on the basis of other sufficient interests other than those of the subject, so long as the subjects' rights have been protected.

In practice, the researchers follow a common agreement that the benefits and risks should be in a favourable ratio. The rules to assess the risk-to-benefit ratio have already been mentioned. Also, the assessment of research should be very accurate and precise. "There should first be a determination of the validity of the presuppositions of the research; then the nature, probability, and magnitude of risk should be distinguished with as much clarity as possible. The method of ascertaining risks should be explicit, especially where there is no alternative to the use of such vague categories as small or slight risk" [32].

Selection of subjects: The principle of justice provides the moral requirement for the selection of subjects for a research study. The selection is to be made not only based on individual justice but based on social justice also. Individual justice rules out the possibility of selecting some participants, who are in favour of the investigator; for beneficial research and *undesirable* participants for risky research thus ensuring fairness in the selection procedure. Social justice provides an order of preference in the selection of the class of subjects; that is in determining which class (e.g., adult or children) is suitable for a particular study. Any injustice that arises from social, racial, sexual, and cultural biases should not influence the distribution of burden among the participating subjects. Special care is to be taken when selecting a special class of subjects such as institutionalized ones because "they are already burdened in many ways by their infirmities and environments".

A researcher should take precautions against the injustice that results from the involvement of vulnerable subjects. Vulnerable subjects are often selected by researchers because they are readily available. Examples of such groups are racial minorities, the economically disadvantaged, the very sick, prisoners, institutionalized, etc. Such subjects should be protected against the danger of being involved in research solely for administrative convenience. This is important because any harm resulting from the experiment can easily be manipulated as a result of their illness or socioeconomic condition.

9.7 COSTS OF RESEARCH MISCONDUCT

The consequence of research misconduct affects not only the individual but it has a ripple effect. It will spoil the reputation of the organization where the research was conducted and at large the society may lose faith in scientific research. The most common type of consequences that research misconduct can produce include [33]:

- Individual costs
- Brand costs
- Capital costs
- Human costs

Below is a brief discussion of these consequences.

Individual costs: Any research misconduct has a direct effect on the individual researcher. The integrity of the researcher will be questioned. Such misconduct leads to "loss of their professional career or reputation, lawsuits, revoked awards, designations, titles and academic degrees". There are a number of such incidents where the individuals lost their career, position, awards, etc. (2012 iThenticate Report, www.ithenticate.com).

Brand costs: Brand cost is one of the long-term damages caused by research misconduct. The term *brand* represents a kind of fame that the universities, publications, and journals all have. The brand plays an important role in attracting the students to the universities, increasing the subscription of journals, increasing the funding from various funding agencies, etc. An individual researcher is always a part and a representation of an organization. Any research misconduct in which the researcher has an involvement definitely spoils the reputation of the institution. Similarly, the reputation of a journal will be affected if it publishes erroneous reports continuously.

Capital costs: A case of research misconduct will lead to a capital loss to the institution where the research was conducted. This loss includes not only the amount sanctioned to the research project but money spent in the investigation of the case, the amount spent in repairing any equipment, and many other expenses related to the research.

Human costs: Human costs resulting from research misconduct may not get much public attention. Even though a report was retracted because of misconduct, it might have influenced many researchers. The data from many retracted papers may already be used by some other researchers. For example "*The Journal of Medical Ethics* published a study in 2011 that examined 180 retracted papers which contained research linked to 28,000 patients. The study found that 6573 of those patients were treated utilizing the data and research from the retracted papers" (cited in 2012 iThenticate Report) [34]. Sometimes the falsified data may be used by the researchers even after the retraction of the paper. This is because the removal of such papers from all sources on the internet is not easy and may be available for a further long period. The researcher uses such papers and data, very often, unknowingly. This then degrades the quality of further research.

9.8 DEALING WITH RESEARCH MISCONDUCT

Scientific misconduct is one of the major problems faced by the scientific community. Many people, institutions, universities, and other research organizations have provided various definitions of scientific misconduct concentrating mainly on plagiarism, falsification, and fabrication of data. However, I could find a comprehensive definition of scientific misconduct on the website of the Welcome Trust, U.K. According to the website, the scientific misconduct is:

"The fabrication, falsification, plagiarism or deception in proposing, carrying out or reporting results of research or deliberate, dangerous or negligent deviations from accepted practices in carrying out research. It includes failure to follow established protocols if this failure results in unreasonable risk or harm to humans, other vertebrates, or the environment and facilitating of misconduct in research by collusion in, or concealment of, such actions by others. It also includes intentional, unauthorized use, disclosure, or removal of, or damage to, research-related property of another, including apparatus, materials, writings, data, hardware or software, or any other substances or devices used in or produced by the conduct of research. It does not include honest error or honest differences in the design, execution, interpretation, or judgment in evaluating research methods or results or misconduct unrelated to the research process. Similarly, it does not include poor research unless this encompasses the intention to deceive" [35].

Scientific misconduct can be mainly classified into three:

1) *Misconduct that distorts scientific knowledge*: This distorts the entire scientific knowledge and thus puts the scientific community and the society as a whole at risk. Examples are falsification and fabrication of non-existent data.
2) *Misconduct that misleads the scientific community*: Plagiarism; unjustified authorship; duplicate publication; deliberate false evaluation of projects and results.
3) *Questionable research practices*: Such activities generate doubt on the seriousness of the research. They include sloppy handling of data; division

of the results, with publication in several different journals solely for the purpose of increasing the list of publications.

For responsible conduct of research, a researcher should maintain the integrity of scientific research and take precautions to avoid any kind of scientific misconduct at any stage of the research work. For the responsible conduct of scientific research Emilio Bossi (2010) suggested the following practices be followed by a researcher [36].

- In the planning phase, the feasibility of the research plan must be checked, so that later no one will be tempted to distort data in order to complete the project.
- The roles of the individuals involved must be clearly defined, to avoid interpersonal conflicts.
- The right of authorship and the order in which the various authors are listed in publications must be clarified at as early a stage as possible.
- Conflicts of interest must be presented honestly and transparently; if such conflicts exist, even a faultless project can lose its entire value.
- The handling of data and materials must also be clearly defined: who can be informed of the results in the course of a project, and when?
- Agreements between researchers and sponsors, other research groups, and other instances must be documented.
- During the execution of a project, the procedure must be accurately and reliably documented.
- Data and materials must be stored in such a way that they cannot be lost or manipulated. There must be an open exchange of information within the research group.
- Publication of the results of the research must be unbiased and complete. The deliberate withholding of data is considered a falsification.
- Presentation of the results in separate publications for the purpose of increasing the number of titles published is to be avoided.

Scientific misconduct may occur unknowingly and sometimes deliberately. Plagiarism Occurs because of the unawareness of the researcher about what plagiarism is. However, falsification and fabrication of data result from a deliberate action from the researcher. All credible allegations of research misconduct should be investigated, and that corrective actions should be undertaken if the investigation makes a positive finding. Many articles that deal with scientific misconduct and research ethics suggest the formation of the Research Ethics Committee at the institutional level as well as the appointment of a research Ombudsman.

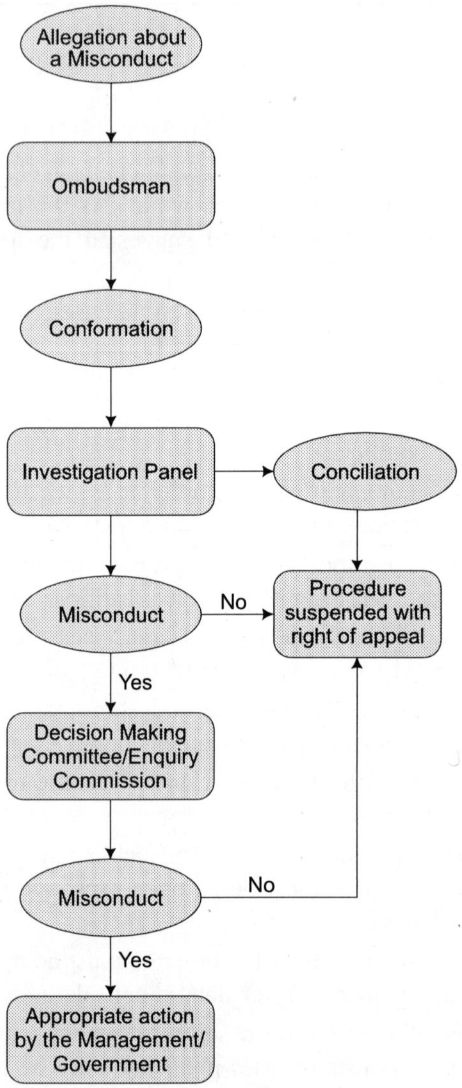

Fig. 9.1 Flowchart depicting the various processes involved in the investigation of scientific misconduct

The duty of such committees is to investigate the scientific misconduct within the organization and suggest necessary action. In addition, the committee is responsible to create awareness about scientific misconduct among the researchers through educative programmes. The principal investigator of any research team "should be aware of the appropriate channels for reporting possible misconduct in science. Leaders of research teams are responsible for insuring that scientists under their direction are familiar with various ethical standards and that they follow them" [37]. Such committees should also be constituted by professional societies and funding agencies with wider jurisdiction than an institutional committee. There should also be international committees on research ethics which constituted and managed by scientific societies or by governments (David B. Resnik, 1998, p. 156).

The ombudsman with responsibility for investigational procedures is appointed by the institution usually for a fixed time period. The ombudsman initiates the investigational procedures on suspected scientific misconduct if the allegation is found to be valid on the basis of an initial evaluation. Then an expert committee will be appointed by the ombudsman. The committee may include internal and external persons as members. "The internal panelists should be included because of their knowledge of the local situation, and the external ones for their broader view of the situation and to facilitate acceptance of the final decision" [38]. The committee then makes an investigation on the issue and reports the findings to the ombudsman. If the report confirms that scientific misconduct has occurred, then the institution appoints a decision-making committee. This committee then makes decisions based on the report of the investigation panel and suggests the higher authority, in most cases the management, to initiate appropriate measures (Emilio Bossi, 2010).

The investigation process should be transparent. The suspected individual should get the opportunity to explain his stand. Therefore for fair conduct of the investigation, the investigation procedure should meet the following basic requirements (Emilio Bossi, 2010).

- Hearing of the individual accused.
- Confidentiality for all persons involved in the procedure.
- Protection of the person making the allegation.
- Any persons who may be biased must not be involved in the procedure.
- The individual procedural steps must be documented in writing, with the possibility of appeal.

These basic requirements ensure that the investigation is not a one-sided process. The accused individual should get a chance for an appeal before the committees. Also, the principles, rules, and procedures related to scientific misconduct should be clearly defined and well publicized. They should include the definitions of misconduct and the steps for receiving and processing allegations. This will be helpful in promoting fairness in the investigation process and ensures that the process will not be seen as arbitrary or deliberately targeted against any individual. In other words, the scientific misconduct investigations themselves satisfy the highest levels of integrity and accuracy. Fairness and credibility in such an investigation are crucial because it is easy to spoil the reputation of a scientist but it is difficult to restore the damaged reputation. A flowchart showing the various processes involved in the investigation of suspected scientific misconduct is shown below.

9.9 RESEARCH ETHICS COMMITTEES

Research ethics committees have an important role in ensuring ethical standards and integrity in scientific research. Such committees are established to prevent the misconduct in scientific research. Research ethics committees are often appointed by the institutions which are responsible for the ethical conduct of scientific research. A research ethics committee can be considered "a group of people appointed to review research proposals to assess formally if the research is ethical" [39]. The research must be performed according to ethical standards and should consider the rights, dignity, and safety of the participating people. The members of an ethical committee can be from various fields of expertise. Scientists, professionals, legal experts, and ethical experts are usually included in an ethics committee.

At an institutional level, the ethics committee is especially important. The main roles of a research ethics committee are given below.

- To determine the research policy of the institution.
- To provide ethical education to researchers and other members of the institution.
- To suggest how to use the available resources including scientific data.
- To protect the rights and dignity of the individual participating in the research both as a researcher and as the subject.
- To determine the ethical standards to be observed in a particular research programme. This is especially important in clinical research.
- To evaluate the research process.
- To point out any scientific misconduct noticed.
- To help the enquiry committee in the case of proven scientific misconduct.

The main objective of any research ethics committee is to ensure the ethical practice of scientific research by maintaining the integrity and preventing scientific fraud. There are many reasons why a scientific ethic committee is to be constituted. A few of them are listed below [40].

- To accept the ethics committee's approval as evidence that research is performed according to the ethical principles.
- Journals' requirement of ethics committee's approval for publication of scientific articles.
- To provide performing of international research protocols according to the ethical principles those are projected in the general research protocol and to certificate this.

- To examine the usage of financial support from local and/or foreign foundations for scientific researches according to the aim of the research and ethical principles.
- To protect the subject and researcher's rights and to counsel if any ethical problems occur by consideration of both of them.
- To retain some researches those do not have any scientific support in journals and to cause vital problems.

The responsibilities of ethics research committees largely depend on the discipline in which the research is being conducted. For example, in clinical research, the committee is more concentrated on ensuring the safety and well-being of the participants. In disciplines like social sciences, ethics committees are more concerned about plagiarism, falsification, or fabrication.

The majority of the institutions appoint members of the research ethics committee selected randomly from various areas of expertise. One disadvantage associated with this is that many such members may not be sufficiently equipped with the knowledge of ethical conduct of scientific research. Institutions usually do not conduct any orientation programmes or ongoing training programmes for the members of the ethical committee. Such training programmes are essential in many areas of research such as medicine. The ethics committee itself has some ethical obligations. A member of the committee should never allow a conflict of interest to act in making a decision. If any such conflict of interest is present then the member should declare it before the other members of the committee and should refrain from the decision making.

In ethics committees, the members usually work on an honorary basis. They are not paid for their work. They are members of the ethics committee because of their personal interest in the research process. This reduces the possibility of the decision being influenced by any conflict of interest. This idea is not again problem-free; some members may not maintain enthusiasm and interest in their unpaid contribution for a long period. Therefore many articles on research ethics committees suggested the formation of a "small but professionalized committee of highly qualified and trained experts who are adequately paid for their contribution…." [41].

Research ethics committees at various levels including local, national, and international insure transparency in scientific research. Regarding the role of research ethics committees, the Helsinki Declaration says that "the research protocol must be submitted for consideration, comment, guidance, and approval to a research ethics committee before the trial begins. This committee must be independent of the researcher, the sponsor, and any other undue influence". Thus, right from the proposal for a research, the entire research work is continuously monitored by the ethics committees.

REVIEW QUESTIONS

1. Define research ethics.
2. What are the principles of scientific research?
3. Write a note on research ethics.
4. Why ethics is important in research?
5. What are the social elements that influence scientific misconduct?
6. Explain self-regulation.
7. Discuss the values and principles of ethics.
8. Discuss the various ethical principles involved in scientific research.
9. Explain copyright.
10. Write a note on patents and trademarks.
11. Discuss various intellectual property rights.
12. Mention some agreements on intellectual property rights.
13. What is royalty?
14. What is a stem cell case?
15. Give an account of the Baltimore case.
16. What are the various scientific misconducts?
17. What do you mean by plagiarism?
18. Write an essay on plagiarism.
19. Mention some common forms of plagiarism.
20. What are the possible causes of plagiarism?
21. Explain how plagiarism can be avoided.
22. Distinguish between summary and paraphrase.
23. Mention some tips for writing a good summary.
24. What is quotation? What is its importance?
25. What are the different methods for detecting plagiarism?
26. What are the responsibilities of an author in view of plagiarism?
27. Discuss the various tools for checking plagiarism.
28. What kind of exemptions are given by UGC in plagiarism checking?
29. What do you mean by misuse of privileged information?
30. Explain how data is misused in scientific research.
31. What is the efficient use of data and its importance in scientific research?

32. Write a note on the authorship and publication issues.
33. Give an account of undesirable authorships.
34. Discuss various codes of ethics.
35. Discuss the Nuremberg codes of ethics.
36. Explain the Belmont principles.
37. What are the requirements for applying ethical principles in scientific research?
38. Discuss various costs of scientific misconduct.
39. What are the main classifications of scientific misconduct?
40. Explain the possible practices for avoiding scientific misconduct.
41. Explain the various stages involved in the investigation of scientific misconduct.
42. Explain the role of research ethics committees.
43. Why the research ethics committee is important in an institution?
44. Discuss the functions of research ethics committees.

REFERENCES

[1] Adil E. Shamoo and David B. Resnik, "*Responsible Conduct of Research*", Oxford University Press, 2009, p. 14.

[2] David B Resnik, "*Philosophical Foundations of Scientific Ethics*", p. 4.

[3] Panel on Scientific Responsibility and the Conduct of Research. Committee on Science, Engineering, and Public Policy. National Academy of Sciences, National Academy of Engineering, Institute of Medicine. "*Responsible Science: Ensuring the Integrity of the Research Process*", Volume 1 (Washington, D.C.: National Academy Press, 1992), p. 1.

[4] National Committee for Research Ethics in Norway, "*Guidelines for Research Ethics in Science and Technology*", 2008, p. 41, www.etikkom.no/English/Publications/NENTguidelines.

[5] Mark Frankel, "*Professional Codes: Why, How, and with What Impact?*", 1989, Journal of Business Ethics, 8: 109-115.

[6] Tuohy, Carolyn, and Wolfson, Alan: 1977, "*The Political Economy of Professionalism: A Perspective in Four Aspects of Professionalism*", Consumer Research Council, Ottawa.

[7] "*Guidelines for Research Ethics in Science and Technology*", 2008, pp. 11-19.

[8] Patricia A. Bolton, "*Scientific Ethics*", 2002.

[9] Adil E. Shamoo and David B. Resnik, pp. 28-29.

[10] George C Thomas., p. 576-579.

[11] www.assets.publishing.service.gov.uk/government/uploads/system/uploads/attachment_data/627956/IP-Rights-in-India.pdf

[12] Bogner, Alexander and Menz, Wolfgang. "*Science Crime: The Korean Cloning Scandal and the Role of Ethics*", October 2006, Science and Public Policy. 33.8, 585-589.

[13] Kevles, Daniel J. "*The Baltimore Case: A Trial of Politics, Science, and Character*", 1998, New York: W.W. Norton.

[14] 2012 iThenticate Report, "*True Costs of Research Misconduct*", www.ithenticate.com

[15] Van Noorden, Richard. "*Science publishing: The trouble with retractions*", Nature. October 5, 2011. http://www.nature.com/news/2011/111005/full/478026a.html

[16] Centre for Bioethics, University of Minnesota, "*A Guide to Research Ethics*", 2003, p. 11.

[17] "*How to Avoid Plagiarism: The Scourge of the Academe*", p. 3, www.kgsupport.com

[18] Department of Sociology, Western Washington University, "*The Student's Guide to Avoiding Plagiarism*", p. 1.

[19] "*How to Avoid Plagiarism: The Scourge of the Academe*", p. 5.

[20] University of Malaya, "*How to Avoid Plagiarism: A Handbook for Postgraduate Students*", p. 10.

[21] Asim M. El Tahir Ali, Hussam M. Dahwa Abdulla, and V´aclav Sn´aˇsel, "*Overview and Comparison of Plagiarism Detection Tools*", Dateso 2011, pp. 161–172.

[22] https://www.ugc.ac.in/pdfnews/7771545_academic-integrity-Regulation2018

[23] University of Pittsburgh, "*Guidelines for Responsible Conduct of Research*", March 2011, p.4, www.pitt.edu

[24] Ibid

[25] Organisation for Economic Co-operation and Development, "*OECD Principles and Guidelines for Access to Research Data from Public Funding*", p. 10.

[26] https://www.insaindia.res.in/pdf/Ethics_Book.pdf

[27] "*Trials of War Criminals Before the Nuremberg Military Tribunals*" Under Control Council Law no. 10, Vol. II, October 1946-April 1949, pp. 181-182.

[28] Janet Zimmerman, "*The Belmont Report: An Ethical Framework for Protecting Research Subjects*", The Monitor, Summer 1997.

[29] The National Commission for the Protection of Human Subjects of Biomedical and Behavioral Research, "*The Belmont Report: Ethical Principles and Guidelines for the Protection of Human Subjects of Research*", April 18, 1979, p. 5.

[30] Ibid

[31] Ibid p. 8

[32] Ibid

[33] 2012 iThenticate Report

[34] Grant R Steen, "*Retractions in the medical literature: how many patients are put at risk by flawed research*?" Journal of Medical Ethics. May 17, 2011.

[35] http://www.wellcome.ac.uk/doc%5Fwtd002756.html

[36] Emilio Bossi, "*Scientific integrity, misconduct in science*", Swiss Med. Wkly. 2010; 140(13–14):183–186

[37] David B Resnik, "*The Ethics of Science: An Introduction*", 1998, Routledge London, p. 156.

[38] Emilio Bossi

[39] "*Governance arrangements for research ethics committees*", May 2011, p. 6, www.dh.gov.uk

[40] Elif ATICI MD, "*Situation of Research Ethics Committees in Turkey*", JISHIM 2006, 5, 43-46.

[41] Christiane Druml, M. Wolzt, J. Pleiner and E. A. Singer, "*Research ethics committees in Europe: trials and tribulations*", Intensive Care Med (2009) 35:1636–1640

Suggested Readings

1. Sana Loue, "*Text Book of Research Ethics: Theory and Practice*", 2002, Kluwer Academic Publishers.
2. David B Resnik, "*The Ethics of Science: An Introduction*", 1998, Routledge London.
3. Adil E. Shamoo and David B. Resnik, "*Responsible Conduct of Research*", 2009, Oxford University Press.
4. Judy E. Stern and Deni Elliott, "*The Ethics of Scientific Research*", 1997, University Press of New England, Hanover.
5. Paul Oliver, "*The Student's Guide to Research Ethics*", 2003, Open University Press, McGraw-Hill Education.

Chapter 10

Publication Ethics

Scientific research? Only when not at the cost of ethics—and first of all, those of the researchers themselves.

— **Aleksandr Solzhenitsyn**

Ethics is knowing the difference between what you have a right to do and what is right to do.

— **Potter Stewart**

10.1 INTRODUCTION

There is no denying that research affects every area of human society and that the type of study conducted today will determine the world's future. The research activities and then the following publication of the research findings must be based on some ethical principles. Researchers, students, and the general public can all suffer serious consequences as a result of unethical study practices. The proper study design, study execution, data gathering, data analysis, and publication are only a few of the many coordinated procedures and processes that go into academic research. Even though going through all of these procedures and ending up with a publication can be an amazing experience, one should be mindful of the ethical code of behaviour that binds researchers at every level. Codes, regulations, and principles are crucial and helpful, but like any set of guidelines, they may not apply in all circumstances, frequently contradict, and require a great deal of interpretation. Therefore, it is crucial for researchers to learn how to decipher, evaluate, and implement different research guidelines as well as how to form judgments and behave morally in diverse circumstances.

The necessity for the existence of various norms of research ethics is to address research and publication misconduct. The fabrication or falsification of data, plagiarism, problematic data presentation or analysis, failure to obtain informed consent from the subject or ethical approval from the Research Ethics Committee, inappropriate claims of authorship, duplicate publication, and undisclosed conflicts of interest are just a few examples of research misconduct. A detailed discussion of various research and publication misconduct and various norms of research and publication ethics are given in this chapter.

10.2 IMPORTANCE OF ETHICAL NORMS

There are several reasons why it is crucial to follow ethical standards when doing research [1].

- The goals of research, such as knowledge, truth, and error prevention, are promoted by norms. For instance, laws against creating, manipulating, or presenting research results incorrectly encourage the truth and reduce inaccuracy.
- Ethics encourage the values that are vital to collaborative work, such as trust, accountability, mutual respect, and fairness, as research frequently includes a lot of cooperation and coordination among many different people in various fields and institutions. Most scientists want to be acknowledged for their work and do not want their concepts to be misappropriated or revealed too soon.
- Many ethical rules make it possible for the public to hold researchers accountable.
- Public support for research is bolstered by ethical standards in the field. If people can trust the calibre and integrity of the research, they are more likely to support it.
- Numerous research norms support a range of other crucial moral and social values, including civic duty, respect for others, human rights, animal care, legal observance, and public health and safety.

There are many practices that are considered unethical. David B. Resnik pointed out the following practices that are considered unethical [1].

- Publishing the same paper in two different journals without telling the editors
- Submitting the same paper to different journals without telling the editors
- Not informing a collaborator of your intent to file a patent in order to make sure that you are the sole inventor
- Including a colleague as an author on a paper in return for a favour even though the colleague did not make a serious contribution to the paper
- Discussing with your colleagues confidential data from a paper that you are reviewing for a journal
- Using data, ideas, or methods you learn about while reviewing a grant or a paper without permission
- Trimming outliers from a dataset without discussing your reasons on paper
- Using an inappropriate statistical technique in order to enhance the significance of your research
- Bypassing the peer review process and announcing your results through a press conference without giving peers adequate information to review your work

- Conducting a review of the literature that fails to acknowledge the contributions of other people in the field or relevant prior work
- Stretching the truth on a grant application in order to convince reviewers that your project will make a significant contribution to the field
- Stretching the truth on a job application or curriculum vita
- Giving the same research project to two graduate students in order to see who can do it the fastest
- Overworking, neglecting, or exploiting graduate or post-doctoral students
- Failing to keep good research records
- Failing to maintain research data for a reasonable period of time
- Making derogatory comments and personal attacks in your review of the author's submission
- Promising a student, a better grade for sexual favours
- Using a racist epithet in the laboratory
- Making significant deviations from the research protocol approved by your institution's Animal Care and Use Committee or Institutional Review Board for Human Subjects Research without telling the committee or the board
- Not reporting an adverse event in a human research experiment
- Wasting animals in research
- Exposing students and staff to biological risks in violation of your institution's biosafety rules
- Sabotaging someone's work
- Stealing supplies, books, or data
- Rigging an experiment so you know how it will turn out
- Making unauthorized copies of data, papers, or computer programs
- Deliberately overestimating the clinical significance of a new drug in order to obtain economic benefits

These kinds of activities are strictly unethical and sometimes they may invite legal actions. Most universities and government institutions issue proper guidelines and norms for the ethical conduct of research activities. It is to be noted that many of the activities mentioned above may not be specified as unethical in norms issued by the universities or governments; however, they are still unethical practices. Misconduct can have a significant influence on science and society even if it only happens occasionally because it can undermine research integrity, decrease public confidence in science, and waste time and money.

10.3 GENERAL RESPONSIBILITIES OF AUTHORS

Researchers are typically thought of as playing a crucial role since research aims to increase the body of knowledge in society. Authors of research articles have several responsibilities throughout the research process until the publication of the article and sometimes even after that. During research and while preparing the research article or manuscript, the following responsibilities must be kept in mind by the authors [2].

- *Declaration of Originality*: Almost all journals demand that the authors should offer a declaration attesting to their uniqueness before it is taken into consideration. Some journals frequently ask the authors to submit copies of papers from relevant research to the one they are reviewing.
- *Proper Record-keeping and Maintenance of Data*: Researchers are required to keep daily logbooks to track their activities and findings. Negligence and carelessness related mistakes ought to be avoided. Even after the data have been published, the results, together with any supporting images or data outputs, must be kept on file for a sufficient amount of time to adequately address any ensuing questions, doubts, etc.
- *Avoiding Plagiarism*: Plagiarism is the unjustified use of another person's thoughts, methods, conclusions, or words (including those gleaned from a private examination of their research proposals or manuscripts) without giving due credit. A proper citation must be given whenever published material is used in a manuscript. To utilize someone else's data, proper written authorization from the copyright holder must be obtained and kept on file.
- *Honesty, Objectivity, and Integrity*: In all scientific communications, authors must adhere to honesty, objectivity, and integrity while avoiding bias in the design of experiments, data analysis, data interpretation, and reporting of data, results, methods, and procedures. Falsifying, fabricating, or misrepresenting data is unethical and should never be used.
- *Conflict of Interest*: Any conflict of interest; personal or financial, that might affect the research should be disclosed by the authors. Additionally, researchers should avoid bias in grant writing, grant approval, peer review, hiring decisions, and other areas of study. Any grant or subsidy for research expenses obtained from a for-profit or non-profit organization, a government agency, or a research support group must be properly disclosed by the authors. These disclosures aid the reader in understanding the authors' relationships with various for-profit organizations that could be interested in the data reported in the study. Therefore, the conflicts of interest can be classified as *intellectual conflicts of interest* and *financial conflicts of interest*. Situations where widespread knowledge may clash with what is reported are included in definitions of intellectual conflict of interest. For instance, most authors quote references to support their work, yet some authors may do so incorrectly or with inadequate evidence for their argument. A financial relationship between the author(s) and a for-profit organization raises a financial conflict of interest. People who

accept such assistance should take care that their acts are not seen as product promotion [3].

- *Respect for Intellectual Property*: It is responsible to the authors to respect copyrights, patents, and other kinds of intellectual property. In 2018 Chaddah [4] pointed out that if you have accessed someone else's unpublished data or research plans (as a reviewer, editor, lab visitor, lecture audience member, etc.), you should not exploit their ideas for your work without their consent because doing so is considered "concept plagiarism."
- *Copyright Agreement*: Authors are usually paid when publishing an article, although required to assign copyright to the publication of their study in the journal. In a legal document known as an assignment of copyright details, the authors provide the publisher with a few rights. Typically, these contracts permit the authors to keep some ownership of the content. Writers must read and comprehend the copyright agreement text very carefully. A link or a copy of the original text in the original language must be explicitly recognized and submitted to any publications that use translated versions of quoted texts.
- *Disclosures*: Authors must adhere to the guidelines for journal submission by being truthful and impartial. Disclosures regarding the originality of the content and the real contributions made by each author to the study are required. Correct disclosure of financial assistance and conflicts of interest is required.
- *Consent for Reuse of Published Material*: Sometimes authors need to reuse information and/or photos that they or others have already published. Reuse may require formal permission from the publisher and/or original writers, depending on the copyright restrictions of the relevant content; in contrast, when there are no copyright restrictions, reuse may be permitted without formal permission.
- *Confidentiality*: Every exchange of information between authors and the journal must be kept private. Throughout the peer review and publication processes, all correspondence regarding the paper should be sent to the designated corresponding author. Authors should adhere to the journal's communication guidelines with outside peer reviewers.
- *Responsible Publication*: Instead of just advancing a person's career, the main goal of a research publication is to spread new knowledge and improve scholarship. Anything that is published therefore indicates author's accountability.
- *Public Announcement of New Results*: In general, a piece of work that has been submitted and accepted for publication shouldn't be announced to the public until it has been formally published. You should adhere to the standards established by the applicable journal for this purpose.

- *Preprints and Prior Publication*: Internet distribution and conference proceedings may be viewed as past publications, which could jeopardise the originality of the paper.

Although these guidelines were made particularly for main research journals, they might also be applicable to reviews and other professional publications. Besides these ethical considerations, there are many other responsibilities for the author. Some of them are listed below.

- The authors are responsible to maintain accurate records of the data related to their submitted paper by utilizing the journal's interface correctly and supplying or grant access to these data within a fair amount of time. Depositing data in a proper repository or storage site for sharing and future use by others is appropriate when permitted by the employer, funding agency, and any other parties with an interest.
- Be mindful that the article proposals should contain original material that has not been published in other journals; papers presented at international conferences may be approved following the usual journal's review procedure.
- The authors should make sure the manuscript is not already being used or accepted for publication elsewhere. When parts of the content are similar to previously published or submitted content, must credit those sources.
- The authors should certify that the submitted paper contains only original work, and to acknowledge and credit any content that has been copied from other sources. To get authorization to use any content from external sources for reproduction.
- The authors must accept that they must address the reviewers' critical observations, opinions, and recommendations in a special statement that is accessible via the journal interface.
- If a substantial error in their publication is found, contact the journal editor or publisher as soon as possible. Assisting the editor and publisher when it is judged is essential to issue an erratum, addendum, or corrigendum notice, or to retract the paper.
- After a paper is accepted, the author shouldn't edit it. The author is required to explicitly communicate the nature of the desired change to the editor if there is a strong cause to do so. Any such suggested changes must ultimately be approved by the editor.
- Any inaccuracies that authors find after their manuscript has been published must be reported to the editor. When an author finds a substantial error or inaccuracy in their own published work, they have a responsibility to contact the journal's editor as soon as possible and work with them to retract or correct the publication.

10.4 SUBMISSION GUIDELINES FOR AUTHORS

The publication of an article in a peer-reviewed academic journal, which is the formal component of the scholarly communication system, has several goals beyond straightforward communication. It is a component in the growth of a well-organized and reputed network of knowledge. Before submitting a research article to a journal for publication, should follow some basic principles or guidelines. Many journals provide proper guidelines for the authors for the submission of their articles. Specific requirements may vary from journal to journal. However, basic requirements remain the same. General criteria for manuscript acceptance are almost similar for all journals. Some of them are [5]:

- Suitability for the journal's subject scope
- Scientific soundness, originality, salience, and timeliness of the information
- Compliance with the standards of medical ethics governing research conducted on human and animal subjects
- Compliance with specific research reporting protocols
- Coherence of the research design and methodology

Apart from these, Eun Seong Hwang et al. [6] pointed out the following submission guidelines for the authors of a research article. These guidelines almost completely address the basic issues related to research and publication ethics.

Basic responsibilities of authors

Authors are responsible for the information contained in their own research papers as well as the methodology and findings. Every author must be aware of the following basic responsibilities.

- Authors should thoroughly review their own research articles to ensure that calculations, data presentation techniques, result analysis, etc., are accurate. Authors are responsible for ensuring that their research complied with all applicable laws and ethical standards.
- Peer review, editing, and publication requirements for academic publications must be adhered to by authors.
- Authors must submit unique submissions, and they cannot submit the same work to multiple academic journals.
- The author must immediately notify the journal editor and discuss any errors found after the work has been submitted or published, even if they are minor follow-up actions.
- The author must contribute any relevant data, rules, instructions, software, lab notes, or other information as soon as the editor asks for it.
- Even after the publication of the manuscript, the author must cooperate fully with any requests made by the editor (or the readers); in addition, any errors or problems discovered regarding the manuscript must be resolved by cooperating with the editor.

Ensuring the objectivity, relevance, and transparency of the paper

The scientific research must be carried out using a sufficient amount of academic evidence, and the results must be analyzed logically in order for the researcher to ensure objectivity, relevance, and reproducibility of the research stated in the paper. The following points must be noted:

- To enable future researchers to repeat and reproduce the same findings, the author's suggested research methodology and analysis approach must be made clear in the publication.
- Researchers must faithfully document the process and outcomes and keep the data in an organized manner. The researcher must be able to provide the proof if the reviewers or journal editor asks for it at any stage throughout the peer review.
- The researcher must make explicit any limitations in the study if there are any.
- All conflicts of interest, both financial and non-financial, must be disclosed by researchers so that the journal editor, reviewers, and readers are fully informed.
- The funding sources for the research must be made public. The publication must clearly state whether the funding institution provided assistance with the experiment's design, execution, analysis, interpretation, translation, or report in addition to financial support. Any person who assisted with the execution, statistical analysis, or conceptualization of the experiment, as well as the source from which materials and equipment were obtained, must be identified.

First publication of original content

Results of research studies should be reported in an honest and accurate manner by all researchers. The originality of the research findings is very important. In this context, the following aspects are to be kept in mind.

- It is best to avoid extensive data change because it can be interpreted as fabrication.
- Researchers should be especially careful not to indicate a biased result when using digital photographs and other images like SEM, TEM XRD, etc. It might not be evident in some situations if the manipulation was allowed or compliant with rules. If these questions are posed, the author must explain their thought process and provide justifications in the document so that the reviewers or readers can evaluate the work properly.
- When producing a research paper, the findings shouldn't be purposefully ignored just because they are challenging to understand, inconsistent with other data, or don't support the hypothesis or the study's conclusion.
- The authors must use their own language to write all expressions utilized in the article.

- Additionally, the document must follow all publication and copyright regulations. Before using a source that is copyright protected, permission from the author of the source must be acquired.
- Authorship must be extended to any researchers who are thought to have contributed significantly to the planning, execution, and writing of the study as well as the manuscript.
- The approval, permission, and enrollment requirements must be made public for any research involving human experimentation or animal testing. The study subjects should be as fully protected as feasible when writing manuscripts.

10.5 ETHICAL ISSUES IN PUBLICATION

In this section, we will discuss some basic ethical issues related to the publication of research articles. The sole driving force behind early scientific inquiry was a pure curiosity to learn more about a particular topic. The way science and society interact nowadays has changed. Due to the high expense of the material and human resources needed for current research, it is dependent on funding from the government or privately owned firms with very specified objectives. In this situation, researchers must compete for funds in order to carry out their studies and publish the findings as soon as possible in order to enhance their scientific standing in the field of inquiry. The urge to publish has caused some writers to compromise the integrity of their study, which can occasionally lead to scientific fraud. In this regard, we must keep in mind that the publication process is founded on veracity, accuracy, and scientific integrity [7]. The editors of various journals meet regularly to discuss initiatives to further these goals, such as the use of particular software programs to detect plagiarism or the disclosure of conflicts of interest by authors, reviewers, and editors. These initiatives aim to ensure transparency in the publication of articles and combat scientific fraud.

Some important ethical considerations in the publication are given below [8-10]

Duplicate/Multiple Submission

The simultaneous submission of a study report to multiple journals is unethical and against the law. Authors must properly inform and request the editor of the journal in which their article is under consideration that their manuscript is withdrawn from further consideration by that journal if they intend to submit it to another journal while it is being considered elsewhere. The withdrawal request must have the support of all coauthors, and this support must be made known to the editor of the journal that is considering the study. The authors may only submit their paper to another publication after receiving communication from the original journal recognizing the withdrawal.

Publication of the same article in multiple journals

Duplicating or wilfully concealing the sources of work in new publications by reorganizing or reformulating existing publications is unethical and constitutes plagiarism. It is also bad to divide a given study into multiple publications because the value of the study could be lost due to the fragmented material in each.

Suggesting Potential Reviewers

Certain journals request that authors submit the names of potential reviewers as well as, in some situations, people who they might not want to serve as reviewers due to potential conflicts of interest. To provide names of potential reviewers based on a buddy or acquaintance who may increase the likelihood of acceptance is unethical. To suggest false names as reviewers with self-directed addresses so that the authors also serve as reviewers is utterly immoral.

Fabrication or Falsification

Falsification is the manipulation of data or experimental procedures to generate a desired conclusion or to avoid an ambiguous or unexplained result. Fabrication of data is the recording of false data when none exists. The genuine research conducted by other authors and researchers is weakened by this kind of misbehaviour. It not only consumes up the limited research resources but also wastes the time of other researchers who have planned further studies based on studies with fake data. The conclusions reached as a result of such investigations may have an impact on future research practice, and when such studies are made public, the public's confidence in scientific research is undermined [10]. Some example of fabrication or falsification are:

- Artificially creating data when it should be collected from an actual experiment
- Unauthorized altering or falsification of data, documents, images, music, art, or other work
- Unauthorized omission of data, information, or results in documents, reports, and presentations
- Hiding data, results, or information using inappropriate scales, magnification, and representation in charts, graphs, and other forms of representation
- Falsifying information pertaining to the subjects participating in an experiment
- Falsely recruiting subjects for experiments without revealing the purpose of the experiments or receiving institutional approval for involving subjects in the experiment
- Fabricating sources of information
- Unauthorized impersonation of another person to complete an academic activity

- Unauthorized use of another individual's computer login ID and password
- Unapproved deviation from a predetermined experimental procedure

Sabotage

Sabotage is the act of interfering with or destroying someone else's work in order to prevent that person from successfully completing an academic task [10]. Some examples are:

- Destroying another person's work (including documents, design, data, artwork, etc.)
- Not contributing to a collaborative effort adequately when required to do so
- Withholding information when it should be shared with others in a collaborative project
- Revealing confidential data about another person's project or procedures to others
- Disturbing the setup or calibration of equipment arranged by another person
- Colluding with others to falsely accuse others of academic dishonesty
- Installing viruses, spyware, or other damaging software on another person's computer
- Stealing another person's property, such as computers, experimental setups, data, or designs, to prevent that person from completing his or her academic activities
- Destroying books and materials in the library, laboratory, or similar places to prevent other students from completing their work successfully

Correction of errors in published articles

After an article has been published, the author or authors may discover certain mistakes in the data or their interpretation as a result of a mistake. If a mistake was unintentionally made when gathering, analyzing, or interpreting data, all authors of the work, if it had more than one author, should be made aware of it. The error, as well as its nature and any effects it had on the conclusions drawn, should be notified to the journal's editor. In order to correct the earlier error, authors are free to publish either an erratum or a new manuscript in the same journal, in conjunction with the editor.

Withdrawal of a Published Paper

An intentional act of misconduct, such as data fabrication, plagiarism, etc., on the part of authors in a published work, may be noticed or suspected by an author, editor, or reader. To prove the claimed misconduct, this needs to be properly probed, especially by the journal. In the event that a problem is discovered, all authors must be informed. The author(s) may then choose to retract or withdraw the article, or the journal may publish a retraction notice for the previously published paper along with appropriate justification. When such misconduct is discovered, the responsible institution or institutions must deal with it in accordance with the rules and laws that were in effect at the time.

To address all of these ethical issues related to the publication of a research article and to give special relevance to the ethical aspect of publishing scientific articles most scientific journals issue proper guidelines. Some of them are [7]:

- All of the authors have given their permission for the corresponding author to submit and publish the article that was submitted for review.
- Without leaving anyone out, all of the authors have made significant contributions to the article, and each author's work is noted.
- The paper is a unique work that hasn't been published before and isn't being considered by another publication at the same time (presentations to scientific conferences are excluded from this definition).
- The article does not contain any unpublished material copied from other authors without their consent.
- Whether or not the data are from the same authors, all references to data from earlier studies that are part of this article have been made. A publication must always be cited in an article that is submitted for review if it is a sub-analysis of previously released project results.
- If the article satisfies all of the aforementioned requirements, it will be maintained on file in the journal and be regarded as a valid publication.
- The authors must notify the journal right away if any of the aforementioned requirements are not followed so that the article can be removed from publication.

10.6 ROLE AND RESPONSIBILITIES OF EDITORS

Ethics must also be observed by the editors and reviewers. The writers' most important body of research has been submitted. It is expected of the reviewers and editors to uphold confidentiality and not use another person's words or ideas without permission. They ought to deliver the reviews on schedule. Any type of conflict of interest, including financial, personal, and other types, should be disclosed. Only when the reviewer feels competent to do so should the review be conducted. Members of the editorial team should not misuse data and should protect the confidentiality of any research that is submitted. A journal usually issues necessary guidelines for the editors and reviewers of the research article.

A chief editor, subeditors, and editorial board members are appointed by the majority of journals. Most of the time, editorial positions in journals are honorary, thus the work is done more for the love of the job and the status attached to it. Editors and their decisions are crucial in assuring the calibre of published work and the journal's overall reputation. Their duties include receiving timely and thorough peer reviews of the submission, checking the text's linguistic properties, assuring that the data and any supporting materials are accurate and sufficient, formatting the text properly for publication, receiving timely proofreading corrections, and finally publishing with high-quality results as per the schedule.

10.6.1 Role of Editors

The role of an editor in a journal can be summarized as follows:

- The editor is in charge of ensuring that the journal's intellectual content is of a high standard overall and is reviewed thoroughly, fairly, and promptly.
- The editor is in charge of choosing papers that make fresh, original, and significant advances to knowledge as well as upholding the journal's objective and scope.
- The editor chooses the editorial board members, reviewers, associate editors, and any other editorial roles necessary to create material for the publication.
- The editor chairs the editorial board and is responsible for directing its activities. For these special issues, the editor may appoint guest editors, request manuscripts, and write editorials.
- In addition to creating an annual report for the publisher, the editor looks for possibilities to host an annual meeting of the editorial board.

10.6.2 Ethical Responsibilities of Editors

In 2016, James Galipeau et al. [11] suggested the following general guidelines for the editors. Even though the aimed at the editors of biomedical journals, the principles are equally applicable to any journal editors.

- In order to fulfil his or her duties effectively and with authority, a journal editor must be academically qualified in the journal's designated field and enjoy editorial work.
- Before taking the position of chief editor or editor of a journal, the applicant should research not just the duties involved, but also the publications and its publishers' standards of excellence.
- All manuscripts submitted for publication should be given fair treatment by editors, who should evaluate each one on the basis of its own merits without taking the author's race, gender, religious affiliation, ethnic background, citizenship, or political philosophies into account(s)
- In order to prevent editing work from being delayed or postponed, all editors must agree to commit the necessary time to fulfil their editorial responsibilities.

- Since editors are required to understand the nature of research and authors' expectations, an active researcher may perform better in this role.
- All submissions should be submitted by the editor to check for plagiarism, and the quality of illustration materials.
- More papers are submitted to most publications than can be published. Because submission is obviously of poor quality or falls beyond the journal's mandate, editors frequently have the option of simply rejecting or returning it.
- The editor should carefully choose peer reviewers based on their subject-matter knowledge and professional experience.
- When a single- or double-blind review method is used, the review process must be kept confidential. Early on in the single- or double-blind review procedure, appropriate secrecy must be kept, even when the reviewers' names and remarks are eventually made public.
- Many journals ask authors to recommend potential reviewers, but some also request names of people they might not want to serve as reviewers due to potential conflicts of interest. In each scenario, the editor must use his or her own expertise, judgment, and knowledge to agree or disagree with the authors or take other actions. The editor shall reject the submission and notify the author's institution of ethical misbehaviour if the reviewer's name given by the author turns out to be fictitious.
- The reviewers' comments on the manuscript must be received by the editors in a timely manner. The majority of online submission tools utilized by various publications offer automatic reviewer reminders. Unnecessary delays can harm the author and the journal's reputation.
- Instead of only serving as the middleman between the two parties, the editor should directly assess the reviewers' reports and authors' responses.
- The manuscript should be handled by another member of the editorial team in a way that prevents the editor-cum-author in this case from having any access to the review process if an editor also happens to be an author in a submission to the same journal that uses a blind or double-blind review process.
- The editor's duties continue after publication, particularly when issues with priority, plagiarism, unethical data manipulation, etc., come up. In these situations, proactive and knowledgeable action and decision-making are required.
- The dates of the initial and revised submissions, if appropriate, and the date of acceptance should be included with the published papers. Ahead of print internet publication should also be attempted as soon as the manuscript is accepted.
- Unpublished information, arguments, or interpretations disclosed in a submitted manuscript should not be used in an editor's own research except with the consent of the author [12].

- One of the other editors of the editor's journal will be in charge of handling the editorial duties and authority for any submission written by the editor and submitted to the editor's journal. Real or perceived conflicts of interest should be avoided by editors.

10.6.3 General Responsibilities of Editors

Apart from these ethical responsibilities, an editor of a journal has many other duties and responsibilities for making the journal popular among the scientific community and in general improving the standards of the journal. Some of the additional responsibilities are given below:

- The editor consults with the editorial board, analyses the editorial mission and scope of the journal on a regular basis, and provides recommendations.
- The editor checks to see that the papers released follow the editorial mission.
- The editor is in charge of selecting the papers for publication as well as deciding whether to publish any thematic issues of the magazine. If requested, the editor may choose one or more guest editors for a thematic issue.
- In addition to the journal's normal issues, the editor may suggest to the publisher that a special issue be issued. Such a request will need to be approved or rejected by the publisher.
- The editor will allocate papers to reviewers who are qualified to assess them, keep an eye on the calibre of the reviews written, and replace editors whose evaluations are insufficient or late.
- When reviewers' assessments of work differ, the editor will make the final decision regarding whether or not to accept it.
- The editor will make sure that the review process is finished on time and that authors receive helpful input regarding their papers.
- The editor makes sure that the manuscripts are full, with all the necessary references, illustrations, and author contact details.
- The editor submits the manuscripts for each issue on time according to the schedule agreed upon with the publisher.
- Any ethical transgressions committed by authors and reviewers will be found, looked into, and addressed by the editor.
- The editor is responsible for ensuring that contributors are handled fairly, politely, objectively, and honestly.
- The editor needs to be equipped to handle mistakes and accusations of wrongdoing.
- The editor will try to speak about the goals and principles of the journal at conferences and other gatherings, encouraging potential authors to think about submitting papers to the journal.

- The editor will assist the publisher in promoting the journal by pointing out venues, gatherings, and people who should receive promotional materials.

10.7 ROLE AND RESPONSIBILITIES OF REVIEWERS

The primary method used to assess the calibre of research is peer review. Peer-reviewed papers serve as the foundation for the majority of funding decisions in science and the academic development of scientists. The quality of the peer-reviewing process and the calibre of the editorial board are considered the main influencers on a journal's reputation, Journal Impact Factor (JIF), and status in the field due to the growth in the number of scientific articles published each year. Scientific referees or reviewers, who often provide their time and expertise without compensation, are essential to scientific publications that publish peer-reviewed manuscripts. Reviewers assess articles submitted to journals according to their specifications, predetermined standards, and the calibre, thoroughness, and accuracy of the research presented. They offer criticism, give suggestions for enhancements, and advise the editor on whether to approve, reject, or ask for modifications to the article. The editor is always in charge of making the final choice, but reviews are crucial to the process. Editors select reviewers whose knowledge most closely fits the manuscript's topic and extend invitations to them to examine the work, sometimes with the help of electronic databases of reviewers maintained by the offices of their publication.

10.7.1 Role of Reviewers

The Elsevier journal specifies the following roles of a reviewer [13].

- Reviewers ensure the rigorous standards of the scientific process by taking part in the peer-review system.
- Reviewers uphold the integrity of the journal by identifying invalid research, and helping to maintain the quality of the journal.
- Reviewers fulfill a sense of obligation to the community and their own area of research.
- Reviewers establish relationships with reputable colleagues and their affiliated journals, and increase their opportunities to join an editorial board.
- Reviewers can help prevent ethical breaches by identifying plagiarism, research fraud, and other problems by dint of their familiarity with the subject area.
- Reviewers reciprocate professional courtesy, as authors and reviewers are often interchangeable roles – as a reviewer, researchers "repay" the same consideration they receive as authors.

10.7.2 Responsibilities of Reviewer Towards the Author, Editor, and Readers

The responsibilities of a reviewer can be broadly classified as responsibilities towards the author and responsibilities towards the editor of the journal.

The responsibilities of a reviewer towards an author are [14]:

- Giving timely written, objective, and constructive criticism on the work's scholarly merits and scientific value, together with the supporting evidence for the reviewer's judgement
- Evaluating the work's composition, scientific integrity, creativity, and appeal to the journal's readers as well as its clarity, conciseness, and relevance
- Staying away from personal remarks or criticism
- Maintaining the privacy of the review process by refraining from discussing or revealing any information from the reviewed paper with outside parties.

The responsibilities of a reviewer towards the editor of the journal are [14-16]:

- Notifying the editor as soon as possible if unable to review in time and, if possible, giving the names of substitute reviewers.
- Notifying the editor of any potential conflicts of interest, whether real or imagined, and rejecting to review if there are any.
- Obeying the editor's written instructions regarding the requirements of the journal for the review's scope, topic, and quality.
- Provide a thorough, unbiased, constructive, and instructive critique of the work that has been submitted, which may include any further materials that the author has offered to the journal.
- Evaluate the work's originality, scientific merit, and scope; suggest improvements; and, if required, recommend acceptance or rejection using the editor's preferred rating scale.
- Noting any ethical issues, such as any violations of accepted standards for the ethical treatment of animal or human subjects, or any significant similarity between the manuscript under review and any paper that has already been published or any manuscript that is concurrently being submitted to another journal that the reviewer may be aware of.
- Avoiding making direct contact with authors.

Finally, the responsibilities of a reviewer towards the readers are:

- Ensuring that the procedures and results are sufficiently described so that the reader may assess the study's scientific validity and perform a replication
- Ensuring that every important scientific paper is cited in the article

10.7.3 Ethical Responsibilities of Reviewers

Apart from the general responsibilities towards the author, editor, and readers, there are many ethical concerns to which the reviewers must adhere. Some of them are given below [17-20]:

- A reviewer should accept the assigned task only if they are sufficiently qualified and skilled in the subject area, and they should adhere to the journal's deadline for submission of comments. Peer review should be accepted as a responsibility that comes with being a professional.
- Any possibility of a conflict of interest should be immediately reported to the editor.
- Reviewers should be aware that they are or may have been written as well as active researchers, therefore they should give authors enough "room" to communicate their interpretation of the evidence, especially if it differs significantly from the prevailing viewpoints.
- Reviewers should look for originality in the issue or questions being addressed as well as some uniqueness in the findings that allow for advancement in knowledge using appropriate techniques, resources, experimental designs, etc.
- In addition, reviewers should look into any potential unethical actions taken by authors and report them to the editor or journal.
- Reviewers should be aware of the fact that the authors chose to conduct the specified study in a certain environment, moved forward in the manner they did for personal reasons, and prepared the paper in the manner they did.
- It is not appropriate for reviewers to attempt to impose their own favoured theories, hypotheses, experimental designs, etc. Reviews should be constructive in their criticism of the work and manuscript so that, even if it is rejected, the authors can still use the reviewer's remarks and recommendations to improve their future research output.
- Since the reviewer was given access to the article as part of the prepublication review, the reviewer should not utilize the information for his or her personal or another person's benefit. Without the journal's approval, it should not be done to involve another person in the review process.
- The editor anticipates a truthful and objective evaluation of the article under review's advantages and disadvantages. It is customary for reviewers to submit both confidential comments to the editor and more in-depth comments to the authors.
- Re-reviews of a changed version should be accepted by reviewers upon request. In general, they should refrain from bringing up fresh issues unless they are related to the amended material in order to be rejected.

- According to the journal's policy, the review process must be kept confidential whether it is single- or double-blind, and this obligation endures even if the manuscript under review is published or rejected.

The Council of Science Editors (CSE), an international membership organization for editorial professionals publishing in the sciences classified the ethical responsibilities of the reviewers under the following categories [14].

- ***Confidentiality***: Unless absolutely required and with the editor's permission, material under review should not be discussed or shared with anybody outside the review process. Material submitted for peer review should be treated confidentially and with care to protect the author's identity and creative output because it is a privileged communication. Reviewers shouldn't keep copies of the papers they receive, and they shouldn't utilize the information they learn about them for anything other than peer review.
- ***Constructive critique***: Reviewer comments should highlight the positive features of the content under consideration, constructively point out any problematic parts, and list any modifications that are required. Anything less prevents the author from understanding the flaws in the work that has been submitted. Editors and authors should be able to comprehend the reviewer's reasoning behind the judgement by the way the reviewer presents and supports the review. The reviewer must make sure that any observation or claim that has been made before is supported by an appropriate citation and must notify the editor right once they become aware of a duplicate publication.
- ***Competence***: Reviewers who are aware of their limited knowledge of the subject matter of the manuscript owe it to the editor to be honest about it. Reviewers need not be experts in every area of a manuscript's content, but they should only undertake a task if they have the necessary knowledge to make a reliable judgement. Without the necessary experience, a reviewer runs the risk of recommending either the acceptance of a submission with serious flaws or the rejection of a paper with value. The reviewer ought to decline the review in such circumstances.
- ***Impartiality and integrity***: Reviews should be based on facts that have been considered objectively and impartially without the influence of personal or professional bias. Without regard to the authors' race, ethnicity, sex, gender, religion, or citizenship, reviewers should base all comments exclusively on the paper's scientific merit, creativity, and writing quality as well as its relevance to the journal's scope and mission. A reviewer should make every effort to prevent even the appearance of making use of information received through the review process. A reviewer should not use the information available through the privileged communication of peer review for scientific, financial, personal, or other advantages.

- ***Disclosure of conflict of interest***: The review process should, to the greatest extent practicable, be set up to reduce any real or perceived prejudice on the side of the reviewer. Reviewers should decline to serve as reviewers if they have any conflicts of interest that would prevent them from providing an objective evaluation, or they should disclose these conflicts to the editor and inquire about the best way to resolve them.
- ***Timeliness and responsiveness***: Reviewers are accountable for taking prompt action, following the guidelines to the letter, and submitting their reviews on time. The review procedure will be harmed if this is not done. To the best of one's ability, the review should be finished in the allotted time. Reviewers should instantly reject to do reviews if they cannot complete them by the deadline.

10.8 GUIDELINES FOR EDITORS AND REVIEWERS

An international panel of subject-matter authorities makes up the editorial board. Each board member holds a position inside an organization that conducts research or education. The peer review procedure and the journal's content are under the control of the editors, who are authorities in their disciplines. Their responsibility is to oversee the peer review of manuscripts, provide recommendations regarding the acceptance or rejection of a paper, and solicit submissions of the highest calibre. An editor's main duty will be to revise manuscripts after they have undergone peer review. If peer reviewers request revisions, the editor will make sure they have been made. The editor will also check that manuscripts are free of plagiarism and other forms of scientific misconduct and that they are grammatically and stylistically consistent, readable, and readable. This will guarantee that the articles that are published have the necessary level of scientific rigour [21].

10.8.1 Guidelines for Editors

While choosing reviewers and making decisions, the editors must keep the following points in mind [22].

- Editors need to make sure that the right reviewers are chosen for submissions.
- In many journals, the default setting on manuscript central is set to three reviewers, thus editors should ideally choose at least two reviewers to provide a report. They should also make sure that, unless there is a compelling reason otherwise, no reviewer is recommended by the paper's authors.
- Reviewers that repeatedly produce rude, subpar, or late evaluations should no longer be used by editors.
- Editors should seek out possible new reviewers from a variety of places like bibliographic databases.

- Even if a choice appears to follow naturally from reviewers' remarks, editors should still provide written feedback to writers regarding that decision. In this scenario, one or two sentences summarizing the reviewers' views is adequate.
- Editors should monitor the performance of peer reviewers and take steps to ensure this is of a high standard.
- Reviewers should be urged by editors to address any ethical concerns and potential research and publication misconduct like unethical research design, inappropriate data manipulation, and presentation that are brought up by submissions, the originality of entries, as well as to watch out for plagiarism and duplication of work.
- A manuscript's importance, originality, and clarity, as well as the study's validity and relevance to the journal's scope, should be considered by editors when recommending whether to accept or reject a paper for publication.
- Unless there are significant issues with the submission, editors shouldn't overturn decisions to accept submissions.
- Unless there are significant issues, new editors shouldn't reverse the choices taken by the prior editor to publish contributions.

10.8.2 Guidelines for Reviewers

The requirements for double blind reviews are essential for assuring the calibre of scholarly journals, and they are a major factor in the editors' choice to publish a research paper. Every journal anticipates that each of the reviewers will provide a thorough, impartial evaluation of the papers in a timely manner. The following are some important guidelines for the reviewers.

- The reviewer is responsible for protecting the privacy of the research paper and any correspondence with the journal and other reviewers. Without the editorial board's prior approval, don't give the manuscript to any third parties.
- The reviewer's remarks on the article should be respectful and constructive. It is assumed that the reviewers are capable of offering a knowledgeable assessment of the manuscript. Try to identify the article's advantages as well as its shortcomings.
- Reviewers should analyze the article based on following criteria [23]:
    - ♦ Significance of the study
    - ♦ Meaningful ideas and clarity of arguments
    - ♦ Consciences of the manuscript
    - ♦ Coherence between titles, abstract, keywords, introduction, and findings
    - ♦ The logical structure of the article

    - ♦ Appropriateness of statistical analysis, tables, and figures
    - ♦ Relevance with regard to the journal
    - ♦ Standards of citation
    - ♦ Layout of the article
- Reviewers should make sure that your suggestions are accurate and fair since they will be forwarded to the appropriate authors. Be straightforward and precise as well so that authors can grasp them. It is requested that the reviewer should refrain from criticizing the author directly for his work.
- A reviewer should avoid all kinds of improper activities like [20]:
    - ♦ Presenting the facts in a review in the wrong manner.
    - ♦ Delaying the review process unnecessarily.
    - ♦ Criticizing a competitor's work unfairly.
    - ♦ Breaching the secrecy of the review.
    - ♦ Suggesting changes that appear to support the reviewer's own work.
    - ♦ Misusing confidential information for personal motives.
    - ♦ Stealing ideas or texts from a manuscript under review.
    - ♦ Including personal criticism of the author.
    - ♦ Failing to reveal a conflict of interest would have shunned the reviewer from the procedure.

In summary, a reviewer, by taking part in the peer review process, upholds the requirements of the scientific method and keeps the journal's quality high by pointing out dubious research. A reviewer builds connections with prestigious colleagues and the publications they publish in, by helping them to increase their opportunity to serve on editorial boards. It is essentially the reviewer who identifies plagiarism, research fraud, and other problems to stop unethical behaviour.

10.9 RESPONSIBILITIES OF PUBLISHERS

The research outcome reaches the readers through the published articles. In every country, there are numerous publishers who publish research articles. As a stand-alone volume, conference proceedings are occasionally published by individuals or academic institutions. The publisher gives the journal the necessary resources, including a portal for submission and publication, tools for detecting plagiarism, DOI numbers, hosting, and content preservation. The publisher must take reasonable measures to safeguard editorial independence. Therefore, some ethical principles are to be followed by the publishers of research articles [24-26].

- It is not appropriate to see the dissemination of research results as a simple business activity because it involves a public responsibility.
- The publication schedule and frequency must be kept to as stated.

- Publishers shouldn't try to boost journals' citations or impact factors artificially.
- Publishers are not allowed to impede the editorial board's and editors' academic freedom when they are reviewing papers submitted for publication.
- Publishers are required to give editors the infrastructure assistance they need to perform their duties and obligations.
- The journal's editorial, review, ethical, and publication practices must be transparent and open to public inspection. It is important to be clear about the author's standards and any fees the author may be required to pay.
- Any fees due from authors should only be paid once the work has been accepted unless the journal charges a fee for submission.
- It is absolutely unethical to publish in "predatory" or "fake" journals that would quickly publish "anything" for a fee without doing adequate peer reviews.
- The responsibilities of the publisher include:
  - ♦ Safeguard editorial independence
  - ♦ Collaborate to set industry best practice
  - ♦ Provide editors with technical, procedural and legal support.
  - ♦ Educate researchers on publishing ethics.

The general responsibilities of the publisher of a research journal are [27]:

- Create the policies required to support journal editors' editing processes, particularly those pertaining to the editors' independence, research ethics, authorship, transparency, and integrity, peer review, the editorial team's function, appeals, and complaints.
- Make the journal's policies known to everyone participating in the editing process.
- Review and update the journal's policies frequently to reflect any prospective new developments and editor recommendations.
- Promote the accuracy of scholarly work and the integrity of journal's articles.
- Support any agencies (such as governing authorities, institutions, or grant funders) in charge of looking into alleged research misconduct and publication process misbehaviour and, when possible, assist in the resolution of these situations.
- Post any necessary changes, clarifications, and retractions pertaining to the articles that have already been published.
- Publish the academic material as soon as possible.

10.10 FOLLOW-UP MEASURES FOR PUBLISHED MANUSCRIPTS

The publication process is not complete just by making the article available to the public. After the publication also, the author and the editor should keep an eye on the published article so that they may get responses from the readers. Follow-up measures for the published articles are required when the following situation arises [28].

Corrections: The following circumstances should prompt journal editors to think about publishing an erratum or corrigendum to correct the mistake:

- when mistakes are made in the editing process,
- when small errors in calculations or descriptions are discovered after publication, and
- when it is necessary to edit the list of authors, such as by removing authors who do not meet the requirements for authorship or by adding authors who deserve to be on the list.

Expression of Concern: An expression of concern can be used to clarify the situation to readers and take care of any potential negative impacts if a question presented raises serious doubts about the manuscript's veracity but has not yet been proven beyond a reasonable doubt. The expression of concern can be published in the following circumstances.

- when there is a suspicion of research misconduct but no investigative committee has been established or the outcome of the investigation is unsure,
- when the integrity of the investigation into the misconduct is in doubt,
- when the authors or leading experts in the field disagree on the research,
- when the findings of the research could be harmful to readers, and
- when it has been confirmed that a manuscript that was not authorized was submitted as a duplicate.

The editor should publish the "expression of concern" as soon as feasible to lessen the effect of the problematic article. When a decision is taken, the notification should be made in the online database and published as soon as possible in the printed edition of the journal. By clearly stating the author and title at the beginning of the expression of concern, the editor should make sure that the article in question can be easily confirmed; the parts and contents in question should be explained in as much detail as possible.

Retraction: "Retraction is a process which announces to readers that the publication cannot be relied upon, and thus should be ignored". When articles are published without first ensuring their legitimacy, they are retracted. This is done to uphold publication ethics, and the author shouldn't view it as a punishment. Due to research misconduct like falsification, fabrication, and plagiarism, publications are retracted. Retractions, however, are also used when there have been sincere errors, including arithmetic or experimental errors. The editor should retract the published article within the following circumstances.

- Publications are retracted when the findings cannot be accepted as the consequence of unethical research practices or sincere mistakes like mathematical or experimental errors.
- When research is proven to have been unethically conducted or to have involved in plagiarism, retractions are also employed.
- The second duplicated paper should be retracted when a duplicate publication is confirmed without clear prior disclosure or the appropriate authority.

Issuing a corrigendum can be considered:

- If only a small fraction of the paper contains an honest mistake.
- If the text's legitimacy may be upheld after repair.
- If only the author has been altered and not the content.
- If plagiarism has been confirmed in only a very small portion such as one or two sentences of the article.

The retraction notice should be published as soon as possible after the decision has been reached, both online and in the initial printed version of the article. This is done to lessen the potential harm to other researchers who might squander their time and effort or come to the wrong conclusions as a result of the false publication.

REVIEW QUESTIONS

1. Why ethics is important in scientific publishing?
2. Give some reasons why ethical standards are to be followed in research.
3. Mention possible unethical practices in research.
4. Give a detailed account of the ethical responsibilities of an author.
5. What is plagiarism?
6. Explain the conflict of interest.
7. What are intellectual and financial conflicts of interest?
8. What are the general responsibilities of an author?
9. Mention some general criteria for the acceptance of a research paper.
10. What are the basic responsibilities of authors?
11. Discuss the publication guidelines for the authors.

12. Discuss the possible ethical issues in publication.
13. What is multiple submission?
14. Explain fabrication and falsification.
15. Give some examples of falsification.
16. What is sabotage? Give examples.
17. Discuss major publication guidelines given by most journals.
18. What are the roles of editors?
19. What are the ethical responsibilities of editors?
20. What are the general responsibilities of editors?
21. What are the roles of reviewers?
22. What are the responsibilities of reviewers?
23. What are the responsibilities of a reviewer towards the author?
24. What are the responsibilities of a reviewer towards the editor?
25. What are the responsibilities of a reviewer towards the reader?
26. Discuss the ethical responsibilities of a reviewer.
27. Discuss the various categories of responsibilities of reviewers as per the guidelines of the Council of Science Editors.
28. Write the general guidelines for the editors.
29. Write the general guidelines for the reviewers.
30. What are the criteria based on which the reviewer analyzes an article?
31. What are the possible improper activities of a reviewer?
32. Make a note of the various responsibilities of a publisher.
33. Discuss the follow-up measures for a published manuscript.
34. What is the "expression of concern"? Explain.
35. What is retraction? Explain in detail.

REFERENCES

[1] David B. Resnik, "*What is ethics in research & why is it important*". National Institute of Environmental Health Science, 1-10 (2011).

[2] Muralidhar, Amith Ghosh and Sanghvi "*Ethics in Science Education, Research and Governance*: Indian National Science Academy (2019).

[3] Cynthia R. King, Ethical Issues in Writing and Publishing, *Clinical Journal of Oncology Nursing*, 5(3) (2001).

[4] P. Chaddah "*Ethics in competitive research: Do not get scooped; do not get plagiarized*", Pothy.com, (2018).

[5] https://www3.paho.org/journal/sites/default/files/201812/Instructions%20for%20 Authors_EN_170701_0.pdf

[6] Eun Seong Hwang et al. "*Manual for research and publication ethics in science and engineering*". Korean Federation of Science and Technology Societies, (2016).
[7] Pablo Avanzas et al. "*Ethical Considerations in the Publication of Scientific Articles*", 64(5): 427-429 (2011).
[8] Muralidhar et al. pp. 72-73
[9] Anil K Jain. "*Ethical issues in scientific publication*", Indian.J. Orthop. 44(3) (2010).
[10] https://www.niu.edu/academic-integrity/students/cheating/fabrication-orfalsification.shtml
[11] James Galipeau et al. "*A scoping review of competencies for scientific editors of biomedical journals*", BMC Medicine (2016) 14:16.
[12] https://academy.pubs.asha.org/erc/editor/
[13] https://www.elsevier.com/reviewers/role.
[14] http://publicationethics.org/files/u2/05A_CoI_Submitted.pdf
[15] http://publicationethics.org/files/u2/01A_Redundant_Submitted.pdf
[16] https://publicationethics.org/files/plagiarism%20A.pdf
[17] S. C. Lakhotia "*Peer review: then and now*", Current Science 105: 745 (2013).
[18] A. Moore, "*What's in a peer review report?*". BioEssays, 35: 77 (2013).
[19] Muralidhar et al. pp. 79-81
[20] https://www.councilscienceeditors.org/resource-library/editorial-policies/publication-ethics/2-3-reviewer-roles-and-responsibilities/
[21] https://hsssjournal.com/guidelines-for-editors-and-reviewers/
[22] https://academic.oup.com/journals/guidelines/
[23] https://www.mdpi.com/reviewers
[24] Muralidhar et al. pp. 81-82
[25] S. C. Lakhotia, "*The fraud of open access publishing*", Proc. Indian Natl Sci. Acad . 83: 33-36 (2017).
[26] Patwardhan et al. "*A critical analysis of the 'UGC-approved list of journals*", Current Science 114: 199-1303 (2018).
[27] https://expertjournals.com/responsibilities-publisher-relationship-journal-editors/
[28] Eun Seong Hwang et al. pp. 75-78.

Suggested Readings

1. C. Neal Stewart Jr. "Research Ethics for Scientists", (2011), John Wiley and Sons.
2. Sana Loue, "Text Book of Research Ethics: Theory and Practice", (2002) Kluwer Academic Publishers.
3. Ana Smith Iltis (Ed.), "Research Ethics" (2006) Routledge, New York.
4. Kristin Shrader-Frechette, "Ethics of Scientific Research" (2000) Rowman & Littlefield Publishers.
5. D'Angelo, John, "Ethics in Science: Ethical Misconduct in Scientific Research" (2012) CRC Press, London.
6. Peter Pruzan, "Research Methodology: The Aims, Practices and Ethics of Science", (2016), Springer International Publishing.

Index

S